Applied Mathematics and Parallel Computing

Herbert Fischer, Bruno Riedmüller
Stefan Schäffler (Editors)

Applied Mathematics and Parallel Computing

Festschrift
for
Klaus Ritter

With 39 Figures

Physica-Verlag
A Springer-Verlag Company

Dr. Herbert Fischer
Dr. Bruno Riedmüller
Priv.-Doz. Dr. Stefan Schäffler
Institut für Angewandte Mathematik und Statistik
Technische Universität München
Arcisstraße 21
D-80333 München
Germany

ISBN-13: 978-3-642-99791-4

Die Deutsche Bibliothek – CIP-Einheitsaufnahme
Applied mathematics and parallel computing: Festschrift for Klaus Ritter/Herbert Fischer... (ed.). – Heidelberg: Physica-Verl., 1996
ISBN-13: 978-3-642-99791-4 e-ISBN-13: 978-3-642-99789-1
DOI: 10.1007/978-3-642-99789-1
NE: Fischer, Herbert [Hrsg.]; Ritter, Klaus: Festschrift

Softcover reprint of the hardcover 1st edition 1996

SPIN 10536516 88/2202-5 4 3 2 1 0 – Printed on acid-free paper

Preface

The authors of this Festschrift prepared these papers to honour and express their friendship to Klaus Ritter on the occasion of his sixtieth birthday. Because of Ritter's many friends and his international reputation among mathematicians, finding contributors was easy. In fact, constraints on the size of the book required us to limit the number of papers.

Klaus Ritter has done important work in a variety of areas, especially in various applications of linear and nonlinear optimization and also in connection with statistics and parallel computing. For the latter we have to mention Ritter's development of transputer workstation hardware. The wide scope of his research is reflected by the breadth of the contributions in this Festschrift.

After several years of scientific research in the U.S., Klaus Ritter was appointed as full professor at the University of Stuttgart. Since then, his name has become inextricably connected with the regularly scheduled conferences on optimization in Oberwolfach. In 1981 he became full professor of Applied Mathematics and Mathematical Statistics at the Technical University of Munich. In addition to his university teaching duties, he has made the activity of applying mathematical methods to problems of industry to be centrally important.

The editors wish to thank all authors of this Festschrift for their contributions. Our TEX-nicians Margit Stanglmeir and Christian Hertneck were particularly helpful in the final stage of the book. We are also grateful to Dr. Werner A. Müller from Physica-Verlag for his efficient and expeditious production of this book.

All the authors wish Klaus Ritter many more years of fruitful scientific work with friends and colleagues. At the same time, the editors also wish him much time with his family.

Munich, February 1996

Herbert Fischer Bruno Riedmüller Stefan Schäffler

Contents

Informatics and the Internal Necessity for the Mathematization of the Sciences

N. Apostolatos, Athens

Abstract: In this paper some general thoughts about the significant role of Informatics in the future, the necessity for a strict foundation of Informatics and the Mathematization of Sciences are presented.

1 Mathematization

The question What is Mathematics? is a question of fundamental character. Mathematics has a unique universality. When we abstract the structures which we find in any branch of science we do really only Mathematics. So Mathematics is Queen and Servant of Science in accordance with the title of [9]. My belief that Mathematics is not a simple scientific branch but the unique and decisive means for the formulation and solving of all problems independent of field, is continually strengthened. And so the necessity for the mathematization of the sciences is evident.

Mathematics, if classified in the field of sciences, could be characterised as the science of sciences. However, if we wish to be more precise, we need to classify Mathematics in the field of languages. Mathematics is the most precise language available to people today.

According to Gauß: *"Die Mathematik ist die Königin der Wissenschaften, und die Zahlentheorie ist die Königin der Mathematik."*

We would say: *Mathematics is the language of all sciences (in a strict sense of the concept of science) as well as the language of all languages.*

A science is considered more advanced when a more precise language is used for the formulation and processing of its concepts. The level of a science is, therefore, determined by the extent of the use of the mathematical language by this science.

Any lack of clearness or inaccuracy or doubt accredited to the definitions and processing of the concepts of various scientific fields, is substantially due to the fact that we have not as yet succeeded in using exclusively the mathematical language at these points.

The lack of mathematization in the various scientific fields constitutes a barrier for the decisive application of modern methods of Informatics. A barrier which will be surpassed only with drastic changes to the structure of these fields, changes that will enable their systematic mathematization.

The 1940s end with the establishment of the field of computers. It would not be an exaggeration to say that our century will close with the name "Century of Computers" or, more appropriately, "Century of Informatics". Therefore, language, with the wider meaning of the term, constitutes the predominant element in our life. The necessity for a strict definition of a language led us to a mathematization of it. Mathematics becames the basic instrument for this purpose. Mathematics, in reality, is a super-language because not only is it a language but also substantially constitutes the only language for the strict definition of the other languages.

The division of Mathematics into sub-fields has been made in such a way as to make it practically impossible to work in one of them continuously without using the others as well. That is, the various sub-fields are not disjoint to each other.

In periods of intensive philosophical thought, such as this of ancient Greece, mathematicians, and not only, have discussed on the nature of Mathematics. In a sense, Mathematics is unique in the intellectual development. We must not forget that Euclid's *Elements* are not only the most used mathematical book in world's history, but also this work has had the greater influence on scientific thought. The strict axiomatic foundation, and so the preparation for the mathematization, is an internal necessity for any scientific branch.

The introduction of infinity in Mathematics was a decisive turning point in the history of Mathematics. And while the finite seems to us attainable, many times in practice it leads us to insuperable impasses. Let us try to conceive the magnitude of the finite number:

$$10^{10^{10^{10}}}$$

In order to, therefore, estimate "how great" this number is, we will attempt to "describe" it by using other magnitudes. This number, of course, is, as we say in Mathematics, finite, it is, nonetheless, inconceivably great. Let us imagine the whole of the universe as a sphere with such a radius that it would take ten billion years for the light to cover it with its unimaginable speed. Let us accept now that we have the ability to register a digit on some material matter which occupies a mass such as that of the smallest atom of matter. Then, within the whole of the universe, it would be impossible to place anything other than a "minute" part of the digits of this number, so minute that we could say that "nearly all" digits of this number have been excluded from the universe.

The enormous development of computers made it possible for us to use enormous amounts of information, however always within a limited space whose power is close to zero compared with the above number. This need to be within accessible finite spaces was what led to the development and foundation of Interval Mathematics. The Interval Arithmetic is a necessity for the computation of safe solutions. Without a relative theory (e.g. Arithmetic of sets) the strict foundation of Numerical Mathematics would be impossible.

2 Mathematical Foundations of Informatics

The subject of Informatics is immense. Mathematicians are regarded as the substantial creators of the current branch of Informatics. In the oncoming years, the mathematical language will play the decisive role for the on steady basis foundation of Informatics.

A significant example supporting the thesis that every founded progress within the field of Informatics cannot but rely on the possibilities offered by the unlimited and continuously expanding mathematical language, is the modern functional programming.

Characteristic, in this case, is the following paragraph taken from the paper of J. Backus: "Can Programming Be Liberated from the Von Neumann Style? A Functional Style and its Algebra of Programs." ([8] p. 614).

> *"Associated with the functional style of programming is an algebra of programs whose variables range over programs and whose operations are combining forms. This algebra can be used to transform programs and to solve equations whose "uknowns" are programs in much the same way one transforms equations in high school algebra."*

It is an undisputable fact that today we live in an exceptional phase of human history. The catalytic effects of modern technology which are characterized by the surprising and unique speed of the development of Informatics, determine the future development and mark the frame in which human beings will move in during the following decades.

For the first time in the human history, the capital, the workforce and the raw materials do not constitute the basic kernel of development. In the years to come, knowledge, the renewal of human activity and the information will constitute the dominant elements for future development. The Informatics, as we have emphasised previously in an other place, will have a decisive effect on the form of the future community, since, prior to this, it will have caused successive, essential changes in the political systems, with the major ones existing today, approaching each other. Thus, new political systems will result which will be able to meet the demands of the dynamic conditions created by the continuously developing of Informatics.

A political system cannot be static; essentially, a correct political system cannot be anything else but a "dynamic" algorithm leading to a solution which, nonetheless, will have to readapt itself continuously to its respective tendencies, thus causing a continuous readaptation of this algorithm.

Since the question arises from time to time, whether or not a computer can think, this question depends on what is meant by "to think". Thus, let us begin with the following

Definition. *The term "to think" means the process of interaction between informations and the rules of processing informations for the creation of new informations and new rules of processing informations.*

Therefore, an electronic computer, just like a person, can think. However, if we were to add to this definition that the above interaction has to be made independently without external influence, then we would have to accept that one electronic computer does not think, and, simultaneously, a person also does not think, because here, too, external factors (although

we do not know how) influence the way a person thinks. That a person has a conscience of its existence is definitely regarded as a matter of fact today, but we cannot say the same for the computer.
According to René Descartes: *"Cogito, ergo sum"*, here "thinking" is something that "exist" a priori. So since the copmputer has not the conscience of its "thinking" it's not possible to say: "I am", that, of course can say a person.

The continuously lowering costs of the hardware and software of Informatics, which simultaneously is developing rapidly, opens up avenues for the enrolment of continuously more individuals to the process of learning. And in this way, althoug the process of learning is undergoing a revolutionary redefining, simultaneously a new human force is created which is entering a new form of progressive renaissance.

24 years ago, in our book "Digital Computers" ([2]), we were stating:

> *"In the near future the person who completely ignores some basic topics related to the computer science will have the feeling that he lives in a foreign world with serious effects upon his neuropsychological state and his production within the society."*

In 1980, the President of ACM, P.J. Denning, in an ACM President's Letter "The U.S. Productivity in Crisis" ([11] p. 618), writes:

> *" Computers are now everywhere - in cars, in watches and pens, in light switches and in calculators, in radios and in ovens, in cash registers and in banks, and in personal computers - they have invaded the province of Everyman. The ability to understand this tecnology, and use it wisely, will be a prerequisite of society in 1990 and beyond."*

Today, in the mid 1990s, the above constitutes a reality, which, he who does not take into account due to ignorance, is left outside place and time together with all the frightening results for himself and whatever is influenced by his actions.

That knowledge increases with geometric progression has been extensively emphasised and continues to be so. Simultaneously any biological improvement in a person's ability to think is essentially restricted. How then will a person be able to command this knowledge when such a plethora is offered to him and indeed when some of this knowledge can be used destructively upon himself?

This immense problem cannot be confronted except through a revolutionary reformation of the educational system. The various educational systems lack the capability of a dynamic readaptation, something which is primarily necessary, especially with the rapidly developing technology of our times.

The only language which within its foundations and development has followed logical rules, is Mathematics, and it plays and still plays and will continue to do so, a protagonistic role in future developments. It is necessary, therefore, for conventional languages, before the forth coming developments, to adapt themselves in accordance with the dominant logic. That is to acquire a unified orthological structure with the assistance of the super-language of Mathematics. In the near future, the gradual creation of a logical syntax, equipped with logical-dynamic rules which will allow it to adapt itself to the respective demands, seems to be the natural course of development.

3 Complexity

0.4cm The criterion of performance is the basic one for valueing algorithms. So algorithms that perform significantly quickly a class of problems than any previously algorithm for the same class of problems, is a valuable scientific contribution. Otherwise for classes of problems for which there is not an algorithm to solve them, the only objective is to find a such algorithm.
During the recent years, significant progress has been made in the parallel systems and in the parallel algorithms. This, together with the assistance of modern computers, will increase the speed of processing of problems, significantly. Today we have made it possible to have parallel systems with many processors. Numerical models, such as fractal structure and chaotic behavior, have disclosed noticeable complexity. However, greatly complicated problems cannot be dealt with successfully by depending only on the increase of the speed of computer systems which is derived either from the general improvement or the further use of any parallelism.
On these occasions we need to develop new revolutionary algorithms and this will be attained in two ways:

(a) With the assistance of the existing Mathematics and its normal development.

(*b*) With the assistance of revolutionary development in Mathematics, similar to the creation of Infinitisimal Calculus. With this assistance revolutionary algorithms will be developed that will definitely exploit whatever progress in computer media, but foremostly will have classes more greater speed than the speed of algorithms existing today, for the solution of corresponding problems.

At present, we only have possibility (a). At this stage, I will mention a simple example in order to illustrate what is meant by faster algorithm ([7] p. 375).

Problem

The number 9801 has the following property:

$$(98+01)^2 = 9801$$

We are searching to find all the numbers with an even number of digits:

$$y_1 y_2 \cdots y_i y_{i+1} \cdots y_{2i}$$

so that for all $i = 1(1)n$ hold:

$$(y_1 y_2 \cdots y_i + y_{i+1} y_{i+2} \cdots y_{2i})^2 = y_1 y_2 \cdots y_i y_{i+1} \cdots y_{2i}$$

If we work typically, we will have to examine all the numbers with $2, 4, \ldots,$ $2n$ digits, in other words we will examine:

$$90 + 9000 + 900000 + \ldots + 9 * 10^{2n-1} = 9090 \ldots 90 \ldots 90$$

numbers.

With the algorithm we are giving in [7], it will be necessary to examine only:

$$\left[\frac{25 * 10^{2n-2}}{10^n - 1}\right] + 1$$

numbers.

A number which is by far much smaller. So e.g. for $n = 4$, using the typical method 90909090 numbers will need to be examined, whereas with the giving, only 2501.

4 Future Developments

Due to the leaping increase of human knowledge and the limits of the human brain, it is obvious that the way of teaching must aim not on the accumulation of knowledge but on the development of the capacities of learning that will enable, with the relevant methodology, access to the sources of knowledge.
In this way, every member of tomorrow's society, will be involved in the process of learning for life.

Certainly the recent political developments in the various countries of our planet depend closely upon the revolution of Informatics. Along side with whatever rapid reorganisation of the politico-economic and social structures, Informatics will have a decisive influence on the structure of human thought.
It may appear strange if one were to maintain that the great problems preoccupying people today such as environmental pollution, traffic congestion and, above all energy consumption, will find a solution through Informatics. And this because substantially these problems and many others, are created and worsened by the unavoidable increase in the movement of individuals on Earth's restricted space. Movement with the revolutionary conquests of Informatics, not only will become less essential with the passage of time, but will become at one stage and after, negative for the individual's very productivity within society.
This circumstantial "immobility" that will be imposed on individuals will have enormous effects on the very biological and intellectual development of individuals, resulting with also the complete readaptation of the socio-political systems.

And the speedup of the computers today, as well as those of the future, will be too weak to confront the enormous calculating problems, if effective algorithms are not discovered. It's precisely the need for finding a way to overcome limited possibilities of any technological development that will provide the basic motive for a new period in the development of the mathematical language and a systematic utilization of its applications. Such as the revolutionary development of Mathematics as the creation of the Infinitesimal Calculus, which was based on the fundamental work of Newton and Leibnitz.

In a time of rapid increase of human knowledge, the inability of direct estimation causes, in many cases, unnecessary strain on the thinking person. We believe that in the near future a strong attempt will be made for a unified form of human knowledge with the simulataneous removal of the non useful parts. The specific need to confront this acute problem will provide us with new ways of separating and classifying this knowledge, something which will become possible with the suitable development of Informatics.

At this point it is perhaps relevant to refer to something we have put forward at the beginning of our last book ([7]):

> *"The universality, the endurance in time, the uniqueness and, above all, the precision, are basic properties of the mathematical language. Thus, in the attempt for a continuously more unified expression of human knowledge, something which is imposed due to its increase with geometric progress, mathematics will be the field which, itself, continuously unified will constitute the basic instrument for this purpose."*

References

[1] Apostolatos, N., Kulisch, U.: *Grundlagen einer Maschinenintervallarithmetik. Computing 2, 89-104, (1967).*

[2] Apostolatos, N.: *Digital Computers. Book (Greek). Athens 1972.*

[3] Apostolatos, N.: *A proposal for the renaming of the scientific branch of Computer Science. The Computer Journal. Vol. 32, No 3, p. 282, 1989.*

[4] Apostolatos, N.: *The Magic of Mathematics. (Greek). TO BHMA, Athens 25-26.3.1989.*

[5] Apostolatos, N.: *The Informatics determines the Structure of all Sciences (Greek). Mathematical Review (Greek Mathematical Society) 37, 1-10, (1990).*

[6] Apostolatos, N.: *On a unified concept of Mathematics. Computing Suppl. 9, 1-10 (1993).*

[7] Apostolatos, N.: *Programming - Applications. Book (Greek). Athens 1995.*

[8] Backus, J.: *Can Programming Be Liberated from the von Neumann Style? A Functional Style and Its Algebra of Programs. Comm. of ACM 21, 613-641 (1978).*

[9] Bell, E. T.: *Mathematics, Queen and Servant of Science. McGraw-Hill, New York 1951.*

[10] Courant, R., Robbins, H.: *Was ist Mathematik? Sringer-Verlag, Berlin 1962.*

[11] Denning, P. J.: *U.S. Productivity in Crisis. Comm. of the ACM 23, 617-619 (1980).*

[12] Nash, S. G.: *A History of Scientific Computing. ACM Press - Addison Wesley, 1990.*

[13] Strassen, V.: *Algebraic Complexity Theory. In Handbook of Theoretical Computer Science. Vol. A, Algorithms and Complexity, 633-672. Edided by Jan van Leeuwen. Elsevier Science Publ. B. V. 1990.*

[14] Wilder, R. L.: *Introduction to the Foundations of Mathematics. John Wiley, 1956.*

A New Semi–infinite Programming Method for Nonlinear Approximation

Miroslav D. Ašić, Department of Mathematics,
The Ohio State University
Vera V. Kovačević-Vujčić, Faculty of Organizational Sciences,
University of Belgrade

Abstract. The paper presents a new method for solving approximation problems based on the ideas of semi-infinite programming. The nonlinear problem of generalized rational approximation is replaced by a sequence of linear semi-infinite programming problems which are solved approximately with the increasing precision. It is shown that the sequence of approximate solutions converges to the solution of the initial problem.

Key Words: Nonlinear approximation, Semi-infinite programming, Discretization, Linear programming.

1 Introduction

This paper considers the following nonlinear approximation problem: Let $T \subseteq R^d$ be a compact set and let f, $g_1, \ldots, g_m$, $h_1, \ldots, h_n$ be continuous real valued functions on T. Approximate f by a generalized rational function

$$F(t, x, y) = (x_1 g_1(t) + \cdots + x_m g_m(t))/(y_1 h_1(t) + \cdots + y_n h_n(t))$$

in the sense of Chebyshev, i.e. find $\bar{x}$ and $\bar{y}$ as a solution to the problem

$$(1) \qquad \min_{x,y} \|f - F\| = \min_{x,y} \{ \max_{t \in T} |f(t) - F(t, x, y)| \}$$

If $T = [\alpha, \beta]$, $g_i(t) = t^{i-1}, i = 1, \ldots, m$, $h_j(t) = t^{j-1}$, $j = 1, \ldots, n$, (1) is the standard approximation problem by a rational function

$$R(t, x, y) = (x_1 + x_2 t + \cdots + x_m t^{m-1})/(y_1 + y_2 t + \cdots + y_n t^{n-1})$$

and it can be shown that the best approximation always exists (see e.g. [7] and [9]). Under the additional condition that the numerator and the denominator have no common zeros and at least one of them is of a full degree ($\bar{x}_m \neq 0$ or $\bar{y}_n \neq 0$), the function of the best approximation is unique and the optimal parameters satisfy the alternation criterion (see e.g. [8]). Since for $\lambda \neq 0$ we have $R(t, \lambda x, \lambda y) = R(t, x, y)$ it is clear that the optimal solution is not unique in the parameter space. Uniqueness can be achieved by adding additional normalizing conditions, such as $R(\alpha, x, y) = 1$.
In the case of generalized rational approximation the existence of the best approximation cannot be guaranteed since common zeros of the numerator and the denominator cannot always be eliminated. In order to overcome this difficulty, some additional constraints are usually added to (1), e.g. (c.f. [8])

$$(2) \qquad 1 \leq y_1 h_1(t) + \cdots + y_n h_n(t), \quad t \in T$$

There are several different approaches to solving nonlinear approximation problems (see [8] for the detailed overview). Direct discretization techniques lead to nonlinear programming problems, while refinement of the discretization usually requires a global maximization problem over the index set to be solved at each iteration. The other possible approach is to replace the nonlinear approximation problem by a sequence of linear approximation problems which are again solved using discretization techniques. We require that the k-th linear problem be solved up to the precision 2^{-k}, and we show that a satisfactory approximate solution can be obtained in such a way that no maximization over the index set T is required. We prove that the sequence of approximate solutions converges to the solution of the initial problem.

2 Description of the method

Problem (1) with the additional requirement (2) can be reformulated as the following nonlinear semi-infinite programming problem:

$$\min\ z$$

$$(3)\qquad z \geq |f(t) - \frac{x_1 g_1(t) + \cdots + x_m g_m(t)}{y_1 h_1(t) + \cdots + y_n h_n(t)}|, \quad \forall t \in T$$

$$y_1 h_1(t) + \cdots + y_n h_n(t) \geq 1, \quad \forall t \in T$$

The method which will be described below solves the problem (3) using iteratively two phases: In the first phase of the k-th iteration z is fixed to the value z_k and the auxiliary problem of linear semi–inifinite programming is formulated and solved with the precision 2^{-k}. Let x^k, y^k be the components of the approximate solution obtained in the first phase. In the second phase (x, y) is set to (x^k, y^k) and z_k is modified. A similar idea is used in [10] to solve a simpler problem of rational approximation on a finite set T.

Algorithm 1.

Step 0: Find numbers z_0 and M such that the system

$$z_0 \geq |\ f(t) - \frac{x_1 g_1(t) + \cdots + x_m g_m(t)}{y_1 h_1(t) + \cdots + y_n h_n(t)}\ |, \quad \forall t \in T$$

$$M \geq y_1 h_1(t) + \cdots + y_n h_n(t) \geq 1, \quad \forall t \in T$$

has a solution. Set $k = 0$.

Step 1: Find an approximate solution (x^k, y^k, w_k) of the following auxiliary semi–infinite programming problem:

$$\min\ w$$

$$(4)\quad w \geq (|f(t) - \frac{x_1 g_1(t) + \cdots + x_m g_m(t)}{y_1 h_1(t) + \cdots + y_n h_n(t)}| - z_k)(y_1 h_1(t) + \cdots + y_n h_n(t)), \ \forall t \in T$$

$$M \geq y_1 h_1(t) + \cdots + y_n h_n(t) \geq 1\ , \quad \forall t \in T$$

which satisfies the condition $w_k - \bar{w}_k \leq 2^{-k}$, where $\bar{w}_k$ is the optimal objective function value.

Step 2: If $w_k \geq 0$ set $z_{k+1} = z_k$, replace M by $M + 1$ and go to Step 1. Otherwise go to Step 3.

Step 3: Find an approximate solution z_{k+1} of the following auxiliary semi–infinite programming problem

$$\min z$$

$$z \geq \left| f(t) - \frac{x_1^k g_1(t) + \cdots + x_m^k g_m(t)}{y_1^k h_1(t) + \cdots + y_n^k h_n(t)} \right|, \forall t \in T \tag{5}$$

such that $z_{k+1} \leq z_k$ and $z_{k+1} - \bar{z}_{k+1} \leq 2^{-k}$, where $\bar{z}_{k+1}$ is the optimal value of the objective function z. Replace k by $k+1$ and go to Step 1. □

Let us note that the problem in Step 1 is in fact linear semi–infinite programming problem. Indeed, the problem (4) can be rewritten in the following equivalent form:

$$\min w$$

$$(6)\quad \begin{aligned} w &\geq (f(t) - z_k)(y_1 h_1(t) + \cdots + y_n h_n(t)) - (x_1 g_1(t) + \cdots + x_m g_m(t)), \ \forall t \in T \\ w &\geq (-f(t) - z_k)(y_1 h_1(t) + \cdots + y_n h_n(t)) + (x_1 g_1(t) + \cdots + x_m g_m(t)), \ \forall t \in T \end{aligned}$$

$$M \geq y_1 h_1(t) + \cdots + y_n h_n(t) \geq 1, \ \forall t \in T$$

The linear semi–infinite programming problem in Step 3 can alternatively be formulated as a global optimization problem (over the set T) with nondifferentiable objective function.

The details about the actual implementation of Algorithm 1 will be given in the next section. In the remaining part of this section we investigate some convergence properties of the algorithm.

Let us first note that Algorithm 1 is well defined. Indeed, z_0 in Step 0 can be determined by taking

$$z_0 \geq \max_{t \in T} |f(t)|$$

To pick a suitable M take a linear combination of $h_1(t), \ldots, h_n(t)$ which is positive on T and let M be greater than the ratio of its maximum and minimum values. The problem in Step 1 has an approximate solution since it is feasible and $w \geq -z_k M$. Similarly, the problem in Step 3 has an approximate solution since the feasible set is nonempty and $z \geq 0$. It remains to show that in Step 3 z_{k+1} can be chosen to satisfy $z_{k+1} \leq z_k$ and $z_{k+1} - \bar{z}_{k+1} \leq 2^{-k}$. Since Step 3 is executed only in the case $w_k < 0$ and (x^k, y^k, w_k) is a solution to (4) we obtain

$$\left| f(t) - \frac{x_1^k g_1(t) + \cdots + x_m^k g_m(t)}{y_1^k h_1(t) + \cdots + y_n^k h_n(t)} \right| - z_k < 0, \ \forall t \in T$$

so that $\bar{z}_{k+1} < z_k$. Hence z_{k+1} can be chosen to satisfy both $z_{k+1} \leq z_k$ and $z_{k+1} - \bar{z}_{k+1} \leq 2^{-k}$as specified in Step 3.
From Steps 2 and 3 it follows that the sequence (z_k) is nondecreasing. The next theorem shows that it converges to the optimal value of problem (3).

Theorem 1. Assume that the functions $f, g_1, \dots, g_m,\; h_1, \dots, h_n$ in (1) are continuous on the compact set T and that the best approximation

$$F(t, \hat{x}, \hat{y}) = \frac{\hat{x}_1 g_1(t) + \cdots + \hat{x}_m g_m(t)}{\hat{y}_1 h_1(t) + \cdots + \hat{y}_n h_n(t)}$$

satisfying (2) exists. Let

$$\max_{t \in T} |f(t) - F(t, \hat{x}, \hat{y})| = \eta$$

and let $((x^k, y^k, z_k))$ be a sequence generated by Algorithm 1. Then $z_k \to \eta$, $k \to \infty$ and each cluster point $(\bar{x}, \bar{y})$ of the sequence $((x^k, y^k))$ yields the coefficients of a best approximating function.

Proof: Since the sequence (z_k) is nondecreasing and $z_k \geq \eta$ it follows that it converges. Suppose that $z_k \to \eta + \delta$, where $\delta > 0$. Then $\bar{w}_k$ can be nonnegative only a finite number of times. Indeed, after a finite number of iterations in which $w_k \geq 0$ we will have

$$\hat{y}_1 h_1(t) + \cdots + \hat{y}_n h_n(t) \leq M, \quad \forall t \in T \tag{7}$$

since M is increased each time. If (7) is satisfied we have, for each $t \in T$,

$$\begin{aligned}(|f(t) - \frac{\hat{x}_1 g_1(t) + \cdots + \hat{x}_m g_m(t)}{\hat{y}_1 h_1(t) + \cdots + \hat{y}_n h_n(t)}| - z_k)\;(\hat{y}_1 h_1(t) + \cdots + \hat{y}_n h_n(t)) &\leq \\ \leq (\eta - (\eta + \delta)) = -\delta &\end{aligned}$$

so that $\bar{w}_k \leq -\delta$. If k is large enough (say, $k \geq k_0$) to satisfy $2^{-k} < \delta/2$ we will also have $w_k < -\delta/2$, which means that z_{k+1} will be determined according to the rules in Step 3.
To estimate the actual decrease in the value of z note that, from (4) it follows that for $k \geq k_0$ and each $t \in T$

$$\begin{aligned}&-\delta/2 > w_k \geq \\ &\geq (|f(t) - \frac{x_1^k g_1(t) + \cdots + x_m^k g_m(t)}{y_1^k h_1(t) + \cdots + y_n^k h_n(t)}| - z_k)(y_1^k h_1(t) + \cdots + y_n^k h_n(t)) \geq \\ &\geq (|f(t) - \frac{x_1^k g_1(t) + \cdots + x_m^k g_m(t)}{y_1^k h_1(t) + \cdots + y_n^k h_n(t)}| - z_k)\; M\end{aligned}$$

and hence for each $t \in T$

$$(|f(t) - \frac{x_1^k g_1(t) + \cdots + x_m^k g_m(t)}{y_1^k h_1(t) + \cdots + y_n^k h_n(t)}| \leq z_k - \frac{\delta}{2M}$$

Therefore, $z_{k+1} \leq z_k - \delta/(2M) + 2^{-k}$, $k \geq k_0$. For $k \geq k_1 \geq k_0$ we will have $2^{-k} < \delta/(4M)$, so that $z_{k+1} \leq z_k - \delta/(4M)$, $k \geq k_1$. This contradicts the convergence of the sequence (z_k).
If $(\bar{x}, \bar{y})$ is a cluster point of $((x^k, y^k))$ the function $F(t, \bar{x}, \bar{y})$ is a function of the best approximation since $z_k \to \eta, k \to \infty$. □

It is worth mentioning that this theorem shows that a nonlinear problem of the type (1) (with possibly many local optima) can be replaced by a sequence of linear auxiliary problems which need to be solved only approximately. Furthermore, the sequence obtained by Algorithm 1 finds a *global* solution to the original nonlinear optimization problem.

3 A linear semi–infinite programming method

In this section we shall propose a method which solves a linear semi–infinite programming problem approximately, with a given precision 2^{-k}. The method can be used for the implementation of Steps 1 and 3 in Algorithm 1. It belongs to a class introduced in [2], [3], [4]. Application of these methods to linear approximation problems has been considered in [1], [5] and the related ideas have been used in global optimization [6]. For the sake of simplicity we shall assume that $T \subseteq R^d$ is a product of intervals, although the method, with slight modifications can be applied for the arbitrary compact index set.
Consider a linear semi–infinite programming problem

(8) $$\min_{u \in U} c^T u, \; U = \{ \, u \in R^r | \; a(t)^T u \leq b(t), \; \forall t \in T \, \}$$

where $a : T \to R^r$, $b : T \to R^r$, $c \in R^r$. In the sequel we shall use the following:

Assumption 1. (i) $T = [p_1, q_1] \times \cdots \times [p_d, q_d]$.
(ii) The Slater's condition is fulfilled, i.e. there exists a point $\bar{u} \in R^r$ such that $a(t)^T \bar{u} \leq b(t) - \beta$ for all $t \in T$ and some $\beta > 0$.
(iii) The level set $\bar{U} = \{u \in U | c^T u \leq c^T \bar{u}\}$ is bounded. Let A be a lower bound for the objective function value, i.e. $c^T u \geq A$ for $u \in U$.

(iv) The following (uniform) Lipschitz condition is satisfied:

$$|a(t')^T u - b(t') - a(t'')^T u + b(t'')| \leq L\|t' - t''\|,\ t', t'' \in T,\ u \in \bar{U}$$

□

In the definition of Algorithm 2 we will use a sequence of uniform nets $M_j \subseteq R^d$, defined by:

$$M_j = \{(p_1 + k_1 h_1^j, \dots, p_d + k_d h_d^j) |\ k_i \in \{0, \dots, 2^j m_i\},\ i = 1, \dots, d\}$$

where $h_i^j = (q_i - p_i)/2^j m_i$, and m_i are suitably chosen positive integers (see Step 0).

In Step 2 of Algorithm 2 we use a piecewise linear function f which is defined as follows: For given $v \in R^r$, $t \in T$ and $h_1 > 0, \dots, h_d > 0$ let

$$f(s) = a(t)^T v - b(t) + \sum_{i=1}^{d} \bar{A}_i (s_i - t_i) + \sum_{i=1}^{d} \bar{B}_i |s_i - t_i| \tag{9}$$

where $\bar{A}_1, \dots, \bar{A}_d,\ \ \bar{B}_1 > 0, \dots, \bar{B}_d > 0$ are such that

$$f(s) \geq a(s)^T v - b(s) \text{ for } |s_i - t_i| \leq h_i, \quad i = 1, \dots, d$$

Such a function can usually be obtained using a Taylor expansion. For example, if a and b are twice continuously differentiable functions, the coefficients $\bar{A}_i$ can be chosen as the first partial derivatives of the constraint functions with respect to the i-th variable at the point t, while $\bar{B}_1, \dots, \bar{B}_d$ can be obtained by rounding the remainder terms so that $\bar{B}_i \to 0$ when $h_1 \to 0, \dots, h_d \to 0$.

Algorithm 2.

Input parameters: β, L, A as in Assumption 1, a positive integer k.

Step 0: Find $m_1, \dots, m_d$ such that

$$\max_{1 \leq i \leq d} \frac{q_i - p_i}{m_i} \sqrt{d} \leq \beta / L$$

Step 1: Solve the linear programming problem

$$\begin{aligned} & \min\ c^T u \\ \text{s.t.}\quad & a(t)^T u \leq b(t) - \beta,\ t \in M_0 \end{aligned}$$

and let v^0 be its solution. Let $J(k)$ be the smallest even positive integer such that $2^{-J(k)/2}(c^T v^0 - A) \leq 2^{-k}$. Set $C = M_0$.

Step 2: Solve the linear programming problem

$$\min c^T u$$
s.t.
$$a(t)^T u \leq b(t) - \beta/2^{J(k)},\ t \in C$$

and let v be its solution. Set $j = 0,\ E_0 = M_0$.

Step 3: If $j = J(k)$ set $v(k) = v,\ C(k) = C$ and STOP. Otherwise, for each $t \in E_j$ determine a function f of the type (9) satisfying

$$f(s) \geq a(s)^T v - b(s),\ |s_i - t_i| \leq h_i^j/2,\ i = 1, \ldots, d$$

Let E'_{j+1} be the set of points $t \in E_j$ for which the corresponding function f is greater than or equal to $-\beta/2^{J(k)}$ at some extreme point of the set $[t_1 - h_1^j/2, t_1 + h_1^j/2] \times \cdots \times [t_d - h_d^j/2, t_d + h_d^j/2]$. For each $t \in E'_{j+1}$ let $N(t) = \{\ (t_1 + \kappa_1 h_1^j/2, \ldots, t_d + \kappa_d h_d^j/2),\ \kappa_i \in \{-1, 0, 1\}\ \} \cap T$ and set

$$E_{j+1} = \bigcup_{t \in E'_{j+1}} N(t)$$

Replace j by $j + 1$.

Step 4: If $a(t)^T v \leq b(t) - \beta/2^{J(k)}$ for $t \in E_j$ go to Step 3. Otherwise replace C by $C \cup \{t \in E_j |\ a(t)^T v > b(t) - \beta/2^{J(k)}\}$ and go to Step 2. □

Basically, if parameters β, L and A are fixed, Algorithm 2 accepts a positive integer k and generates as its output an even positive integer $J(k)$, a discretization $C(k)$ of the set T and an approximate solution $v(k)$.

Theorem 2: Let Assumption 1 be satisfied. Let for a given positive integer $k, J(k)$ and $v(k)$ be defined as in Steps 1 and 3 of Algorithm 2, respectively. Then

(i) $$a(t)^T v(k) \leq b(t) - \beta/2^{J(k)},\ t \in M_{J(k)}$$
(ii) If $\bar{v}$ is an optimal solution of the problem (8) then

$$c^T v(k) - c^T \bar{v} \leq 2^{-k}$$

Proof: (i) Let us first note that the set

$$U_{J(k)} = \{u \in R^r | a(t)^T u \leq b(t) - \beta/2^{J(k)}, t \in M_{J(k)}\}$$

is nonempty and contained in U. Indeed, by Slater's condition there exists a point $\bar{u}$ such that $a(t)^T\bar{u} \le b(t) - \beta$ for each $t \in T$. Therefore $\bar{u} \in U_{J(k)}$. Let $u \in U_{J(k)}$ and $t \in T$ be arbitrary. Take $\bar{t} \in M_{J(k)}$ such that

$$\|t - \bar{t}\| \le \max_{1 \le i \le d} \{(q_i - p_i)/m_i\} \sqrt{d}\, 2^{-J(k)-1}$$

Then $|a(t)^T u - b(t) - a(\bar{t})^T u + b(\bar{t})| \le L\|t - \bar{t}\| \le (\beta/L)2^{-J(k)-1} L \le \beta 2^{-J(k)}$. Since for $\bar{t} \in M_{J(k)}$ we have $a(\bar{t})^T u \le b(\bar{t}) - \beta/2^{J(k)}$ it follows that $a(t)^T u \le b(t)$. Hence, $u \in U$.
Similarly, one can prove that the set

$$U_0 = \{u \in R^r \mid a(t)^T u \le b(t) - \beta,\ t \in M_0\}$$

is nonempty and $U_0 \subseteq U$. The proof is almost identical. Therefore, the problem in Step 1 is solvable since it is feasible and $c^T u \ge A$, $u \in U_0 \subseteq U$. That also implies that the problem

$$\text{(10)} \qquad \text{s.t.} \quad \begin{array}{c} \min\ c^T u \\ a(t)^T u \le b(t) - \beta,\ t \in C \end{array}$$

has a solution for each set C such that $M_0 \subseteq C \subseteq M_{J(k)}$. Hence, the dual of (10) has a nonempty feasible set. Since the dual of the problem in Step 2 has the same feasible set it follows that the problem in Step 2 is solvable as well.
The problem in Step 2 may be solved several times during the execution of Algorithm 2. The point $v(k)$ is obtained when it is solved for the last time. The sets E_j used below are generated in Step 3 with $v = v(k)$. Let us note that when $v = v(k)$ the jump Step 4 $\rightarrow$Step 2 cannot occur since it would lead to another solution of the problem in Step 2. Hence,

$$\text{(11)} \qquad a(t)^T v(k) \le b(t) - \beta/2^{J(k)},\ t \in E_j,\ j = 0, 1, \ldots J(k)$$

Suppose that $v(k) \notin U_{J(k)}$. Then for some $\bar{t} \in M_{J(k)}$ we have

$$\text{(12)} \qquad a(\bar{t})^T v(k) > b(\bar{t}) - \beta/2^{J(k)}$$

Let t^j be the point in M_j closest to $\bar{t}$, where $j = 0, 1, \ldots, J(k)$. Since $t^0 \in M_0 = E_0$, the function f in Step 3 will satisfy $f(t) \ge -\beta/2^{J(k)}$ at some extreme point of the set $[t_1^0 - h_1^0/2, t_1^0 + h_1^0/2] \times \cdots \times [t_d^0 - h_d^0/2, t_d^0 + h_d^0/2]$. Indeed, the convexity of f would otherwise imply $f(t) \le -\beta/2^{-J(k)}$ for each t in this set, including the point $\bar{t}$, i.e.

$$a(\bar{t})^T v(k) \leq b(\bar{t}) - \beta/2^{J(k)}$$

which would contradict (12). Therefore, $t^0 \in E_1'$ which means that $N(t^0) \subseteq E_1$. It follows directly from the definition of t^0 and t^1 that $t^1 \in E_1'$, so that $t^1 \in E_1$. Proceeding by induction we conclude that $t^j \in E_j$, $j = 0, 1, \ldots, J(k)$ and since $t^{J(k)} = \bar{t}$, by (11) we have

$$a(\bar{t})^T v(k) \leq b(\bar{t}) - \beta/2^{J(k)}$$

contradicting (12).

(ii) Let us first note that (i) implies that for each k, $v(k)$ is a solution to the problem

$$\min_{u \in U_{J(k)}} c^T u$$

Let us show that $c^T v(k+1) - c^T \bar{v} \leq (c^T v(k) - c^T \bar{v})/2$. We first show that

$$a(t)^T v(k) \leq b(t) - \beta/2^{J(k)+1}, \ \forall t \in T \tag{13}$$

Indeed, let $t \in T$ be arbitrary and choose $\bar{t} \in M_{J(k)}$ such that

$$\|t - \bar{t}\| \leq \max_{1 \leq i \leq d} \left(\frac{q_i - p_i}{m_i} \right) \sqrt{d}\, 2^{-J(k)-1}$$

We have

$$|a(t)^T v(k) - b(t) - a(\bar{t})^T v(k) + b(\bar{t})| \leq L\|t - \bar{t}\| \leq L(\beta/L)2^{-J(k)-1} = \\ = \beta 2^{-J(k)-1}$$

Since $a(\bar{t})^T v(k) - b(\bar{t}) \leq -\beta/2^{J(k)}$ (13) follows.
Next, let us show that

$$(v(k) + \bar{v})/2 \in U_{J(k+1)}$$

As $M_{J(k)+2} \subseteq T$, (13) implies that for all $t \in M_{J(k)+2}$ we have

$$a(t)^T((v(k) + \bar{v})/2) = a(t)^T v(k)/2 + a(t)^T \bar{v}/2 \leq \\ \leq b(t)/2 - \beta/2^{J(k)+2} + b(t)/2 = b(t) - \beta/2^{J(k)+2}$$

The last expression can be rewritten as $b(t) - \beta/2^{J(k+1)}$ since it is clear from Step 1 of Algorithm 2 that $J(k+1) = J(k) + 2$. This means that $(v(k) + \bar{v})/2 \in U_{J(k+1)}$. Finally,

$$c^T v(k+1) - c^T \bar{v} \leq c^T(v(k) + \bar{v})/2 - c^T \bar{v} = (c^T v(k) - c^T \bar{v})/2$$

which implies

$$c^T v(k) - c^T \bar{v} \leq 2^{-J(k)/2}(c^T v^0 - c^T \bar{v}) \leq 2^{-J(k)/2}(c^T v^0 - A) \leq 2^{-k}$$

□

The computational complexity of the algorithms above can be estimated using the results from the authors' earlier papers (see [2], [3]). Specifically, the cardinality of the set of effectively used grid points grows only linearly with k, under some suitable, relatively mild conditions (see [3] for more details). In the conclusion we would like to point out that the method proposed here solves a nonlinear problem using only approximate solutions to linear auxiliary subproblems of moderate size. The potentially troublesome global optimization over the index set is avoided, since Theorem 1 guarantees that the obtained solutions are *globally* optimal.

4 References

[1] Ašić , M.D. and Kovačević-Vujčić, V.V., A semiinfinite programming method and its application to boundary value problems, ZAMM 66, 1986, 403-405.

[2] Ašić, M.D. and Kovačević-Vujčić, V.V., Computational complexity of some semi-infinite programming methods, in: System Modeling and Optimization (A.Prekopa, B.Strazicky, J.Szelezsan, eds.), Springer-Verlag, Berlin, 1986, 34-42.

[3] Ašić, M.D. and Kovačević-Vujčić, V.V., Linear semiinfinite programming problem: A discretization method with linearly growing number of points, in: Proceedings of 17. Jahrestagung "Mathematische Optimierung" (K.Lommatzsch, ed.), Seminarbericht 85, Humboldt Univerzitat zu Berlin, 1986, 1-10.

[4] Ašić, M.D. and Kovačević-Vujčić, V.V., An interior semi-infinite programming method, JOTA, Vol 59, No 3, 1988, 353-367.

[5] Ašić, M.D. and Kovačević-Vujčić, V.V., An implementation of a semi-infinite programming method to Chebyshev approximation problems, in: Numerical Methods and Approximation Theory III (G.V.Milovanović, ed.), Niš, 1988, 111-119.

[6] Ašić, M.D. and Kovačević-Vujčić, V.V., An implicit enumeration

method for global optimization problems, Computers & Mathematics with Applications, Vol 21, No 6/7, 1991, 191-202.

[7] Davis, P.J., Interpolation and Approximation, Dover Publications, New York, 1975.

[8] Hettich, R. and Zencke, P., Numerische Methoden der Approximation und semi-infiniten Optimierung, Teubner, Stuttgart, 1982.

[9] Newman, D.J., Approximation with Rational Functions, Regional Conference Series in Mathematics, No 41, AMS, Providence, 1979.

[10] Zukhovitskiy, S.I. and Avdeeva, L.I., Linear and Convex Programming (in Russian), Nauka, Moskva, 1967.

Exhibition Organized by Klaus Ritter on the Occasion of the 125th Anniversary of the Technical University of Munich

Ludwig Barnerßoi

Institut für Angewandte Mathematik und Statistik
Technische Universität München

Since November 1981, Professor Ritter has been serving as Chair of Applied Mathematics and Mathematical Statistics at the Technical University of Munich. Recognizing the value of methods of applied mathematics and statistics for investigating and solving many practical problems, he complements his research and teaching by maintaining close contact with industry. The complexity of industrial applications often demand parallel computer architectures. Prof. Ritter has the theoretical knowledge for solving problems with parallelism and he has the needed hardware expertise for his coordination in establishing of transputer workstations in the department.

His industrial cooperation and leadership has resulted in the successful completion of numerous projects. For the 125th anniversary celebration of the Technical University, Prof. Ritter organized a large exhibition that displayed 17 of his current projects.

This overview of the projects highlights the multi-faceted research activities of Professor Ritter.

I. Statistics

1. Transputer Based Analysis and Prediction of Time Series

Authors: Dandl, Hogger, Ritter, Rottenegger, Schäffler, Schöne, Schultz, Sturm

ARMA(p,q) models are generally used for modeling time looping data.

Looking at the ML-estimation of the unknown parameters, a very complex optimization problem arises, which can be solved efficiently by means of parallel automatic differentiation and transputer networks. Numerical simulations at the department of Prof. Ritter verify this assertion.

2. Projection Pursuit

Author: Schlee

Often a large number of special marks are observed in the field of statistics, e.g. a lot of data of a single person is collected for a census. Graphical methods are used to get first impressions of the given data. High dimensional samples are treated by projections. This is especially necessary if the size of the samples is too small in comparison with the dimension. The aim of the "Projection Pursuit" project is to develop an interactive computer program for the graphical representation of high dimensional samples.

3. Cluster Analysis

Authors: Ambrosch, Kögler, Ritter, Schlee

In many applications the assumption can be made that the collected data of a sample consist of groups which contain homogeneous elements but differ mutually. Examples are the classification of animals or plants in terms of external characteristics, client profiles at banks or other firms, illness symptoms in medicine, classification of deposits in geology and the like. The estimation of mixed distributions has a lot of advantages. The authors developed a computer program which uses the so called EM-algorithm to visualize the distribution.

4. Chernoff Faces

Authors: Ritter, Schlee, Zimmermann

Statistical analysis is a large branch of mathematics in which - besides a lot of profound theoretical results - the applicability is very important. Plausibility considerations are indeed a necessary part of a correct model. Therefore visual methods were used for the judgement of samples. It is true that difficulties arise from data which have more than three dimensions. Following a suggestion of Chernoff the collected data could be represented as faces where each component of the vector of measured data is responsible for a special facial expression. At a continuous mark this could be the size of the ears and shape of the nose or the number of hair. Discrete marks can be realized through the sex or the temper. There is no limit to the fantasy. The implementation requires a real artist. This method of data analysis is appreciated world wide and is well known as Chernoff Faces.

II. Hardware

1. TransStation: A Flexible Tansputer-Workstation

Authors: Menes, Ritter, Schöne, Schultz
in cooperation with microwi GmbH Augsburg

The transputer workstation "TransStation" which has been developed at the department of Prof. Ritter consists of a variable number of transputers which can be combined to a network. The communication with the user is achieved by a PC-compatible host computer with SCSI-bus and link-adapters. This is a low cost configuration with a high output.
Applications of TransStations are:

- analysis and forecast of time series
- global optimization
- combinatorial optimization
- unconstrained global optimization using stochastic integral equations on transputer networks
- solution of differential equations
- artificial intelligence
- robotics
- Fast Fourier transform

2. Parallel Image Calculation on Transputer Networks

Authors: Menes, Ritter, Schöne, Schultz
in cooperation with microwi GmbH Augsburg

The TransStation is being successfully used in the field of parallel image processing. The linkage of several transputers enables a high output. Examples of special graphic applications are:

- pattern recognition
- ray-tracing
- discrete cosine transformation
- inverse cosine transformation

III. Optimization

1. DATLAB (DATa LABoratory)

Authors: Riedmüller, Ritter, Schröder

The program DATLAB (DATa LABoratory) is a tool for investigating mathematical models which describe the relationship of given points in a plane. Data are read from files, but interactive input is also possible. DATLAB permits modification of the data interactively in a very flexible manner. DATLAB provides various mathematical models to describe such sets of points by mathematical functions and allows the user to compare the respective results. So, the user is able to choose a suitable model in order to establish a well-fitted mathematical description of his input data.

Model 1: Interpolation Model
This model draws a curve as smooth as possible through all the given points. In addition DATLAB offers the possibility to provide lower and upper bounds for the given points. In this case, the interpolation curve is determined by intervals. There are different attempts to get the interpolating function. DATLAB's Model 1.1 offers polynomial interpolation. Model 1.2 uses cubic B-spline interpolation. In both cases, the mathematical models are stated as quadratic programming problems which are solved by suitable algorithms.

Model 2: Regression Model

This models fits regression functions by the method of least squares. The points may be weighted. DATLAB calculates the regression by polynomials or cubic B-splines.

2. PADMOS: Automatic Differentiation and Nonlinear Optimization on PC's

Authors: Greiner, Kredler, Ritter

PADMOS solves general nonlinear programs. Powerful optimization algorithms combined with a convenient input of the problem functions make it a comfortable tool for research applications where several approaches with various model functions are to be compared. The main features are

- no coding of derivatives; computation of gradients, (Hessians) accurate by **automatic differentiation**, cf. section III.4
- objective function, (nonlinear) constraints supplied by text files
- Pascal-like syntax of input functions; detected errors pointed at by a built-in editor for prompt correction
- all functions may involve data from a $(T \times p)$ - matrix
- specific syntax for additive structured data fitting problems requiring only the code for the model function to be fitted
- graphic display of the fitted function (during iteration) versus sampled data points for nonlinear least squares.

The convenient input using automatic differentiation makes the program also attractive for teaching purposes. There are three versions

- full version **PADMOS** for problems with up to 40 (nonlinear) constraints and 15 variables
- demo version **PADFIT**, confined to data fitting problems
- simplified tutorial **PADTUT** for introductory courses in unconstrained nonlinear optimization with algorithms, like steepest descent, conjugate gradients, Ritter's D-method, BFGS, Newton's method, and the trust region approach of More and Sorenson.

Newton's method is made robust by Gill & Murray's Cholesky factorization of the Hessian, and if possible, by directions of negative curvature.

Algorithms for the constrained case, simplifying for linear constraints

- Wilson's method with quadratic solver of Best-Ritter type
- SQP (Han, Powell, Schittkowski)
- Augmented Lagrange
- Robinson's method with an Augmented Lagrangian phase-1
- Active set and Barrier method (only for linear constraints)

3. Global Optimization by Solving Systems of Stochastic Integral Equations

Authors: Ritter, Schäffler, Sturm

A stochastic method is described for solving unconstrained global optimization problems based on a special class of stochastic integral equations. An efficient algorithm is developed using a semi-implicit Euler method for the numerical solution of the stochastic integral equations. Interesting properties and a parallel implementation of the algorithm enable the treatment of problems with a large number of variables.

4. Automatic Differentiation

Authors: Beck, Fischer, Warsitz

The computational solution of many mathematical problems involves derivatives. Programs for computing derivatives may be

- hand-coded
- generated by symbolic manipulation of formulas
- set up via function calls and divided differences
- obtained using automatic differentiation

In practice, the divided differences approach is still the standard approach. But in many cases, derivatives can be computed more accurately by automatic differentiation.

Advantages

- automatic differentiation is not as tedious as hand-coding
- automatic differentiation is more general than symbolic differentiation
- automatic differentiation provides accurate values rather than approximations
- automatic differentiation, where it can be applied, is cheaper than the other methods

Fields of Application

- nonlinear optimization
- nonlinear equations
- parametric equations
- differential equations
- parametric eigenvalue problems
- error estimates
- sensitivity analysis
- time series analysis
- weather-forecast
- nuclear physics
- robotics

IV. Hydrology

1. Computer Aided Control in Sewerage Systems

Authors: Edenhofer, Spielvogel

The complex connections of a sewerage network are described in a mathematical model. It allows calculations on water depth, flow, and valve movements in a sewerage network. These calculations are done by a parallel computer system (e.g. transputer network) in real time and can be displayed graphically on the monitor.

An optimal control of sewerage networks minimizes the sewage overflow to receiving waters and calculates valve movements, which guarantee the predicted use of the water purification plant and exclude capacity crossings.

This control method is superior to previous models of local valve control, because it considers the global situation of the sewerage system including all reservoirs, and admits a fully automatic service of the network. Due to its complexity, it can not be obtained by experience or tests.

Optimal computer aided control of sewerage systems offer a variety of crucial advantages:

- Long-term high efficiency is possible due to fully automatic service and the easing of personnel.
- Minimizing of sewage overflow to receiving waters provides an optimum of environment protection.
- It is possible to predict the consequences of new residential and commercial zone connections to the sewerage network.
- The failure of parts of the sewerage network equipment can be simulated.
- By simulation of extreme rain showers (worst case scenario) it is possible to localize bottlenecks in the network.
- Assistance to make decisions in the planning of new water purification plants or the expansion of already existing ones (e.g. ideal dimensioning of the reservoir, etc.).

2. Optimal Control of a Series of Power Stations

Authors: Edenhofer, Rempter, Ritter,
M. Ulbrich, S. Ulbrich

The aim is to get optimal flow regulation of several river power stations for high water situations in real time. The new mathematical model considers all power stations simultaneously contrary to previous models. In order to simulate big rivers, a one-dimensional model turns out to be sufficiently exact. The simulation was performed on transputer workstations at the department of Prof. Ritter.

3. Surface Irrigation and Ground Water Mound

Authors: **TU München** Edenhofer, Haucke, Schmitz, Schmöller

Vereinigte Arabische Emirate Amiri, Desert & Marine Environment Research Center

Extensive irrigation, artificial as well as natural ground water recharge (e.g. by Wadi-flow), can affect the position of the underlying ground water table to a great extent. Under these infiltration strips the rise of the water table forms a so-called ground water mound. This possibly leads to a high water table, including accumulation of salt in the root zone and even on the soil surface.

In this context the computation of the transient mound geometry is a highly important step in the planning of both irrigation projects and ground water replenishment activities.

The problem of the transient location of the phreatic surface is treated with a basically new principle. Neither simplifying assumptions nor linearization of phreatic surface conditions have been employed. The approach avoids cumbersome discretization techniques (FD, FEM) and leads, using conformal mapping and the theory of complex-valued functions, to the first analytical solution of the ground water mound problem. The application of the exact solution to a variety of realistic problems shows that the established theory is a promising tool to solve engineering problems in this field.

4. Water Quality Model

Authors: **TU München** Edenhofer, Ritter

ATV-Arbeitsgruppe Eidner, Kirchesch, Müller, Schreiner, Sperling, Warg

Flowing water is used in various ways, e.g. for drinking water, sewage, shipping and hydraulic power. Water can become very polluted and contaminated so that its ecological function is impaired. Therefore, the water quality has to be brought into accordance with the ecological requirements. A tool for this is numerical simulation. The very complex flow process as well as the physical, chemical and biological processes are mapped onto a mathematical water quality model. This model has been applied to several rivers in Bavaria.

5. Hydrodynamic Numerical Models

Authors:	**TU München**	**Bayer. Landesamt für Umweltschutz**
	Edenhofer	Mitterer
	Hauger	Rothmeier
	Heider	Schwaller
	Hirt	
	Ritter	
	Schmitz	
	Seus	

It is possible today to calculate the flow in brooks, rivers or canals by hydrodynamic numerical models (HN-models) due to the capacity of modern computers. Therefore, it is not necessary to construct expensive physical models. Besides, HN-models allow fast variations of physical constraints. The HN-models have a vast range of applications. A team of mathematicians and engineers developed these models over a period of several years at the Technical University of Munich. The HN-models have already been tested and proved as good approximations of reality in several case studies.

V. Technical Plants

1. Production Control and Engine Assembling

Authors:	**TU München**	**BMW AG**
	Edenhofer	Flemmer
	Ritter	Olbort
	Wolfseder	

A program system for production control of engine assembly has been developed in cooperation with BMW. In a very short time this system produces the data input by monitor reading programs, Lotus-

Symphony macro programming, and interaction with the user. The machine's operational plan and its organization for the next few months will be calculated and tabulated within a few seconds. The system does not work centrally, nevertheless, automatically uses an existing central databank, reacts spontaneously and is very flexible in extreme changes of the production parameters. It can be used wherever there is need of manufacturing plans for different products in the same assembly line.

2. Wheel Weight Scales

Authors: **TU München** Edenhofer, Rempter, Ritter, M. Ulbrich, S. Ulbrich

GFTT Gesellschaft für Transputertechnik Pietzsch, Feld, Wienecke

In the goods transport by rail a scaling system would be desirable which determines the weight of the wagons while crossing.
The "wheel weight scales" project examines the most promising attempt to solve the problem which consists in establishing the weight on the wheel by means of the rail bending. When a wheel is running over a measuring stripe for stretching mode built into the track, a theoretically well-known stress progress is produced. This is then sampled several thousand times per second. In order to be able to determine the mass resting on the wheel despite extremely disturbed single measurements a segment of the model function is reconstructed out of all data available for this wheel. The resulting curve has still several degrees of freedom, one of these degrees being the mass resting on the wheel. Fitting the model function by means of nonlinear least squares the unknown mass is determined.
Experiments and their evaluation confirmed this proceeding very well.

Concavity of the Vector–Valued Functions Occurring in Fuzzy Multiobjective Decision–Making

Fred Alois Behringer

Munich University of Technology, Dept. of Mathematics

Abstract: Werners' approach to fuzzy multiobjective decision–making leads to a crisp substitute problem which is a problem of lexicographically extended max–min optimization. Nikodem has shown that a real–valued function on a convex subset of R^n is concave (convex) if and only if it is midpoint concave (midpoint convex) and quasiconcave (quasiconvex). This paper investigates the question of how much of Nikodem's theorem can be transferred to vector–valued functions, such as arise in Werners' model, where concavity or generalized concavity is taken with respect to the increasing rearrangement ordering. Some intermediate results are derived as well.

Keywords: Concavity of vector–valued functions, lexmaxmin optimization, lexmaxmin ordering, generalized concavity, quasiconcavity, multiobjective decision–making, fuzzy decision–making

1. Introduction

Werners has developed a method of treating multiobjective decision–making problems where the coefficients of the objective functions are noncrisp ([30],[31]). Werners uses Pareto optimization as her basic decision criterion. As fuzziness together with multiobjectivity implies the intersection of fuzzy sets, and since one immediate way of intersecting fuzzy sets is by means of taking the minimum of the membership functions involved, Werners proposes to approach the problem by way of lexicographically extended maxmin optimization (lexmaxmin optimization for short), thereby covering maxmin as well as Pareto optimization. (Lexmaxmin optimal points are both maxmin and Pareto optimal [1],[4],[7].) Werners' model is linear (the objective functions are affine and the feasible set is defined by inequalities containing affine functions). The resulting lexmaxmin optimization problem can therefore be reduced to m consecutive maxmin problems, where m is the number of objectives, and these, in turn, can be put into the form of problems of linear optimization (see Werners [30] and the references given there).

Practical experiences have shown that the decision-maker's behaviour towards acceptance within a fuzzy environment tends to lead to S-shaped membership functions (compare Section 3). S-shaped functions are strictly monotone increasing and hence quasiconcave as well as semistrictly quasiconcave (explicitly quasiconcave). (For these and other technical terms already employed in this Introduction see the next section on preliminaries.) Superimposing explicitly quasiconcave functions on linear (affine) functions preserves explicit quasiconcavity [10]. So Werners' model, with its crisp substitute formulation, immediately leads to a lexmaxmin problem with explicitly quasiconcave objectives. Such problems, too, can be reduced to a finite set of maxmin problems under some mild conditions ([4]). Explicit quasiconcavity is a sufficient condition.

Similar to a great number of investigations on cone efficiency in Pareto optimization (e.g. [20]), it seems natural to consider the objective functions arising from Werners' lexmaxmin approach en bloc, concavity and generalized concavity immediately regarded with respect to the (vector-valued) overall objective function, rather than their individual ingredients, and with the increasing rearrangement order relation as the underlying means of ordering its (vector) values.

Hitherto known with this respect are some quasiconcavity results ([10]). Also known are some results on functions which take their values from a connected quasiorder (transitive and connected) ([9]). As the increasing rearrangement order relation is transitive and connected, these results apply to the vector-valued objective functions arising from Werners' approach, too.

The present paper focusses on the (almost immediate) fact that a vector-valued function is concave with respect to the increasing rearrangement ordering if its components are concave with respect to the usual ordering of reals.

For real-valued functions, there is a vast literature on the interconnections between midpoint concavity (Jensen concavity) and concavity (e.g. [26]). Verifying midpoint concavity is generally easier than verifying concavity. Nikodem has drawn attention to the fact that a real-valued midpoint concave function is concave if and only if it is quasiconcave [23]. (Nikodem employs convexity rather than concavity. We adopt the concavity language in this paper since the overall goal in Werners' approach to fuzzy multiobjective decision-making is maximization rather than minimization.) Nikodem has formulated his theorem for open domains only. Kominek has shown that the openness condition can be dropped [18]. Quasiconcavity can be replaced by semistrict quasiconcavity [11].

The question treated in this paper is whether Nikodem's result can be transferred to vector-valued functions with respect to the increasing rearrangement ordering. This will be answered affirmatively – to a certain extent, leaving some remainder still waiting for an answer.

The paper is organized as follows. Notation and some definitions are given in Section 2. Section 3 briefly sketches Werners' model. Section 4 is the main body of this paper and contains some results pertaining to

Nikodem's theorem and generalized concave vector-valued functions. Section 5 presents some semilocal extensions, i.e. results where generalized concavity is defined only for a certain range of α-values in inequalities like $f(x\alpha y) \leqslant f(x)\alpha f(y)$. Section 6 briefly points to the fact that results like "local optimality implies global optimality" can be carried over to the vector-valued case discussed in this paper. Section 7 summarizes the results presented in this paper by listing a set of assertions of the type of "*-concavity combined with **-concavity implies ***-concavity".

2. Preliminaries, definitions, and notation

In this paper, vectors are distinguished by superscripts, vector components, by subscripts.

For $u := (u_1,..,u_m)^T \in R^m$ let $(j(1),..,j(m))$ be a rearrangement of $(1,..,m)$ such that $u_{j(1)} \leqslant ... \leqslant u_{j(m)}$. Put $i[u]_k := u_{j(k)}$ for all $k \in \{1,..,m\}$. $i[u] := (i[u]_1,..,i[u]_m)^T$ will be called the *increasing rearrangement* of u. Now let $(j(1),..,j(m))$ be a rearrangement of $(1,..,m)$ such that $u_{j(1)} \geqslant ... \geqslant u_{j(m)}$. Put $d[u]_k := u_{j(k)}$ for all $k \in \{1,..,m\}$. $d[u] := (d[u]_1,..,d[u]_m)^T$ will be called the *decreasing rearrangement* of u.

We define the (binary) relation i< of "u being smaller than w with respect to the increasing rearrangement" by u i< w :⇔ $i[u]$ l< $i[w]$, where l< is the lexicographic order relation ($\exists k$ such that $i[u]_j = i[w]_j$ for all $j<k$ and $i[u]_k < i[w]_k$).

The relation of "u being smaller than w with respect to the decreasing rearrangement" is defined by u d< w :⇔ $d[u]$ l< $d[w]$, where l< is the lexicographic order relation ($\exists k$ such that $d[u]_j = d[w]_j$ for all $j<k$) and $d[u]_k < d[w]_k$).

i< will be called *lexmaxmin smaller* for reasons which will soon become clear; d< will be called *lexminmax smaller.* It goes without saying that u i> w :⇔ w i< u and u d> w :⇔ w d< u.

Contrary to our practice in previous writings, we shall not content ourselves with simply writing $[u]$ for the (decreasing or increasing) rearrangement of u because we should like to keep in line with a forthcoming paper which will contain results where both d⩽ and i⩽ are occuring at the same time.

Clearly, both i< and d< are irreflexive and transitive, and so are i> and d> ([2],[3]).

We define two equivalence relations, i~ and d~, by u i~ w :⇔ $i[u]_j=i[w]_j$ and u d~ w :⇔ $d[u]_j = d[w]_j$ for all j.

Also, u d⩽ w :⇔ (u d< w) v (u d~ w), and u i⩽ w :⇔ (u i< w) v (u i~ w), and we shall occasionally make use of i⩾ and d⩾ whose meaning is obvious.

We shall be using the terms "decreasing rearrangement ordering", "decreasing rearrangement order relation", or "lexminmax order relation" synonymously, meaning $d\preccurlyeq$. Similarly, "increasing rearrangement ordering", "increasing rearrangement order relation", or "lexmaxmin order relation" will all mean $i\preccurlyeq$. Note that generally $i\preccurlyeq \neq d\succcurlyeq$ and $d\preccurlyeq \neq i\succcurlyeq$.

Clearly, both $i\preccurlyeq$ and $d\preccurlyeq$ are connected quasiorders (transitive and connected – and thus reflexive), and so are $i\succcurlyeq$ and $d\succcurlyeq$ ([2],[3]).

Let L be a real linear space. For $x,y \in L$ and $\alpha\in[0,1]$, we write $x\alpha y$ to denote $(1-\alpha)x+\alpha y$. This notation will also appear in the form of $f(x)\alpha f(y)$, meaning $(1-\alpha)f(x)+\alpha f(y)$, where $f(x),f(y) \in R^m$.

For $x,y \in L$, $[x,y]$ is the closed line segment $\{z\in L \mid z=x\alpha y, \alpha\in[0,1]\}$. An open segment will be denoted by (x,y). $[x,y)$ and $(x,y]$ are obvious.

If $(Y,\preccurlyeq)$ is a connected quasiorder , $\sim$ is the symmetric part of $\preccurlyeq$ ($x\sim y :\Leftrightarrow (x\preccurlyeq y)\wedge(y\preccurlyeq x)$ for all $x,y \in Y$) and $<$ is the asymmetric part of $\preccurlyeq$ ($x<y :\Leftrightarrow (x\preccurlyeq y)\wedge(y\not\preccurlyeq x)$ for all $x,y\in Y$). If $(Y,\preccurlyeq)$ is a total order (i.e. $\preccurlyeq$ is transitive, connected, and antisymmetric) and $u,w \in Y^m$, then $u\preccurlyeq w :\Leftrightarrow u_i\preccurlyeq w_i$ for all i and $u<w :\Leftrightarrow (u\preccurlyeq w)\wedge(u\neq w)$.

$X\subseteq L$ will be called *α-convex* if there is $\alpha\in(0,1)$ such that $x\alpha y \in X$ for any pair of points $x,y\in X$. A $\frac{1}{2}$-convex set will be called *midpoint convex*. X will be called *rationally convex* if $x\lambda y \in X$ for any $x,y \in X$ and all $\lambda \in (0,1)\cap Q$.

Let $X\subseteq L$ be α-convex for some $\alpha \in (0,1)\subseteq R$. $f: X\to R$ will be called *α-concave* (with respect to the same α as in X) if $f(x\alpha y) \geqslant f(x)\alpha f(y)$ for all $x,y\in X$.

A function $f: X\to R$ on a midpoint convex (rationally convex, convex) set $X\subseteq L$ will be called *midpoint concave* (*rationally concave*, *concave*) if f is α-concave with respect to $\alpha=\frac{1}{2}$ (any $\alpha \in (0,1)\cap Q$, any $\alpha \in (0,1)\subset R$).

Notice that α is fixed in advance with α-concavity.

Similar concavity definitions apply to $d[f(\bullet)]$ with respect to $l\preccurlyeq$.

In addition to using concavity concepts, we shall also be using quasiconcavity concepts. Here, contrary to our practice in previous papers, we shall adopt the term "semistrict" rather than "strict": Let Y be a set endowed with a connected quasiorder relation (see above). A function $f: X\to Y$ on a convex subset X of a real linear space is *quasiconcave* if $f(x)\succcurlyeq f(y)$ implies $f(x\alpha y)\succcurlyeq f(y)$ for any $x,y \in X$ and all $\alpha\in(0,1)$. $f: X\to Y$ is *semistrictly quasiconcave* if $f(x)>f(y)$ implies $f(x\alpha y)>f(y)$ for any $x,y \in X$ and all $\alpha\in(0,1)$. There are quasiconcave functions which are not semistrictly quasiconcave, and vice versa. Use of the term "semistrict" in place of "strict" seems to have become common practice since the 1981 Congress on Generalized Concavity in Optimization and Economics [27]. The terms (*semistrictly*) *α-quasiconcave*, (*semistrictly*) *midpoint* *quasiconcave*, and (*semistrictly*) *rationally quasiconcave* are clear from the above definition of the corresponding concavity concepts.

A *fuzzy set* μ (on an arbitrary set X) is a function $\mu : X \to [0,1]$, i.e. a set of ordered pairs $(x,\mu(x))$ where $x \in X$. μ is also called a membership function.

3. Werners' model of fuzzy multiobjective decision-making

Following is a brief sketch of Werners' model of fuzzy multiobjective decision-making. This is but one of Werners' suggestions, the one where the constraints remain crisp and only the objectives suffer fuzzification. For details see Werners' book [30].

Starting point is the following linear "vector maximum problem".

(1) $$Cx \to \max, \; Ax \leqslant b, \; x \geqslant 0 \; ,$$

$x \in R^n$, $C \in R^{k,n}$, $A \in R^{m,n}$, $b \in R^m$. Let $C_i x$ be the ith component of Cx. Maximization is meant with respect to the Pareto ordering: $x \in R^n$ is Pareto better than $y \in R^n$ if $C_i x \geqslant C_i y$ for all i and $C_j x > C_j y$ for at least one j. x is Pareto optimal, i.e. optimal in the sense of problem (1), if there is no $y \in R^n$ Pareto better than x. The ith

individual problem is the problem

(2) $$C_i x \to \max, \; Ax \leqslant b, \; x \geqslant 0 \; ,$$

where max is taken in the usual sense. If $\bar{x}^i \in R^n$ is a solution to this problem, then $\bar{c}_i := C_i \bar{x}^i$ is the ith

optimistic value. We shall henceforth assume all optimistic values to exist. A

pessimistic value $\underline{c}_i$ for the ith objective is defined by $\underline{c}_i := \min_j C_i \bar{x}^j$

There can be many, depending on the particular k-tuple of individual optimal solutions $\bar{x}^j$.from which the minimum is taken. For a short discussion with some suggestions see [10]. Let henceforth $\underline{c}_i$ be any such pessimistic value for the ith objective. By

fuzzification we mean the procedure arising from the following situation. With regard to the ith objective, the decision-maker is fully content (which is signalled by $\mu_i(x) = 1$) with a feasible decision $x \in R^n$ leading to the ith objective's maximal value of $\bar{c}_i$. He completely rejects (which is signalled by $\mu_i(x) = 0$) decisions $x \in R^n$ which lead to an individual objective value less than $\underline{c}_i$. Anything in between is dictated by the individual objective value $C_i x$, normalized such that there results a linear dependence, with a membership function value from 0 to 1. For a complete description consult Werners [30]. We thus arrive at the following

fuzzy individual problem

(3) $\mu_i(x) \to \max\ , Ax \leq b\ , x \geq 0\ ,$

where x, C, A, b as in problem (1), and

(4)
$$\mu_i(x) := \begin{cases} 0 & C_i x < \underline{c}_i \\ (C_i x - \underline{c}_i)/(\bar{c}_i - \underline{c}_i) \text{ for} & \underline{c}_i \leq C_i x < \bar{c}_i \\ 1 & \bar{c}_i \leq C_i x\ . \end{cases}$$

Aggregating the individual problems to the overall (i.e. multiobjective) fuzzy decision problem means taking the intersection of the individual fuzzy sets. If this is done by taking the minimum of the individual membership functions, the lower envelope, as the membership function for the whole problem, we arrive at the following

fuzzy problem

(5) $\min \mu_i(x) \to \max\ , Ax \leq b\ , x \geq 0\ .$

Here, the minimum comes from fuzzification. Maximizing the minimum has nothing to do with Pareto optimization. With multiobjective decision problems, however, Pareto optimization is a widely accepted sine-qua-non condition. There is a close connection between Pareto optimality of the above fuzzy problem (5) and the underlying crisp problem (1) ([30],[10]). One way of adding the Pareto aspect to problem (5) is by considering a lexicographically extended version of it instead of only maximizing the minimum. This is what Werners suggests. Doing so, we arrive at the following

lexmaxmin problem

(6) $\mu(x) \to \text{lexmaxmin},\ Ax \leq b\ , x \geq 0\ .$

Here, "lexmaxmin" means: first maximizing the minimum of the μ_i's over the feasible set; then taking the resulting set of optimal decisions and maximizing the minimum but one of the μ_i's over this set, etc. Any optimal decision in the resulting problem is maxmin optimal as well as Pareto optimal. For details see Werners [30] and the references given there.

Adopting the language used in the remainder of this paper, problem (6) consists of taking the lexicographic maximum of $i[\mu(x)]$, or else searching for a maximal element of $\mu(x)$ with respect to the increasing rearrangement order relation $i\leq$.

Problem (6) can be fully reduced to linear optimization. Moreover, there still remains the possibility of reducing problem (6) to a set of consecutive maxmin problems if the individual functions are quasiconcave and semistrictly quasiconcave. In particular, this is the case if $h(\mu(\bullet)) := (h_1(\mu_1(\bullet)),\ldots,h_k(\mu_k(\bullet)))^T$ is used in problem (6) instead of $\mu(\bullet)$, where the h_i's are strictly monotone increasing. S-shaped functions do have this property. Practical experiences reported by Leberling [19], Zimmermann and Zysno [32], and others (see Werners [30]) show that a decision-maker's behaviour towards fuzzy problems usually is of such an "S-shaped" nature. Werners reports at least 12 sources where an S-shaped behaviour is discussed.

4. A Nikodem theorem for vector-valued functions

It is known that the vector-valued function occurring in Werners' fuzzy decision model [30] (as well as in strict descrete Chebyshev approximation [13],[25], the nucleolus problem in n-person game theory [28], and Dresher's weakness exploiting 2-person game [14]) is quasiconcave as well as semistrictly quasiconcave with respect to the increasing rearrangement order relation $\mathrm{i}\!\leqslant$ if these problems are taken in their linear form, as originally posed by the respective authors [10]. It is further known that componentwise (semistrict) quasiconcavity leads to (semistrict) quasiconcavity with respect to $\mathrm{i}\!\leqslant$ [10]. It is also known that $\mathrm{i}\!\leqslant$ is a connected quasiorder (e.g. [10]) so that any result on generalized concavity with respect to connected-quasiorder-valued functions (e.g. [9] and the references given there) apply to the present setting as well. (Some results on generalized concavity from a purely order theoretic view are due to Calzi [12]). The following proposition adds some concavity (rather than only quasiconcavity) considerations. Its proof is almost immediate.

Proposition 1. If a vector-valued function $f: X \to R^m$ on a convex subset X of a real linear space is componentwise concave, then f is also concave with respect to the increasing rearrangement order relation $\mathrm{i}\!\leqslant$. (*Proof:* If $f_j(x\alpha y) \geqslant f_j(y)\alpha f_j(y)$ for all j, then $\mathrm{i}[f(x\alpha y)]\ \mathrm{l}\!\geqslant\ \mathrm{i}[f(x)\alpha f(y)]$. – "Pareto better" implies "lexmaxmin better".)

This also holds with *α-concavity* (the above inequality satisfied for some $\alpha \in (0,1)$ given in advance), *midpoint concavity* ($\alpha = \frac{1}{2}$), and *rational concavity* ($\alpha \in Q \cap (0,1)$).

Generally, concavity is harder to prove than midpoint concavity (Jensen concavity). For real-valued functions, Nikodem's theorem says that f is concave if and only if it is midpoint concave and quasiconcave [23]. Quasiconcavity can be replaced by semistrict quasiconcavity [11]. In a certain sense, the following theorem is a vector-valued formulation of the if part. Some only-if results for vector-valued functions will follow.

Theorem 1. Let $f: X \to R^m$ be a vector-valued function on a convex subset $X \subseteq L$ of a real linear space L. If f is quasiconcave or semistrictly quasiconcave with respect to the lexmaxmin order relation $\mathrm{i}\!\leqslant$ and componentwise midpoint concave (with respect to the usual oder relation in R), then f is concave with respect to i$\leqslant$.

Proof. Make the following assumptions: X is a convex subset of a real linear space; $x, y \in X$; $x \neq y$; $f : X \to R^m$ is quasiconcave or semistrictly quasiconcave with respect to the increasing rearrangement order relation i$\leqslant$; all components $f_i : X \to R$ of f are midpoint concave.

For any $\alpha \in [0,1]$ put $g(\alpha) := f(x\alpha y)$ and $h(\alpha) := f(x)\alpha f(y)$, with their respective components $g_i : [0,1] \to R$ and $h_i : [0,1] \to R$.

From midpoint concavity as was imposed on the f_is it follows that all f_is are rationally concave (e.g. [26]).

To show that f is concave with respect to i$\leqslant$, choose any $\alpha \in (0,1)$, fixed for the rest of this proof, and suppose that

(1) $\quad g(\alpha)$ i$<$ $h(\alpha)$.

We are going to show that this leads to a contradiction.

From (1) it follows that there is k such that

(2) $\quad g_k(\alpha) < h_k(\alpha)$,

since $g_j(\alpha) \geqslant h_j(\alpha)$ for all components would clearly imply $g(\alpha)$ i$\geqslant$ $h(\alpha)$. For any $\delta \in Q \cap (\alpha, 1)$ we have $g_k(\delta) \geqslant h_k(\delta)$ because f_k is rationally concave (see above). h_k is affine. Hence, there is $\eta \in (\alpha, 1)$ and $c > 0$ such that

(3) $\quad g_k(\delta) > g_k(\alpha) + c$ for all $\delta \in Q \cap (\alpha, \eta)$.

For any appropriate choice of δ to be made in the sequel, consider $\beta \in (0, \alpha)$ and $\gamma \in (0,1)$ such that $(x\beta y)\gamma(x\delta y) = x\alpha y$. A few lines of calculation show that $(x\beta y)\gamma(x\delta y) = x(\beta\gamma\delta)y$. (Recall that $\beta\gamma\delta = (1-\gamma)\beta + \gamma\delta$; see the section on notation.) In other words, let us require that

(4) $\quad \beta\gamma\delta = \alpha$.

α is being kept fixed for the rest of this proof (see above). Clearly, for any choice of $\delta \in Q \cap (\alpha, \eta)$, $\beta \in (0, \alpha)$, and $\gamma \in Q \cap (0,1)$, we have

(5) $\quad g_k(\alpha) \geqslant g_k(\beta)\gamma g_k(\delta)$

because g_k is midpoint concave (by assumption) and hence rationally concave (see above). We show that by simply choosing δ close enough to α and giving $\gamma \in Q \cap (0,1)$ a value sufficiently close to 1 but such that (5) remains satisfied, the value of $g_k(\beta)$ must necessarily be smaller than both $g_j(0)$ and $g_j(1)$ for any j. (Recall that $g_j(0) = f_j(x)$ and $g_j(1) = f_j(y)$.) This, then contradicts the assumption that f be quasiconcave or semistrictly quasiconcave (with respect to i$\leqslant$).

Speaking in terms of a cartesian representation, it is immediately clear that the slope of the function $w: [0,1] \to R$ defined by $w(t) := g_k(\beta) t g_k(\delta)$ can be made arbitrarily large if only δ is chosen close enough to α, and γ, close enough to 1, always under the condition that $\beta\gamma\delta = \alpha$, where $\delta \in Q \cap (\alpha, \eta)$, $\gamma \in Q \cap (0,1)$, and $\beta \in (0,\alpha)$.

Analytically speaking, we have the following. From (5) it follows that

(6) $\quad g_k(\beta) \leqslant (1-\gamma)^{-1}(g_k(\alpha) - \gamma g_k(\delta))$.

This, together with (3), entails

(7) $\quad g_k(\beta) \leqslant g_k(\alpha) - \gamma(1-\gamma)^{-1} c$,

where α is predefined and $\gamma \in Q \cap (0,1)$. To ensure that $\beta \in (0,\alpha)$, note that condition (4) entails

(8) $\quad \gamma < \alpha / \delta \Rightarrow \beta > 0$

and

(9) $\quad \alpha < \delta \Rightarrow \beta < \alpha$

(Notice that $\alpha, \gamma \in (0,1)$, and $\alpha < \delta$ is automatically satisfied since $\delta \in (\alpha, 1)$.)

In view of (7),(8), and (9), we see that $g_k(\beta)$ can be made arbitrarily small (including negative values) if only δ is chosen close enough to α, and γ, close to α / δ. The fact that γ and δ must be rational represents no serious restriction. (α can be given any real value such that $\alpha \in (0,1)$, and β, whose value is determined by (4) and an appropriate choice of δ and γ, is not required to be rational.) The proof is complete. □

Remark 1. Quasiconcavity with respect to $i\leqslant$ is strictly weaker than componentwise quasiconcavity in Theorem 1. To see this, let $f_1(x) := 0$ for all $x \in Q$ and $f_1(x) := 1$ for all $x \in R \backslash Q$; moreover, let $f_2(x) := 1$ for all $x \in Q$ and $f_2(x) := 0$ for all $x \in R \backslash Q$. Clearly, $f(\cdot) := (f_1(\cdot), f_2(\cdot))^T$ is quasiconcave with respect to $i\leqslant$, yet neither f_1 nor f_2 is quasiconcave.

In other words, quasiconcavity with respect to $i\leqslant$ does not imply componentwise quasiconcavity. And what is more, it does not even imply quasiconcavity for at least one component.

Interestingly enough, though f is not concave, it is convex with respect to $i\leqslant$, whereas neither f_1 nor f_2 is convex.

Remark 2. The following counterexample shows that the converse of Proposition 1 does not hold. Concavity of f does not imply concavity of all of its components. Let $f_1(x) := -x^2$ and $f_2(x) := x^2$ for all $x \in R$. f is concave with respect to $i\leqslant$. However, f_2 is not concave. f_2 is not even α-concave for any $\alpha \in (0,1)$.

We have found no example with concavity of f and nonconcavity of all of *f*s components. Does concavity of *f* with respect to i⩽ always imply concavity of at least one component f_i? This must remain an open question.

Remark 3. Semistrict quasiconcavity with respect to i⩽ is strictly weaker than componentwise semistrict quasiconcavity in Theorem 1. To see this, let $f_1(x):=x$ for all $x\in Q$ and $f_1(x):=1+x$ for all $x\in R\backslash Q$, moreover, let $f_2(x):=1+x$ for all $x\in Q$ and $f_2(x):=x$ for all $x\in R\backslash Q$. Clearly, $f(\cdot) := (f_1(\cdot), f_2(\cdot))^T$ is semistrictly quasiconcave with respect to i⩽, yet neither f_1 nor f_2 is semistrictly quasiconcave.

In other words, semistrict quasiconcavity with respect to i⩽ does not imply componentwise semistrict quasiconcavity. And what is more, it does not even imply semistrict quasiconcavity for at least one component.

In view of Remark 2, we cannot expect "if" to be replaceable by "only if" in Theorem 1, contrary to the original (real-valued) formulation of Nikodem's theorem. However, we are now going to show that some sort of reduced only-if statements are still valid. For notational convenience, we first state three lemmata, two of which are almost clear from writing them down.

Lemma 1. For any $u, w \in R^m$ and nonnegative $\alpha\in R$, the following hold:

(1) i[u] l= 0 iff u=0
(2) i[αu] l= αi[u]
(3) i[u+w] l⩾ i[u]+i[w]

Proof. Clearly, l= can be replaced by = in (1) and (2). Relations (1) and (2) are clear. (3) was proved by Wang in his book on game theory [29]. Actually, Wang's proof is for d[•] rather than i[•]. The above is a transscription. An n-point extension will be given in a forthcoming paper. The proof is complete. □

Lemma 2. $\alpha > 0 \Rightarrow$ (u i< w $\Leftrightarrow$ αu i< αw) for any $u, w\in R^m$ and $\alpha\in R$.

Proof. Immediate from the definition of i<. □

Remark 4. There is no such relation for $\alpha < 0$, not even if αu i< αw is replaced by αu i> αw. Counterexamples for this and some more relations of a similar nature will be given in a forthcoming paper.

Lemma 3. u l⩾ a and w l> b $\Rightarrow$ u+w l> a+b for any $u, w, a, b \in R^m$.

Proof. If u l= a and w l> b, then there is k such that $u_i = a_i$ and $w_i = b_i$ for all $i<k$, and $u_k = a_k$ and $w_k > b_k$, yielding $u_i+w_i = a_i+b_i$ for all $i<k$, and $u_k+w_k > a_k+b_k$, which means that u+w l> a+b. If, on the other hand, u l> a and w l> b, then there is k such that $u_i = a_i$ and $w_i = b_i$ for all $i<k$, and one of the following relations holds for k: (i) $u_k > a_k$ and $w_k = b_k$, (ii) $u_k > a_k$

and $w_k > b_k$, (iii) $u_k = a_k$ and $w_k > b_k$. Cases (i) to (iii) are exhaustive. They all lead to $u+w \;\text{l}\!> a+b$. The proof is complete. □

The next theorem is well-known from its real-valued appearance.

Theorem 2. Let $f\colon X \to R^m$ be a vector-valued function on a convex subset $X \subseteq L$ of a real linear space L. If f is concave with respect to the lexmaxmin order relation $\text{i}\!\leqslant$, then f is semistrictly quasiconcave with respect to $\text{i}\!\leqslant$.

Proof. Let $x, y \in X$ such that $f(x) \;\text{i}\!> f(y)$. Assume f to be concave with respect to $\text{i}\!\leqslant$. Choose any $\alpha \in (0,1)$. Then,

$\text{i}[f(x)]$	$\text{l}\!> \text{i}[f(y)]$,	by definition because $f(x) \;\text{i}\!> f(y)$;
$\text{i}[f(x\alpha y)]$	$\text{l}\!\geqslant \text{i}[(f(x)\alpha f(y)]$,	because f is concave;
$\text{i}[f(x)\alpha f(y)]$	$\text{l}\!\geqslant \text{i}[f(x)]\alpha \text{i}[f(y)]$,	by (2) and (3) of Lemma 1;
$\text{i}[f(x)]\alpha \text{i}[f(y)]$	$\text{l}\!> \text{i}[f(x)]\alpha \text{i}[f(y)]$,	by Lemma 2 and Lemma 3 because $\text{i}[f(x)] \;\text{l}\!> \text{i}[f(y)]$;
$\text{i}[f(y)]\alpha \text{i}[f(y)]$	$\text{l}\!= \text{i}[f(y)]$,	which is immediate. Hence,
$\text{i}[f(x\alpha y)]$	$\text{l}\!> \text{i}[f(y)]$,	which means that $f(x\alpha y) \;\text{i}\!> f(y)$, by

definition. The proof is complete. □

Corollary. In view of the above proof it is immediately clear that concave and semistrictly quasiconcave can be replaced by *-concave and semistrictly *-quasiconcave in the above theorem, where * is any of the following concepts: α ($\alpha \in (0,1)$ predefined), rationally (all $\alpha \in Q \cap (0,1)$), midpoint ($\alpha = \frac{1}{2}$). Convexity of X can be relaxed accordingly.

From the facts known to hold in a real-valued setting, it is not surprising that the above theorem has a nonsemistrict counterpart.

Theorem 3. Let $f\colon X \to R^m$ be a vector-valued function on a convex subset $X \subseteq L$ of a real linear space L. If f is concave with respect to the lexmaxmin order relation $\text{i}\!\leqslant$, then f is quasiconcave with respect to $\text{i}\!\leqslant$.

Proof. Assume f to be concave with respect to $\text{i}\!\leqslant$. Let $x, y \in X$ such that $f(x) \;\text{i}\!\geqslant f(y)$. If $f(x) \;\text{i}\!> f(y)$, we are done since then the semistrict arguments of the proof of the preceding theorem take over. To complete the proof, suppose that $f(x) \;\text{i}\!\sim f(y)$. (Note in the sequel that $\text{l}\!=$ is equivalent to $=$, the componentwise equality of vectors.) Choose any $\alpha \in (0,1)$. Then,

$\text{i}[f(x)]$	$\text{l}\!= \text{i}[f(y)]$,	by definition because $f(x) \;\text{i}\!\sim f(y)$;
$\text{i}[f(x\alpha y)]$	$\text{l}\!\geqslant \text{i}[(f(x)\alpha f(y)]$,	because f is concave;
$\text{i}[f(x)\alpha f(y)]$	$\text{l}\!\geqslant \text{i}[f(x)]\alpha \text{i}[f(y)]$,	by (2) and (3) of Lemma 1;
$\text{i}[f(x)]\alpha \text{i}[f(y)]$	$\text{l}\!= \text{i}[f(y)]\alpha \text{i}[f(y)]$,	since $\text{i}[f(x)] \;\text{l}\!= \text{i}[f(y)]$;
$\text{i}[f(y)]\alpha \text{i}[f(y)]$	$\text{l}\!= \text{i}[f(y)]$,	which is immediate. Hence,
$\text{i}[f(x\alpha y)]$	$\text{l}\!\geqslant \text{i}[f(y)]$,	which means that
$f(x\alpha y)$	$\text{i}\!\geqslant f(y)$,	by definition. The proof is complete. □

Corollary. In view of the above proof it is immediately clear that concave and quasiconcave can be replaced by *-concave and *-quasiconcave in the above theorem, where * is any of the following concepts: α ($\alpha\in(0,1)$ predefined), rationally (all $\alpha \in (0,1)\cap Q$) , midpoint ($\alpha=\frac{1}{2}$) . Convexity of X can be relaxed accordingly.

5. Some semilocal extensions

Local optimality does not always imply global optimality in a (one-objective or multiobjective) optimization problem. Many algorithms yield local optima only. But even in cases where local optimality implies global optimality, such as will be briefly discussed in the next section, a sort of local generalized concavity criterion tends to be easier to apply than a global one. This section is on "local" concavity concepts with regard to lexmaxmin optimization (multiobjective optimization with respect to i$\leqslant$).

The concept of "semilocal" concavity (actually, convexity) was introduced by Ewing in 1977 [15] and further developed by Kaul and Kaur in 1982 [17] and later papers. Kaul and Kaur have investigated generalized concavity concepts such as quasiconcavity with respect to an applicability of the idea of semilocalness. Ewing as well as Kaul and Kaur were employing their concepts with respect to real-valued functions. If quasiconcavity rather than concavity is considered, there is no problem in transfering these concepts to functions which take their values from a connected quasiorder ([9]). As the lexmaxmin order relation i$\leqslant$ discussed in the present paper is just one instance of a connected quasiorder, the results from [9] carry over to the present (i.e. multiobjective) setting. The only question remaining is whether concavity itself (rather than quasiconcavity or the like) fits into the scheme. To a certain extent it does, as can be seen from the sequel. Kaul and Kaur have defined their functions on locally starshaped domains. As this will make little sense if also concave rather than only "locally" generalized concave functions enter the discussion, we shall only be employing convex domains in this and most parts of the next section.

Definitions 1. A vector-valued function $f: X\to R^m$ on a convex subset X of a real linear space will be called *semilocally quasiconcave* with respect to the increasing rearrangement order relation i$\leqslant$ (i.e. the lexmaxmin order relation) on X if corresponding to each pair of points $x,y \in X$ such that $x\neq y$, the relation $f(x)$ i$\geqslant$ $f(y)$ implies that there is $z\in[x,y)$ such that $f(u)$ i$\geqslant$ $f(y)$ for all $u\in(z,y)$. f will be called *semilocally semistrictly quasiconcave* on X if corresponding to each pair of points $x,y \in X$ such that $x\neq y$, the relation $f(x)$ i> $f(y)$ implies that there is $z\in[x,y)$ such that $f(u)$ i> $f(y)$ for all $u\in(z,y)$.

Theorem 4. If a midpoint quasiconcave vector-valued function $f: X\to R^m$ on a convex subset X of a real linear space is semilocally semistrictly quasiconcave, then it is quasiconcave. Here, all concavity concepts are taken with respect to the increasing rearrangement order relation $i\leq$ (i.e. the lexmaxmin order relation).

Proof. Since $i\leq$ is a connected quasiorder, this theorem is a special case of Theorem 7 from [11] where the statement was shown to hold for symmetrically α-quasiconcave (actually, symmetrically α-quasiconvex) functions. The following proof is a slight modification adapted to the present case.

Suppose there is $x,y \in X$, $x \neq y$, such that $f(x) \; i\geq \; f(y)$. (The case where $f(x) \; i\leq \; f(y)$ is obvious by interchanging x and y.) Suppose there is $z\in(x,y)$ such that $f(z) \; i< \; f(y)$. We are going to show that this leads to a contradiction.

Since f is semilocally semistrictly quasiconcave, there is $b\in(z,y)$ such that $f(w) \; i> \; f(z)$ for all $w\in(z,b)$. By the same argument, there is $a\in(x,z)$ such that $f(u) \; i> \; f(z)$ for all $u\in(a,z)$. Hence, there are $u,w \in (a,b)$ such that $z = \frac{u+w}{2}$ and $f(u) \; i> \; f(z)$ as well as $f(w) \; i> \; f(z)$, contradicting midpoint quasiconcavity. The proof is complete. □

In view of the corollary to Theorem 3, we have the following

Corollary. If a midpoint concave vector-valued function $f: X \rightarrow R^m$ on a convex subset X of a real linear space is semilocally semistrictly quasiconcave, then it is quasiconcave. Here, all concavity concepts are taken with respect to $i\leq$.

From a counterexample given in [11], it is clear that there cannot exist a semistrict counterpart to Theorem 4, i.e. there is no theorem stating that semistrict midpoint quasiconcavity, together with semilocal quasiconcavity, implies semistrict quasiconcavity. Clearly, a "one-objective" counterexample suffices for the present case, too. Here is an adaptation which replaces the quasiconvexity concepts from [11] by their quasiconcavity counterparts: $f(x):=-1$ for $x\in\{\frac{1}{4},1\}$, $f(x):=0$ otherwise in $[0,1]$. f is semilocally quasiconcave. f is semistrictly midpoint quasiconcave. However, f is not semistrictly quasiconcave: $f(0) > f(1)$ but $f(\frac{1}{4}) \not> f(1)$.

What is surprising is the fact that there does exist a "semistrict" counterpart to Theorem 4 if we are ready to replace midpoint quasiconcavity by midpoint concavity (see next theorem).

Theorem 5. If a midpoint concave vector-valued function $f: X \rightarrow R^m$ on a convex subset X of a real linear space is semilocally quasiconcave, then it is semistrictly quasiconcave. Here, all concavity concepts are taken with respect to $i\leq$.

Proof. Suppose there is $x,y \in X$, $x \neq y$, such that $f(x) \; i> \; f(y)$. (The case where $f(x) \; i< \; f(y)$ is obvious by interchanging x and y.) Suppose there is

$z \in (x,y)$ such that $f(z)$ i≼ $f(y)$. We are going to show that this leads to a contradiction.

Since f is semilocally quasiconcave, there is $b \in (z,y)$ such that $f(w)$ i≽ $f(z)$ for all $w \in (z,b)$. From midpoint concavity, it follows that f is semistrictly midpoint quasiconcave (corollary to Theorem 2). Since $f(x)$ i> $f(y)$, by assumption, and $f(z)$ i≼ $f(y)$ (see above), we have that $f(z)$ i< $f(x)$. Hence, $f(x\tfrac{1}{2}z)$ i> $f(z)$.

Repeating the same line of arguments, first with respect to $(x\tfrac{1}{2}z,z)$, then with respect to $((x\tfrac{1}{2}z)\tfrac{1}{2}z,z)$ etc., we see that $f(x\alpha z)$ i> $f(z)$ for any $\alpha \in (0,1)$ such that $\alpha = 1-2^{-n}$, $n \in N$. Choosing n such that $\alpha = 1-2^{-n}$ is sufficiently close to 1 and putting $u := x\alpha z$, we can clearly find $w \in (z,b)$ such that $z = u\tfrac{1}{2}w$. By what has been said before, $f(z)$ i< $f(u)$ and $f(z)$ i≼ $f(w)$.

We cannot have $f(z)$ i~ $f(w)$ since this would contradict semistrict midpoint quasiconcavity (a consequence of midpoint concavity). Thus, we must have $f(z)$ i< $f(u)$ and $f(z)$ i< $f(w)$. But this contradicts midpoint quasiconcavity (also a consequence of midpoint concavity). The proof is complete. □

There was no place in the proof of the above theorem where we actually needed midpoint concavity. The only thing needed was the property of being both midpoint quasiconcave and semistrictly midpoint quasiconcave. (Employing Martos' term "explicit" for "semistrict together with nonsemistrict" [22], we might call this *explicitly midpoint quasiconcave.*) So we have the following

Corollary. If an explicitly midpoint quasiconcave vector-valued function $f: X \to R^m$ on a convex subset X of a real linear space is semilocally quasiconcave, then it is semistrictly quasiconcave. Here, all concavity concepts are taken with respect to i≼ .

Summarizing, we have the following

Corollary. An explicitly midpoint quasiconcave vector-valued function $f: X \to R^m$ on a convex subset X of a real linear space is explicitly quasiconcave if and only if it is semilocally explicitly quasiconcave. Here, all concavity concepts are taken with respect to i≼ .

Of course, this too can be given a midpoint concave (rather than explicitly midpoint quasiconcave) form:

Corollary. A midpoint concave vector-valued function $f: X \to R^m$ on a convex subset X of a real linear space is explicitly quasiconcave if and only if it is semilocally explicitly quasiconcave. Here, all concavity concepts are taken with respect to i≼ .

6. Local versus global optimality and related remarks

There are three well-known results which immediately enter one's mind if quasiconcavity or semistrict quasiconcavity is mentioned in connection with optimization:

(1) The feasible set defined by a set of inequalities of the form $g(x) \geqslant 0$, where the components $g_i(\cdot)$ are quasiconcave, is convex.

(2) The set of optimal points in a (scalar) maximization problem is convex if the objective function is quasiconcave.

(3) Any locally optimal point in a (scalar) problem of maximization is also globally optimal if the objective function is semistrictly quasiconcave.

Assertions (1) and (2) are an immediate consequence of the well-known fact that any upper level set of a quasiconcave function is convex. Clearly, this remains valid if the function's range is a connected quasiorder. R^m, together with the rearrangement order relation i$\leqslant$, is a connected quasiorder. So (1) and (2) also apply to Werners' approach to fuzzy multiobjective decision-making - as they do apply to strict discrete Chebyshev approximation, Dresher's weakness exploiting game, and Schmeidler's nucleolus problem in n-person game theory.

Assertion (2), though, is not adding much information to strict Chebyshev approximation or the nucleolus problem as both problems have unique optimal points ([13],[28]). The latter is a consequence of the underlying linear structure. As a matter of fact, injectivity of the (vector-valued) objective function suffices to guarantee uniqueness in any lexmaxmin problem ([4]).

In particular, this injectivity-uniqueness argument applies to Dresher's game ([5],[6],[24]) and to Werners' fuzzy multiobjective decision-making problem.

Assertion (3) also remains valid in a problem where the objective function's range is a connected quasiorder ([8]). So it remains valid for Werners' problem as well as the other lexmaxmin problems mentioned above. (It remains valid for any lexmaxmin problem.) Actually, it remains valid for any connected-quasiorder-valued function which is "uniformly semistrictly quasiconcavelike" (for an explanation see below) [8]. Replacing "uniformly semistrictly quasiconcavelike" by "uniformly quasiconcavelike" still preserves Assertion (3) if optimality is taken in a strict sense [8].

Uniform quasiconcavelikeness (actually, uniform quasiconvexlikeness) was introduced by Hartwig [16]. The assertions from [8] just mentioned were made with respect to a normed real linear space. However, we can do without a norm if optimality is taken in a "semilocal" sense, similar to the way semilocal generalized concavity is employed by Kaul and Kaur [17] and in the present paper.

Much of the proofs from [8] carries over. To keep this paper selfcontained, let us present a complete reformulation, immediately with the aim in mind of finding the most general concavity concept preserving the validity of the theorem in question with both its nonstrict as well as its strict form. The formulation chosen below makes it an obvious consequence of the very definition of the underlying generalized concavity concept. The proof for semistrict quasiconcavity (as reported e.g. in Mangasarian's well-known book [21]), or even concavity, follows from the descending chain of interrelations between the generalized concavity concepts involved. First we need a few definitions.

Definitions 2. Let $(Y,\leqslant)$ be a connected quasiorder (i.e. $\leqslant$ is transitive and connected). Let X be any subset of a real linear space. $\bar{x}\in X$ is a *global maximizer* of a function $f: X\to Y$ if there is no $x\in X$ such that $f(x) > f(\bar{x})$. $\bar{x}\in X$ is a *global strict maximizer* of f if there is no $x\in X$ such that $x\neq\bar{x}$ *and* $f(x) \geqslant f(\bar{x})$. $\bar{x}\in X$ is a *semilocal maximizer* of f if, given any point $x\in X$ such that $x\neq\bar{x}$, there is $z\in(\bar{x},x)$ such that $f(u) > f(\bar{x})$ for no $u \in (\bar{x},z)\cap X$. $\bar{x}\in X$ is a *semilocal strict maximizer* of f if, given any point $x\in X$ such that $x\neq\bar{x}$, there is $z\in(\bar{x},x)$ such that $f(u) \geqslant f(\bar{x})$ for no $u\in(\bar{x},z)\cap X$.

Definitions 3. Let L be a real linear space, $X\subseteq L$ convex, $(Y,\leqslant)$ a connected quasiorder. A function $f: X\to Y$ is

quasiconcavelike,

semistrictly quasiconcavelike

on X if corresponding to each pair of points $x,y\in X$ there is an $\alpha\in(0,1)$ such that

$f(x) \geqslant f(y) \Rightarrow f(x\alpha y) \geqslant f(y)$,

$f(x) > f(y) \Rightarrow f(x\alpha y) > f(y)$.

f is

accumulating quasiconcavelike,

accumulating semistrictly quasiconcavelike

on X if corresponding to each pair of points $x,y\in X$ such that $x\neq y$,

$f(x) \geqslant f(y) \Rightarrow$ for any $z\in[x,y)$ there is $u\in(z,y)$ such that $f(u) \geqslant f(y)$,

$f(x) > f(y) \Rightarrow$ for any $z\in[x,y)$ there is $u\in(z,y)$ such that $f(u) > f(y)$.

f is

uniformly quasiconcavelike,

uniformly semistrictly quasiconcavelike,

on X if there is a $\delta\in(0,\frac{1}{2})$ such that corresponding to each pair of points $x,y\in X$ such that $x\neq y$ there is an $\alpha\in[\delta,1-\delta]$ such that

$f(x) \geqslant f(y) \Rightarrow f(x\alpha y) \geqslant f(y)$,

$f(x) > f(y) \Rightarrow f(x\alpha y) > f(y)$.

f is

α-quasiconcave,

semistrictly α-quasiconcave,

on X if there is an $\alpha\in(0,1)$ such that corresponding to each pair of points $x,y\in X$,

$f(x) \geqslant f(y) \Rightarrow f(x\alpha y) \geqslant f(y)$,

$f(x) > f(y) \Rightarrow f(x\alpha y) > f(y)$.

f is

symmetrically α-quasiconcave,

symmetrically semistrictly α-quasiconcave,

on X if, given any $\alpha\in(0,1)$, f is

α-quasiconcave and $(1-\alpha)$-quasiconcave with respect to the same α, semistrictly α-quasiconcave and semistrictly $(1-\alpha)$-quasiconcave for the same α.

f is

midpoint quasiconcave,

semistrictly midpoint quasiconcave,

on X if it is α-quasiconcave (semistrictly α-quasiconcave) on X with $\alpha=\frac{1}{2}$.

The definition of rational quasiconcavity, quasiconcavity, and concavity, with their semistrict counterparts, was given in an earlier section.

Accumulating (semistrict) quasiconcavelikeness could do without the convexity condition imposed on X. Starshapedness suffices: a subset X of a real linear space is *locally starshaped* at $\bar{x}\in X$ if corresponding to $\bar{x}$ and each $x\in X$, there exists a maximum positive number $a(\bar{x},x)\leqslant 1$ such that $\bar{x}\lambda x \in X$ for any $\lambda \in (0,a(\bar{x},x))$. X is *locally starshaped* if it is locally starshaped at each of its points.

Most of the above generalized concavity concepts can be given a "semilocal" form in the sense of Kaul and Kaur [17]. We have had semilocal (semistrict) quasiconcavity earlier in this paper. Clearly, this can be given a connected-quasiorder form in place of the more restrictive i$\leqslant$.

The next theorem employs accumulating (semistrict) quasiconcavity. An application to any of the generalized concavity concepts taken from the previous definitions is obvious in view of the following chain of interdependences which, together with their semistrict counterparts, were discussed in [9]: quasiconcave $\Rightarrow$ rationally quasiconcave $\Rightarrow$ midpoint quasiconcave $\Rightarrow$ symmetrically α-quasiconcave $\Rightarrow$ α-quasiconcave $\Rightarrow$ uniformly quasiconcavelike $\Rightarrow$ accumulating quasiconcavelike $\Rightarrow$ quasiconcavelike. Also in [9], it was shown that none of these implications can be reversed.

Theorem 6. Let X be a locally starshaped subset of a real linear space. Let $(Y,\leqslant)$ be a connected quasiorder. If $f: X\rightarrow Y$ is semistrictly accumulating quasiconcavelike, then $x\in X$ is a global maximizer of f if it is a semilocal maximizer of f. If f is accumulating quasiconcavelike, then $x \in X$ is a global strict maximizer of f if it is a semilocal strict maximizer of f.

Proof. To prove the nonstrict part, suppose there is $x, \bar{x} \in X$, $x \neq \bar{x}$, such that $f(\bar{x}) < f(x)$ and $\bar{x}$ is a semilocal maximizer of f. Then there is $z \in (\bar{x}, x)$ such that $f(u) \lessdot f(\bar{x})$ for all $u \in (\bar{x}, z)$, contradicting accumulating semistrict quasiconcavelikeness.

To prove the strict part, suppose there is $x, \bar{x} \in X$, $x \neq \bar{x}$, such that $f(\bar{x}) \lessdot f(x)$ and $\bar{x}$ is a semilocal strict maximizer of f. Then there is $z \in (\bar{x}, x)$ such that $f(u) < f(\bar{x})$ for all $u \in (\bar{x}, z)$, contradicting accumulating quasiconcavelikeness. The proof is complete. □

Corollary. Let X be a convex subset of a real linear space. Let $f: X \to R^m$ be concave with respect to i$\lessdot$. Then $x \in X$ is a global (strict) maximizer of f if x is a semilocal (strict) maximizer of f.

A set of intermediate formulations is obvious.

Semistrict quasiconcavelikeness will not suffice. The following function is semistrictly quasiconcavelike and $\bar{x} = \frac{1}{4}$ is a semilocal maximizer. However, $\bar{x}$ is not a global maximizer: $f(x) := 0$ for $x \in [0, \frac{1}{2}]$; $f(x) := 1$ for $x \in (\frac{1}{2}, 1]$.

Quasiconcavelikeness will not suffice. The following function is quasiconcavelike and $\bar{x} = \frac{1}{4}$ is a semilocal strict maximizer. However, $\bar{x}$ is not a global strict maximizer. It is not even a global maximizer in the nonstrict sense: $f(x) := \frac{1}{2}$ for $x = \frac{1}{4}$; $f(x) := 1$ for $x \in (\frac{1}{2}, 1]$; $f(x) := 0$ otherwise in $[0, 1]$.

7. Summary

Following are the results proved in this paper. Any concavity or generalized concavity concept not carrying one of the prefixes "componentwise" or "for at least one component" is to be regarded with respect to the increasing rearrangement order relation i$\lessdot$. ("* $\not\Rightarrow$ **" means "* does not generally imply **" .)

Clear without proof

Componentwise concave	$\Rightarrow$	concave
componentwise rationally concave	$\Rightarrow$	rationally concave
componentwise midpoint concave	$\Rightarrow$	midpoint concave
componentwise α-concave	$\Rightarrow$	α-concave

$\alpha > 0 \Rightarrow (u \text{ i}< w \Leftrightarrow \alpha u \text{ i}< \alpha w)$ for any $u, w \in R^m$ and $\alpha \in R$.

Some proof needed (and given in the body of this paper)

concave $\Rightarrow$ semistrictly quasiconcave

rationally concave ⇒ semistrictly rationally quasiconcave
midpoint concave ⇒ semistrictly midpoint quasiconcave
α–concave ⇒ semistrictly α–quasiconcave
concave ⇒ quasiconcave
rationally concave ⇒ rationally quasiconcave
midpoint concave ⇒ midpoint quasiconcave
α–concave ⇒ α–quasiconcave

semistrictly quasiconcave and componentwise midpoint concave ⇒
⇒ concave
quasiconcave and componentwise midpoint concave ⇒
⇒ concave
midpoint quasiconcave and semilocally semistrictly quasiconcave ⇒
⇒ quasiconcave
midpoint concave and semilocally semistrictly quasiconcave ⇒
⇒ quasiconcave
midpoint concave and semilocally quasiconcave ⇒
⇒ semistrictly quasiconcave
explicitly midpoint quasiconcave and semilocally quasiconcave ⇒
⇒ semistrictly quasiconcave
midpoint concave ⇒ explicitly midpoint quasiconcave ⇒
⇒ (explicitly quasiconcave ⇔ semilocally explicitly quasiconcave)

quasiconcave ⇏ quasiconcave for at least one component
semistrictly quasiconcave ⇏
⇏ semistrictly quasiconcave for at least one component
concave ⇏ componentwise concave
convex ⇏ convex for at least one component
semistrictly midpoint quasiconcave and semilocally quasiconcave ⇏
⇏ semistrictly quasiconcave

i[u] l= 0 iff u=0
i[αu] l= αi[u]
i[u+w] l> i[u]+i[w]
u l> a and w l> b ⇒ $u+w$ l> $a+b$ for any $u,w,a,b \in R^m$ and nonnegative $\alpha \in R$.

References

[1] Behringer, F.A., "Konvexe N-Stufen-Max-Min-Optimierung". *Zeitschrift für Operations Research* 14 (1970), 276–296.

[2] Behringer, F.A., "*Optimale Entscheidungen bei Unsicherheit und Spiele unter Ausnutzung von Fehlern des Gegners*", thesis (Habilitationsschrift), University of Technology, Munich, 1975.

[3] Behringer, F.A., "On Optimal Decisions under Complete Ignorance: A New Criterion Stronger than Both Pareto and Maxmin". *European Journal of Operational Research* 1 (1977), 295–306.

[4] Behringer, F.A., "Lexicographic Quasiconcave Multiobjective Programming". *Zeitschrift für Operations Research* 21 (1977), 103–116.

[5] Behringer, F.A., "Quasikonkav-quasikonvexe Spiele unter Ausnutzung gegnerischer Schwächen und lexikographische Optimierung". *Optimization* 8 (1977), 75–88.

[6] Behringer, F.A., "Lexikographische Optimierung und Zweipersonennullsummenspiele unter Ausnutzung von Fehlern des Gegners". *Zeitschrift für Angewandte Mathematik und Mechanik* 57 (1977), 345–346.

[7] Behringer, F.A., "A Simplex Based Algorithm for the Lexicographically Extended Linear Maxmin Problem". *European Journal of Operational Research* 7 (1981), 274–283.

[8] Behringer, F.A., "Uniformly quasiconvexlike functions". *Optimization* 17 (1986), 19–29.

[9] Behringer, F.A., "A Local Version of Karamardian's Theorem on Lower Semicontinuous Strictly Quasiconvex Functions". *European Journal of Operational Research* 43 (1989), 245–262.

[10] Behringer, F.A., "Lexmaxmin in Fuzzy Multiobjective Decision-Making". *Optimization* 21 (1990), 23–49.

[11] Behringer, F.A., "Convexity is equivalent to midpoint convexity combined with strict quasiconvexity". *Optimization* 24 (1992), 219–228.

[12] Calzi, M.L., "Quasiconcave Functions over Chains". *Optimization* 24 (1992), 15–29.

[13] Descloux, J., "Approximations in L^p and Chebyshev Approximations". *J. Soc. Ind. Appl. Math.* 11 (1963), 1017–1026.

[14] Dresher, M., "*Games of Strategy*". Prentice-Hall, Englewood Cliffs, 1961.

[15] Ewing, G.M., "Sufficient conditions for global minima of suitably convex functionals from variational and control theory". *SIAM Review* 19 (1977), 202–220.

[16] Hartwig, H., "Generalized convexities of lower semicontinuous functions". *Optimization* 16 (1985), 663–668.

[17] Kaul, R.N., and Kaur, S.,
"Generalizations of Convex and Related Functions".
European Journal of Operational Research 9 (1982), 369–377.

[18] Kominek, Z., "A characterization of convex functions in linear spaces".
Zeszyty Naukowe Akademii Górniczo-Hutniczej Im. Stanislawa Staszica (1989), Kraków, 71–74.

[19] Leberling, H., "On finding compromise solutions in multicriteria problems using the fuzzy min-operator.
In: *Fuzzy Sets and Systems* 6 (1981), 105–118.

[20] Luc, D.T., "Theory of Vector Optimization". *Lecture Notes in Economics and Mathematical Systems* 319, Springer-Verlag Berlin, 1989.

[21] Mangasarian, O.L, "*Nonlinear Programming*". McGraw-Hill, New York, 1969.

[22] Martos, B., "*Nonlinear Programming, Theory and Methods*".
Amsterdam, 1975.

[23] Nikodem, K., "On some class of midconvex functions".
Ann. Pol. Math. L (1989), 145–151.

[24] Potters, J.A.M., and Tijs, S.H., "The Nucleolus of a Matrix Game and Other Nucleoli". *Mathematics of Operations Research* 17 (1992), 164–174.

[25] Rice, J.R., "Tchebycheff Approximation in a Compact Metric Space".
Bull. Amer. Math. Soc. 68 (1962), 405–410.

[26] Roberts, A.W., and Varberg, D.E., "*Convex Functions*". New York, 1973.

[27] Schaible, S., and Ziemba W.T. (eds.),
"*Generalized Concavity in Optimization and Economics*". New York, 1981.

[28] Schmeidler, D., "The Nucleolus of a Characteristic Function Game",
SIAM J. Appl. Math. 17 (1969), 1163–1170.

[29] Wang, J., "*The Theory of Games*", Oxford, 1988.

[30] Werners, B., "*Interaktive Entscheidungsunterstützung durch ein flexibles mathematisches Programmierungssystem*". München, 1984.

[31] Werners, B., "Interactive Multiple Objective Programming
Subject to Flexible Constraints".
European Journal of Operational Research 31 (1987), 342–349.

[32] Zimmermann, H.-J., und Zysno,P., "Zugehörigkeitsfunktionen: Modellierung, empirische Bestimmung und Verwendung in Entscheidungsmodellen.
Arbeitsbericht 1. Teil, Projekt der Deutschen Forschungsgemeinschaft,
Zi 104/15-1, 1982.

An Algorithm for the Solution of the Parametric Quadratic Programming Problem*

Michael J. Best
Department of Combinatorics and Optimization
University of Waterloo
Waterloo, Ontario N2L 3G1
Canada

Abstract

We present an "active set" algorithm for the solution of the convex (but not necessarily strictly convex) parametric quadratic programming problem. The optimal solution and associated multipliers are obtained as piece-wise linear functions of the parameter. At the end of each interval, the active set is changed by either adding, deleting, or exchanging a constraint. The method terminates when either the optimal solution has been obtained for all values of the parameter, or, a further increase in the parameter results in either the feasible region being null or the objective function being unbounded from below. The method used to solve the linear equations associated with a particular active set is left unspecified. The parametric algorithm can thus be implemented using the linear equation solving method of any active set quadratic programming algorithm.

Key Words Parametric Quadratic Programming Algorithm, Quadratic Programming, Active Set

*This research was supported by the Natural Sciences and Engineering Research Council of Canada under Grant No. A8189. The author gratefully acknowledges the technical assistance of P.M. Creagen.

1 Introduction

We present an "active set" algorithm for the solution of the parametric quadratic programming problem

$$\min\{(c + tq)'x + \frac{1}{2}x'Cx \mid a_i'x \leq b_i + tp_i, \ i = 1, \ldots, m\} \quad (1.1)$$

where c, q, $a_1, \ldots, a_m$ are n-vectors, C is an (n, n) symmetric positive semi-definite matrix and $b_1, \ldots, b_m$, $p_1, \ldots, p_m$ are scalars. A prime is used to denote transposition. The algorithm solves (1.1) for all values of the parameter t satisfying $t \geq t_0$ where t_0 is the given initial value of the parameter. The algorithm determines finitely many critical parameter values $t_1, t_2, \ldots, t_\nu$ satisfying

$$t_0 < t_1 < \cdots < t_{\nu-1} < t_\nu.$$

The optimal solution and associated multipliers are obtained as linear functions of t for all t between consecutive critical parameter values. The algorithm terminates when either $t_\nu = \infty$, t_ν is finite and either (1.1) has no feasible solution for $t > t_\nu$ or (1.1) is unbounded from below for $t > t_\nu$. The relevant reason for termination is also made explicit.

It has been shown in [1] that the "active set" algorithms of Fletcher [10], Gill and Murray [11], Whinston and van de Panne - Dantzig [19], and Best and Ritter [6] for the solution of a non-parametric quadratic programming problem with C positive semi-definite, all construct identical sequences of points. They differ only in the way they solve linear equations with coefficient matrix

$$H_j = \begin{bmatrix} C & A_j' \\ A_j & 0 \end{bmatrix}$$

where A_j' is a matrix of gradients of active constraints at iteration j. The various algorithms typically utilize a factorization of H_j, submatrices of H_j, or partitions of H_j^{-1}. They specify updating formulae for the new factors when A_j is modified by the addition, deletion, or exchange of a row. The algorithm presented here requires solutions of linear equations of this same form. At the end of each parametric interval, the active set changes by either an addition, deletion or exchange. We leave unspecified the method by which the relevant linear equations are to be solved and in this sense the algorithm is quite general. Using the linear equations solving method associated with a particular quadratic programming algorithm provides a natural extension of that method for the solution of the parametric quadratic programming problem (1.1).

Work on the closely related topic of sensitivity analysis is given by Boot [8] for the case of C being positive definite. Ritter [17] has established properties of optimal solutions for more general types of parametric quadratic programming problems. Grigoriadis and Ritter [12] give a solution procedure for a parametric quadratic programming problem having explicit non-negativity constraints and C semi-definite. Wolfe's quadratic programming algorithm [20] can solve the special parametric problem having $c = 0$ and $p_1 = \ldots = p_m = 0$ in (1.1). Murty [16, page 515] gives an extension of the complementary pivot method to parametric quadratic programming problems. Some extensions of parametric linear complementarity problems have been studied by Cottle [9] and Kaneko [13,14]. Ritter [18] extends Bland's anti-cycling rules for a linear programming problem [7] to a parametric programming problem.

In recent years, interest in the area of mean-variance portfolio analysis has grown considerably. Parametric quadratic programming plays an important role in portfolio analysis, as both a theoretical and a computational tool [4,5]. Practical portfolio analysis problems can be quite large. It is quite common to see problems having several thousands of variables. Also, portfolio models differ depending on what features they emphasize; e.g., transaction costs, variable transaction costs, a factor model or sector constraints. An important contribution of this paper is that it allows special purpose algorithms to be readily derived for these types of problems simply by dealing with their linear equation solving requirements in an efficient manner.

A recent result of Best and Ding [3] uses parametric quadratic programming to find all isolated local minimizers (and some non-isolated local minimizers) for the non-convex QP

$$\min\{c'x + \tfrac{1}{2}x'Cx + x'DQ'x \mid Ax \leq b\} \tag{1.2}$$

where D and Q are (n, k) matrices and C is positive semi-definite. The method proceeds by formulating the parametric QP:

$$\min\{c'x + \tfrac{1}{2}x'Cx + t'Q'x \mid Ax \leq b,\ D'x = t\}, \tag{1.3}$$

where t is a parameter vector in R^k. Letting $R(t)$ denote the set of feasible solutions for (1.3), the derived problem for (1.2) is

$$\min\{f(t) \mid t \in R^k\}, \tag{1.4}$$

where

$$f(t) = \begin{cases} \inf\{c'x + \tfrac{1}{2}x'Cx + t'Q'x \mid x \in R(t)\}, & \text{if } R(t) \neq \phi, \\ +\infty, & \text{otherwise.} \end{cases}$$

It is shown in [3] that the isolated local minimizers of (1.2) and (1.4) are in one to one correspondence. In particular, if t^* is an isolated local minimizer for (1.4) then any optimal solution for (1.3) with $t = t^*$ is a local minimizer for (1.2). A case of particular interest is $k = 1$ so that D and Q are vectors. In that case, $f(t)$ becomes a piece-wise quadratic function of a single variable and its local minima are readily obtained. In addition, (1.3) becomes a parametric QP with a scalar parameter both in the linear part of the objective function and in the right hand-side of the constraints; i.e., precisely the form of our model problem (1.1).

2 Solution for the First Interval: Update of the Active Set

Let C be as in (1.1) and let A be any (k, n) matrix. We define $H(A)$ according to

$$H(A) = \begin{bmatrix} C & A' \\ A & 0 \end{bmatrix}$$

The following lemma is used frequently to assess the situation when A is modified by adding, deleting, or exchanging a row. It is proven in [1] and is restated here for convenience.

Lemma 2.1:
(a) Suppose A has full row rank. Then $H(A)$ is non-singular if and

only if $s'Cs > 0$ for all non-zero s such that $As = 0$.
(b) Suppose A_0 has full row rank, $A_0' = (A_1', d)$ and $H(A_0)$ is non-singular. Then $H(A_1)$ is singular if and only if there exists a unique n-vector s_1 such that $A_1 s_1 = 0$, $Cs_1 = 0$ and $d's_1 = -1$. Furthermore, if $H(A_1)$ is singular, then A_2 has full row rank and $H(A_2)$ is non-singular, where $A_2' = (A_1', e)$, e is any n-vector with $e's_1 \neq 0$ and s_1 is the uniquely determined n-vector of the previous statement.

We begin the solution of (1.1) by solving it for $t = t_0$. Let $x_0(t_0)$ be the optimal solution so obtained and let $J_0 \subseteq \{1, 2, \ldots, m\}$ be the set of indices of constraints active at $x_0(t_0)$. We call J_0 the *active set* for $t = t_0$. Let A_0 be a matrix whose rows are the a_i', all $i \in J_0$.

Assumption 2.1
a) A_0 has full row rank,
b) $s'Cs > 0$ for all $s \neq 0$ with $A_0 s = 0$,
c) $v_0(t_0) > 0$.

Assumptions 2.1(a) and (b), together with Lemma 2.1(a), imply H_0 is non-singular. The Karush-Kuhn-Tucker conditions then assert that $x_0(t_0)$ and $v_0(t_0)$ are uniquely determined solution of the linear equations

$$H_0 \begin{bmatrix} x_0(t_0) \\ v_0(t_0) \end{bmatrix} = \begin{bmatrix} -c \\ b_0 \end{bmatrix} + t_0 \begin{bmatrix} -q \\ p_0 \end{bmatrix}.$$

The full; i.e., m-dimensional, vector of multipliers $u_0(t_0)$ is obtained from J_0 and $v_0(t_0)$ by assigning zero to those components of $u_0(t_0)$ associated with constraints inactive at $x_0(t_0)$ and the appropriately indexed component of $v_0(t_0)$, otherwise.

Now suppose t is increased from t_0. Let $x_0(t)$ denote the optimal solution and associated active constraint multipliers as functions of the parameter t. Provided there are no changes in the active set, $x_0(t)$ and $v_0(t)$ are the uniquely determined solution of the linear equations

$$H_0 \begin{bmatrix} x_0(t) \\ v_0(t) \end{bmatrix} = \begin{bmatrix} -c \\ b_0 \end{bmatrix} + t \begin{bmatrix} -q \\ p_0 \end{bmatrix}. \tag{2.1}$$

The solution can conveniently be obtained by solving the two sets of linear equations:

$$H_0 \begin{bmatrix} h_{10} \\ g_{10} \end{bmatrix} = \begin{bmatrix} -c \\ b_0 \end{bmatrix}, \qquad H_0 \begin{bmatrix} h_{20} \\ g_{20} \end{bmatrix} = \begin{bmatrix} -q \\ p_0 \end{bmatrix},$$

which both have coefficient matrix H_0. Having solved these for h_{10}, h_{20}, g_{20}, (the second subscript refers to the first critical interval $[t_0, t_1]$, t_1 being defined below), the solution of (2.1) is

$$x_0(t) = h_{10} + th_{20}, \tag{2.2}$$

and,

$$v_0(t) = g_{10} + tg_{20},$$

The full vector of multipliers $u_0(t)$ may then be obtained from J_0 and $v_0(t)$ as described above. We write $u_0(t)$ as

$$u_0(t) = w_{10} + tw_{20}. \tag{2.3}$$

The optimal solution for (1.1) and the associated multiplier vector are given by (2.2) and (2.3), respectively, for all $t > t_0$ for which $x_0(t)$ is feasible for (1.1), and, $u_0(t)$ is non-negative. The first restriction implies $t \leq \hat{t}_1$, where

$$\begin{aligned} \hat{t}_1 &= \min\left\{ \frac{b_i - a_i'h_{10}}{a_i'h_{20} - p_i} \mid \text{all } i = 1, \ldots, m \text{ with } a_i'h_{20} > p_i \right\}, \\ &= (b_l - a_l'h_{10})/(a_l'h_{20} - p_l). \end{aligned} \tag{2.4}$$

The second restriction implies $t \leq \tilde{t}_1$, where

$$\begin{aligned} \tilde{t}_1 &= \min\left\{ \frac{-(w_{10})_i}{(w_{20})_i} \mid \text{all } i = 1, \ldots, m \text{ with } (w_{20})_i < 0 \right\}, \\ &= -(w_{10})_k/(w_{20})_k. \end{aligned} \tag{2.5}$$

We have used $(w_{10})_i$ to denote the i-th component of w_{10}. We use the convention $\hat{t}_1 = +\infty$ to mean that $a_i'h_{20} \leq p_i$ for $i = 1, \ldots, m$. In this case, l is undefined in (2.4). However, l is not required for further calculations and thus no difficulty occurs. The analogous convention is used when $(w_{20})_i \geq 0$ for $i = 1, \ldots, m$. Setting $t_1 = \min\{\hat{t}_1, \tilde{t}_1\}$, it follows that the optimal solution is given by (2.2) and the multipliers are given by (2.3) for all t satisfying $t_0 \leq t \leq t_1$. Furthermore, Assumption 2.1(c) implies $\tilde{t}_1 > t_0$, and the inclusion of all indices of constraints active at $x_0(t_0)$ in J_0 implies $\hat{t}_1 > t_0$. Therefore, $t_1 > t_0$.

For $t > \hat{t}_1$, $x_0(t)$ violates constraint l and for $t > \tilde{t}_1, (u_0(t))_k$ becomes negative. This suggests that the analysis be continued by setting $J_1 = J_0 + \{l\}$ if $t_1 = \hat{t}_1$ and $J_1 = J_0 - \{k\}$ if $t_1 = \tilde{t}_1$. A_1 would then be obtained from A_0 by either adding the new row a_l' or deleting the existing row a_k'. Similarly, b_1 and p_1 would be obtained from b_0 and p_0 by either adding the components b_l and p_l, respectively, or deleting the components b_k and p_k, respectively. $H_1 = H(A_1)$ would be obtained accordingly. The optimal solution for the next interval would then be $x_1(t) = h_{11} + th_{21}$ with multiplier vector $u_1(t) = w_{11} + tw_{21}$ for all t with $t_1 \le t \le t_2$, where h_{11}, h_{21}, w_{11},and w_{21} are obtained from the solution of the linear equations

$$H_1 \begin{bmatrix} h_{11} \\ g_{11} \end{bmatrix} = \begin{bmatrix} -c \\ b_1 \end{bmatrix}, \qquad H_1 \begin{bmatrix} h_{21} \\ g_{21} \end{bmatrix} = \begin{bmatrix} -q \\ p_1 \end{bmatrix}, \tag{2.6}$$

$t_2 = \min\ \{\hat{t}_2, \tilde{t}_2\}$, and $\hat{t}_2$ and $\tilde{t}_2$ are computed from formulae analogous to those for $\hat{t}_1$ and $\tilde{t}_1$, respectively. We will justify this intuitive approach in two cases (Cases 1.1 and 2.1). In two other cases (Cases 1.2.1 and 2.2.1), we will show that J_1 must be obtained from J_0 by exchanging the indices of two constraints. Nonetheless, all quantities for the next parametric interval are to be found using (2.6) and are completely specified in terms of J_1. In the two remaining cases, we will show that for $t > t_1$ either (1.1) has no feasible solution (Case 1.2.2) or (1.1) is unbounded from below (Case 2.2.2).

We make the assumption that $\hat{t}_1 \neq \tilde{t}_1$ and that no ties occur in the computation of $\hat{t}_1$ and $\tilde{t}_1$; i.e., the indices l and k of (2.4) and (2.5) are uniquely determined.

Case 1: $t_1 = \hat{t}_1$
The analysis continues according to whether or not a_l is linearly dependent on a_i, all $i \in J_0$. The relevant possibility may be determined by solving the linear equations

$$H_0 \begin{bmatrix} s \\ \xi \end{bmatrix} = \begin{bmatrix} a_l \\ 0 \end{bmatrix}. \tag{2.7}$$

If the solution for (2.7) has $s = 0$, then a_l is indeed linearly dependent on the a_i, all $i \in J_0$ and the components of ξ are the coefficients of the linear combination. If the solution has $s \neq 0$, then because H_0 is non-singular, we may conclude that a_l and a_i, all $i \in J_0$, are linearly independent. We note that solving (2.7) involves little additional work since H_0 is presumed to be available in a suitably factored form in order to obtain h_{10} and h_{20}.

Case 1.1: a_l and a_i, all $i \in J_0$, are linearly dependent
We set $J_1 = J_0 + \{l\}$, form A_1, b_1, p_1 and $H_1 = H(A_1)$ accordingly. We set $x_1(t) = h_{11} + th_{21}$ and $u_1(t) = w_{11} + tw_{21}$, where h_{11}, h_{21}, w_{11} and w_{21} are obtained from the solution of (2.6). Observe that $u_0(t_1) = u_1(t_1)$ and consequently $(u_1(t_1))_l = 0$. In order to show that $t_2 > t_1$, we must show that $(u_1(t))_l$ is strictly increasing in t. However, it is straightforward to show from the defining equations for $x_0(t)$ and $x_1(t)$ that for $t > t_1$,

$$(u_1(t))_l = \frac{(x_1(t) - x_0(t))'C(x_1(t) - x_0(t))}{a_l'x_0(t) - (b_l + tp_l)}$$

Now $A_0(x_1(t) - x_0(t)) = 0$, so from Assumption 2.1(b) the numerator of the above expression is strictly positive for $t > t_1$. By definition of $\hat{t}_1$, constraint l is violated for $t > t_1$. Consequently,

$$(u_1(t))_l > 0 \quad \text{for} \quad t > t_1.$$

Case 1.2: a_l is linearly dependent on a_i, all $i \in J_0$
The vector ξ from the solution of (2.7) satisfies

$$a_l = \sum_{i \in J_0} \xi_i a_i. \tag{2.8}$$

There are two sub cases to be considered.

Case 1.2.1: There is at least one $\xi_i > 0, i \in J_0$
The hypothesis of Case 1.2.1 implies that the multipliers for those constraints active at $x_0(t_1)$ are not uniquely determined. The defining equations for $x_0(t_1)$ assert

$$-c - t_1 q - Cx_0(t_1) = \sum_{i \in J_0} (u_0(t_1))_i a_i. \tag{2.9}$$

Let λ be any non-negative scalar. Multiplying (2.8) by λ and subtracting the result from (2.9) gives

$$-c - t_1 q - Cx_0(t_1) = \lambda a_l + \sum_{i \in J_0} ((u_0(t_1))_i - \lambda \xi_i) a_i.$$

Thus λ and the coefficients of the a_i are also valid multipliers for the active constraints for all λ for which these coefficients remain nonnegative. The largest such value of λ is

$$\begin{aligned} \lambda^* &= \min\left\{ \frac{(u_0(t_1))_i}{\xi_i} \mid \text{all } i \in J_0 \text{ with } \xi_i > 0 \right\}, \\ &= (u_0(t_1))_k / \xi_k, \end{aligned} \tag{2.10}$$

for which the multiplier associated with constraint k is reduced to zero.

We set $J_1 = J_0 + \{l\} - \{k\}$ and obtain the quantities giving the optimal solution and multipliers for the next parametric interval from the solution of (2.6).

The multipliers for the constraints in the new active set for $t = t_1$ are

$$(u_1(t_1))_i = (u_0(t_1))_i - \lambda^* \xi_i, \quad \text{all } i \in J_0,$$

$$(u_1(t_1))_l = \lambda^*, \quad \text{and,} \quad (u_1(t_1))_k = 0.$$

Constraint k, which is active at $x_1(t_1)$, has been deleted from the active set. In order to show that $t_2 > t_1$, we must show that constraint k is inactive at $x_1(t)$ for $t > t_1$. We do this as follows. By definition of t_1, constraint l is violated at $x_0(t)$ for $t > t_1$. Because $A_0 x_0(t) = b_0 + t p_0$ for all t, it follows that

$$b_l + t p_l < \sum_{i \in J_0} \xi_i (b_i + t p_i), \quad \text{for } t > t_1. \tag{2.11}$$

By definition of J_1, we have for all t that

$$a_i' x_1(t) = b_i + t p_i, \quad \text{all } i \in J_0, i \neq k,$$

and

$$a_l' x_1(t) = b_l + t p_l.$$

Therefore,

$$b_l + t p_l = \sum_{i \in J_0} \xi_i (b_i + t p_i) + \xi_k (a_k' x_1(t) - b_k - t p_k).$$

From (2.11) and the fact that $\xi_k > 0$, this implies

$$a_k' x_1(t) < b_k + tp_k, \quad \text{for } t > t_1,$$

as required.

Case 1.2.2: $\xi_i \leq 0$ for all $i \in J_0$

In this case, we claim that there is no feasible solution for (1.1) for $t > t_1$. Suppose, to the contrary, that for some $t > t_1$, there is an x satisfying the constraints of (1.1). In particular, x must satisfy

$$a_i' x \leq b_i + tp_i, \quad \text{all } i \in J_0.$$

Multiplying each such inequality by $\xi_i \leq 0$ and adding gives

$$a_l' x \geq \sum_{i \in J_0} \xi_i (b_i + tp_i).$$

But then from (2.11),

$$a_l' x > b_l + tp_l;$$

i.e., constraint l is violated. The assumption that (1.1) possesses a feasible solution for $t > t_1$ leads to a contradiction and is therefore false.

Case 2: $t_1 = \tilde{t}_1$

Consider the effect of deleting constraint k from the active set. Let $\tilde{A}_1$ be obtained from A_0 by deleting the row containing a_k'. The analysis continues according to whether $H(\tilde{A}_1)$ is non-singular or singular. The relevant possibility can be determined by solving the linear equations

$$H_0 \begin{bmatrix} s_0 \\ \xi_0 \end{bmatrix} = \begin{bmatrix} 0 \\ -e_\rho \end{bmatrix}, \tag{2.12}$$

where e_ρ is a unit vector with one in the same row as in a_k' in A_0. It follows from Lemma 2.1(b) that $H(A_1)$ is a non-singular if and only if $\xi_0 \neq 0$. As in Case 1, we note that solving (2.13) requires little additional work since H_0 is presumed to be available in a suitably factored form.

Case 2.1: $\xi_0 \neq 0$

We set $J_1 = J_0 - \{k\}$. Observe that $x_1(t_1) = x_0(t_1)$ and consequently constraint k is active at $x_1(t_1)$. In order to show that $t_2 > t_1$, we must show that constraint k becomes inactive at $x_1(t)$ for $t > t_1$. As in Case 1.1, it is straightforward to show from the defining equations for $x_0(t)$ and $x_1(t)$ that for $t > t_1$,

$$(a_k' x_1(t) - (b_k + tp_k)) = \frac{(x_1(t) - x_0(t))' C (x_1(t) - x_0(t))}{(u_0(t))_k}.$$

Now $A_1(x_1(t) - x_0(t)) = 0$ and H_1 is non-singular so that from Lemma 2.1(a), the numerator of the above expression is strictly positive for $t > t_1$. By definition of $t_1, (u_0(t))_k < 0$ for $t > t_1$. Consequently,

$$a_k' x_1(t) < b_k + tp_k, \text{ for } t > t_1.$$

Case 2.2: $\xi_0 = 0$
By definition, s_0 satisfies

$$Cs_0 = 0, \qquad a_k' s_0 = -1, \qquad \tilde{A}_1 s_0 = 0.$$

Thus all points $x_0(t_1) + \sigma s_0$ are alternate optimal solutions for (1.1) when $t = t_1$ for all σ such that $x_0(t_1) + \sigma s_0$ is feasible. The largest value of σ is

$$\begin{aligned} \sigma_0 &= \min \left\{ \frac{b_i + t_1 p_i - a_i' x_0(t_1)}{a_i' s_0} \mid \text{ all } i \text{ with } a_i' s_0 > 0 \right\} \\ &= (b_l + t_1 p_l - a_l' x_0(t_1)) / a_l' s_0, \end{aligned} \tag{2.13}$$

where we adopt the convention that $\sigma_0 = \infty$ means $a_i' s_0 \leq 0$ for $i = 1, \ldots, m$. There are two sub cases to be considered.

Case 2.2.1: $\sigma_0 < \infty$
We set $J_1 = J_0 + \{l\} - \{k\}$. Observe that $x_1(t_1) = x_0(t_1) + \sigma_0 s_0, u_1(t_1) = u_0(t_1)$ and constraint k, which was active at $x_0(t_1)$, has become inactive at $x_1(t_1)$. Also, constraint l, which was inactive at $x_0(t_1)$, has become active at $x_1(t_1)$. Since $(u_1(t_1))_l = 0$, we must show that $(u_1(t))_l$ is a strictly increasing function of t in order to establish that $t_2 > t_1$. We do this by first observing that from the defining equations for $x_0(t)$ and the definition of s_0,

$$(u_0(t))_k = (c + tq)' s_0.$$

By definition of t_1 we have

$$(c + tq)'s_0 < 0, \quad \text{for} \quad t > t_1.$$

The defining equations for $x_1(t)$ and the definition of s_0 imply

$$(u_1(t))_l = -\frac{(c + tq)'s_0}{a_l's_0},$$

and thus by definition of σ_0 we have that $(u_1(t))_l$ is a positive, strictly increasing function of t for $t > t_1$.

Case 2.2.2: $\sigma_0 = \infty$
In this case, $x_0(t_1) + \sigma s_0$ is feasible for all $\sigma \geq 0$. As in Case 2.2.1, we have

$$(u_0(t))_k = (c + tq)'s_0,$$

for $t > t_1$. Let $F_t(x)$ denote the objective function for (1.1) evaluated at x for specified t. Because $Cs_0 = 0$, Taylor's series gives

$$\begin{aligned} F_t(x_0(t) + \sigma s_0) &= F_t(x_0(t)) + \sigma(c + tq)'s_0 \\ &\to -\infty, \qquad \text{as } \sigma \to \infty, \end{aligned}$$

demonstrating that (1.1) is unbounded from below for $t > t_1$.

We note that for each of the four cases (1.1,1.2.1, 2.1,2.2.1), it follows from Lemma 2.1 that $H_1 = H(A_1)$ is non-singular.

3 Detailed Formulation of the Algorithm: Examples

Based on the analysis of the previous section, we now formulate the algorithm for the solution of (1.1).

Algorithm PQP

Initialization:
Solve (1.1) for $t = t_0$. Let J_0 denote the set of indices of active constraints and let $H_0 = H(A_0)$. Set $j = 0$.

Step 1: Computation of Optimal Solution, Multipliers and End of the Interval
Solve the linear equations

$$H_j \begin{bmatrix} h_{1j} \\ g_{1j} \end{bmatrix} = \begin{bmatrix} -c \\ b_j \end{bmatrix}, \qquad H_j \begin{bmatrix} h_{2j} \\ g_{2j} \end{bmatrix} = \begin{bmatrix} -q \\ p_j \end{bmatrix},$$

for $h_{1j}, h_{2j}, g_{1j}, g_{2j}$. Form w_{1j}, and w_{2j} from g_{1j} and g_{2j}, respectively. Compute $\hat{t}_{j+1}, l, \tilde{t}_{j+1}, k$ and t_{j+1} according to

$$\begin{aligned} \hat{t}_{j+1} &= \min\left\{ \frac{b_i - a_i' h_{1j}}{a_i' h_{2j} - p_i} \mid \text{ all } i \text{ such that } a_i' h_{2j} > p_i \right\}, \\ &= \frac{b_l - a_l' h_{1j}}{a_l' h_{2j} - p_l}, \\ \tilde{t}_{j+1} &= \min\left\{ -\frac{(w_{1j})_i}{(w_{2j})_i} \mid \text{all } i \text{ such that } (w_{2j})_i < 0 \right\}, \\ &= -\frac{(w_{1j})_k}{(w_{2j})_k}, \\ t_{j+1} &= \min\{\hat{t}_{j+1}, \tilde{t}_{j+1}\}. \end{aligned}$$

Print " the optimal solution is $x_j(t) = h_{1j} + t h_{2j}$ with multiplier vector $u_j(t) = w_{1j} + t w_{2j}$ for all t with $t_j \le t \le t_{j+1}$ ". Go to Step 2.

Step 2: Computation of J_{j+1}
If $t_{j+1} = \infty$, then stop. If $t_{j+1} = \hat{t}_{j+1}$ then go to Step 2.1 and otherwise go to Step 2.2.

Step 2.1: New Active Constraint
Solve the linear equations

$$H_j \begin{bmatrix} s_j \\ \xi_j \end{bmatrix} = \begin{bmatrix} a_l \\ 0 \end{bmatrix},$$

for s_j and ξ_j. If $s_j \neq 0$, then go to Step 2.1.1 and if $s_j = o$, then go to Step 2.1.2.

Step 2.1.1
Set $J_{j+1} = J_j + \{l\}$, form A_{j+1} and H_{j+1}, replace j with $j+1$ and go to Step 1.

Step 2.1.2
If $\xi_j \le 0$, print "no feasible solution for $t > t_{j+1}$ " and stop. Otherwise, compute ρ such that

$$\frac{(g_{1j} + t_{j+1} g_{2j})_\rho}{(\xi_j)_\rho} = \min \left\{ \frac{(g_{1j} + t_{j+1} g_{2j})_i}{(\xi_j)_i} \;\middle|\; \begin{array}{l} \text{all } i \text{ such that} \\ (\xi_j)_i > 0 \end{array} \right\},$$

and let k be the index of the constraint associated with row ρ of A_j. Set $J_{j+1} = J_j + \{l\} - \{k\}$, form A_{j+1} and H_{j+1}, replace j with $j+1$ and go to Step 1.

Step 2.2: Deletion of a Constraint
Solve the linear equations

$$H_j \begin{bmatrix} s_j \\ \xi_j \end{bmatrix} = \begin{bmatrix} 0 \\ -e_\rho \end{bmatrix}$$

for s_j and ξ_j, where e_ρ is a vector of zeros except for component ρ which has value unity, and ρ is the index of the row of A_j associated with constraint k. If $\xi_j \neq 0$ then go to Step 2.2.1 and if $\xi_j = 0$ then go to Step 2.2.2.

Step 2.2.1
Set $J_{j+1} = J_j - \{k\}$, form A_{j+1} and H_{j+1}, replace j with $j+1$ and go to Step 1.

Step 2.2.2
If $a_i' s_j \le 0$ for $i = 1, \ldots, m$ print "problem is unbounded from below for $t > t_{j+1}$" and stop. Otherwise compute σ_j and l according to

$$\begin{aligned} \sigma_j &= \min \left\{ \frac{b_i + t_{j+1} p_i - a_i' x_j(t_{j+1})}{a_i' s_j} \mid \text{all } i \text{ such that } a_i' s_j > 0 \right\}, \\ &= \frac{b_l + t_{j+1} p_l - a_l' x_j(t_{j+1})}{a_l' s_j}. \end{aligned}$$

Set $J_{j+1} = J_j + \{l\} - \{k\}$, form A_{j+1} and H_{j+1}, replace j with $j+1$ and go to Step 1.

The steps of Algorithm PQP are illustrated in the following two examples.

Example 3.1

$$\text{minimize}: \quad -x_2 + t(x_1 + x_2) + 2x_1^2 + 2x_1x_2 + \frac{1}{2}x_2^2$$

$$\begin{array}{rrrcr} \text{subject to}: & -x_1 & & \leq & 0, \\ & -2x_1 & -x_2 & \leq & -2, \\ & & x_2 & \leq & 3. \end{array}$$

We take $t_0 = 0$. For $t = t_0$, the first two constraints are active. With $J_0 = \{1, 2\}$, Step 1 gives

$$x_0(t) = \begin{bmatrix} 0 \\ 2 \end{bmatrix}, \qquad u_0(t) = \begin{bmatrix} 2 \\ 1 \\ 0 \end{bmatrix} + t \begin{bmatrix} -1 \\ 1 \\ 0 \end{bmatrix},$$

$\hat{t} = \infty, \tilde{t} = 2, k = 1$ and $t_1 = 2$. Thus $x_0(t)$ is optimal with multiplier vector $u_0(t)$ for all t with $0 \leq t \leq 2$.

In Step 2, we solve

$$H_0 \begin{bmatrix} s_0 \\ \xi_0 \end{bmatrix} = \begin{bmatrix} 0 \\ 0 \\ -1 \\ 0 \end{bmatrix}$$

to obtain $s_0 = (1, -2)'$ and $\xi_0 = 0$. Control then transfers to Step 2.2.2. Since $\sigma_0 = \infty$, Algorithm PQP terminates with the information that the problem is unbounded from below for $t > 2$.

The geometry of this example is illustrated in Figure 3.1 where the feasible region is shaded. Letting $F_t(x)$ denote the objective function, for $0 \leq t < 2$ the level sets of $F_t(x)$ are parabolas, symmetric about the line $2x_1 + x_2 = \frac{1}{5} - \frac{3}{5}t$. The parabolas are open, up and to the left. They become "thinner" as t approaches 2. For $t = 2$, the level sets become pairs of parallel lines, symmetric about the line $2x_1 + x_2 = -1$ and all points

$$\begin{bmatrix} 0 \\ 2 \end{bmatrix} + \sigma \begin{bmatrix} 1 \\ -2 \end{bmatrix}, \qquad \sigma \geq 0,$$

are alternate optima.

For $t > 2$, the level sets of $F_t(x)$ are again parabolas symmetric about $2x_1 + x_2 = \frac{1}{5} - \frac{3}{5}t$, but now the parabolas open down and to the right and the problem becomes unbounded from below.

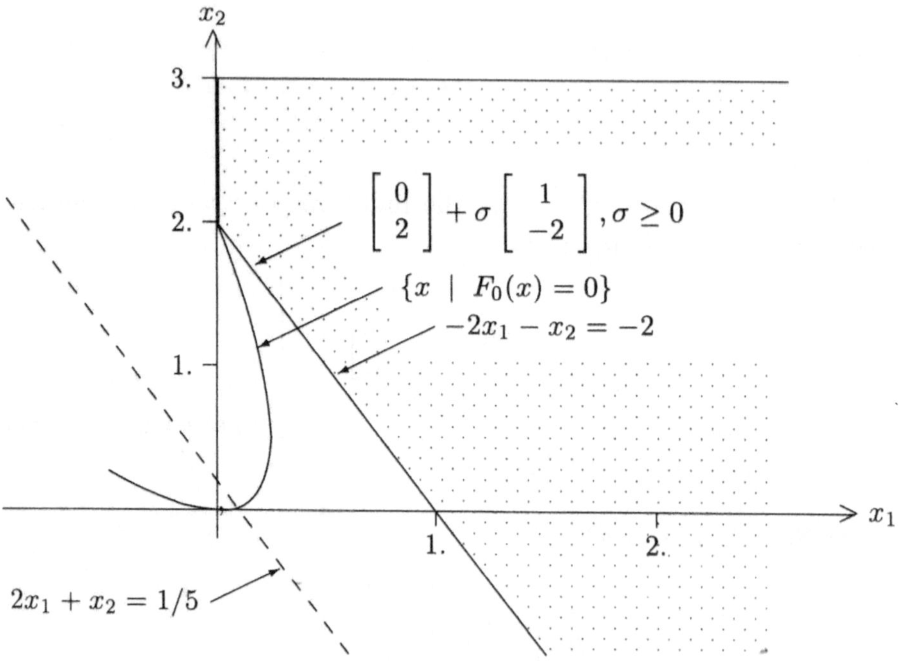

Figure 3.1: Geometry of Example 3.1

Example 3.2
We obtain a second example by replacing the third constraint of Example 3.1 with

$$4x_1 + x_2 \leq 6 - t.$$

Applying Step 1 with $J_0 = \{1, 2\}$ gives

$$x_0(t) = \begin{bmatrix} 0 \\ 2 \end{bmatrix}, \qquad u_0(t) = \begin{bmatrix} 2 \\ 1 \\ 0 \end{bmatrix} + t \begin{bmatrix} -1 \\ 1 \\ 0 \end{bmatrix},$$

$\hat{t} = 4, l = 3, \tilde{t}_1 = 2, k = 1$ and $t_1 = 2$. Thus $x_0(t)$ is optimal with multiplier vector $u_0(t)$ for all t with $0 \leq t \leq 2$.

As in Example 3.1, we obtain $s_0 = (1,2)'$ and $\xi_0 = 0$ in Step 2.2. In Step 2.2.2, we now find $\sigma_0 = 1, l = 3$ and continue by setting $J_1 = \{2,3\}$. We return to Step 1 with $j = 1$ and compute

$$x_1(t) = \begin{bmatrix} 2 \\ -2 \end{bmatrix} + \begin{bmatrix} -1/2 \\ 1 \end{bmatrix}, \qquad u_1(t) = \begin{bmatrix} 0 \\ 0 \\ -1 \end{bmatrix} + t \begin{bmatrix} 0 \\ 3/2 \\ 1/2 \end{bmatrix},$$

$\hat{t}_2 = 4, l = 1, \tilde{t}_2 = \infty$ and $t_2 = 4$. Thus $x_1(t)$ is optimal with multiplier vector $u_1(t)$ for all t with $2 \le t \le 4$, Control now transfers to Step 2.1 to solve the linear equations

$$H_1 \begin{bmatrix} s_1 \\ \xi_1 \end{bmatrix} = \begin{bmatrix} -1 \\ 0 \\ 0 \\ 0 \end{bmatrix}.$$

These have solution $s_1 = 0$ and $\xi_1 = (-1/2 - 1/2)'$. Since $s_1 = 0$, control transfers to Step 2.1.2. Since $\xi_1 \le 0$, Algorithm PQP terminates with the message that the problem has no feasible solution for $t > 4$.

The geometry of Example 3.2 differs from that of Example 3.1 as follows. For $0 \le t \le 2$, the optimal solutions coincide. At $t = 2$, movement to an alternate optima in Example 3.2 results in constraint 3 becoming active. The optimal solution then lies at the intersection of constraints 2 and 3 as t is increased to 4. At $t = 4$, all three constraints are active and the feasible region consists of the single point $(0,2)'$. Increasing t beyond 4 moves the line $4x_1 + x_2 = 6 - t$ up and to the left, and the feasible region is null.

4 Properties of the Algorithm

The properties of Algorithm PQP are established in

Theorem 4.1:
Let Assumption 2.1 be satisfied and let $x_j(t), u_j(t), t_j, j = 1, \ldots,$ be obtained by applying Algorithm PQP to (1.1). Assume that for each j, the indices k, l, and ρ are uniquely determined and that $\hat{t}_j \neq \tilde{t}_j$.

Then Algorithm PQP terminates after $v < \infty$ iterations, and for $j = 1, \ldots, v-1, x_j(t)$ is optimal for (1.1) with multiplier vector $u_j(t)$ for all t with $t_j \leq t \leq t_{j+1}$ and either $t_v = \infty, t_v < \infty$ and (1.1) has no feasible solution for $t > t_v$, or, $t_v < \infty$ and (1.1) is unbounded from below for $t > t_v$.

Proof:

Other than termination after finitely many steps, the theorem follows directly from Cases 1.2, 1.2.1, 1.2.2, 2.2 2.2.1, and 2.2.2 discussed in Section 2. Finite termination is established by showing that no active set is ever repeated, as follows. Suppose, to the contrary, that there are iterations γ, δ with $\delta > \gamma$ and $J_\gamma = J_\delta$. By construction, both $x_\gamma(t)$ and $x_\delta(t)$ are optimal solutions for

$$\min \left\{ (c + tq)'x + \frac{1}{2}x'Cx \mid a_i'x \leq b_i = tp_i, \text{ all } i \in J_\gamma \right\},$$

so that necessarily $t_\gamma = t_\delta$ and $t_{\gamma+1} = t_{\delta+1}$. But by construction, each $t_{j+1} > t_j$. Thus $\gamma = \delta$ and this contradiction establishes that indeed each active set is uniquely determined. Termination after finitely many steps now follows from the fact that there are only finitely many distinct subsets of $\{1, \ldots, m\}$.

Dedication

This paper is dedicated to Klaus Ritter, friend, colleague and mentor, on the occasion of his sixtieth birthday.

References

[1] M.J. Best, "Equivalence of some quadratic programming algorithms", *Mathematical Programming*, 30 (1984) 71-87.

[2] M.J. Best and N. Chakravarti, "An $O(n^2)$ active set method for solving a certain parametric quadratic program", *Journal of Optimization Theory and Applications*, Vol. 72, No. 2 (1992) 213-224.

[3] M. J. Best and B. Ding, "Global and local quadratic minimization", Research Report CORR 95-25, Department of Combinatorics and Optimization, University of Waterloo, 1995.

[4] M.J. Best and R.R. Grauer, "Sensitivity analysis for mean-variance portfolio problems", *Management Science*, Vol. 30, No. 8 (1991) 980-989.

[5] M.J. Best and R.R. Grauer, "The analytics of sensitivity analysis for mean-variance portfolio problems", *International Review of Financial Analysis*, Vol. 1, No. 1 (1992) 17-37.

[6] M.J. Best and K. Ritter, "An effective algorithm for quadratic minimization problems", *Zeitschrift fuer Operations Research*, Vol. 32, No. 5 (1988) 271-297.

[7] G.R. Bland, "New finite pivoting rules for the simplex method", *Mathematics of Operations Research*, 2 (1977) 103-107.

[8] J.C.G. Boot, "On sensitivity analysis in convex quadratic programming problems", *Operations Research*, 11 (1963) 771-786.

[9] R.W. Cottle, "Monotone solutions of the parametric linear complementarity problem", *Mathematical Programming*, 3 (1972) 210-224.

[10] R. Fletcher, "A general quadratic programming algorithm", *J. Inst. Maths Applics,*, 7 (1971) 76-91.

[11] P.E. Gill and W. Murray, "Numerically stable methods for quadratic programming", *Mathematical Programming*, 14(1978) 349-372.

[12] M.D. Grigoriadis and K. Ritter, "A parametric method for semi definite quadratic programs", *SIAM J. Control*, 7(No. 4) (1969) 559-577.

[13] I. Kaneko, "Isotone solutions of parametric linear complementarity problems", *Mathematical Programming*, 12(1977) 48-59.

[14] I. Kaneko, "A parametric linear complementarity problem involving derivatives", *Mathematical Programming*, 15(1978) 146-154.

[15] O.L. Mangasarian, *Nonlinear Programming*, McGraw-Hill, Inc., 1969.

[16] K. Murty, *Linear and Combinatorial Programming*, J. Wiley and Sons, Inc., 1976.

[17] K. Ritter, "A method for solving nonlinear maximum-problems depending on parameters", (translated by M. Meyer), *Naval Research Logistics Quarterly*, 14(1967) 147-162.

[18] K. Ritter, "On parametric linear and quadratic programming problems", MRC Technical Summary Report No. 2197, University of Wisconsin, 1981.

[19] C. van de Panne and A. Whinston, "The symmetric formulation of the simplex method for quadratic programming", *Econometrica*, 37(1969) 507-527.

[20] P. Wolfe, "The simplex method for quadratic programming", *Econometrica*, 27(1959) 382-298.

Optimal and Asymptotically Optimal Equi-partition of Rectangular Domains via Stripe Decomposition*

Ioannis T. Christou[†] Robert R. Meyer[†]

Abstract

We present an efficient method for assigning any number of processors to tasks associated with the cells of a rectangular uniform grid. Load balancing equi-partition constraints are observed while approximately minimizing the total perimeter of the partition, which corresponds to the amount of interprocessor communication. This method is based upon decomposition of the grid into stripes of "optimal" height. We prove that under some mild assumptions, as the problem size grows large in all parameters, the error bound associated with this feasible solution approaches zero. We also present computational results from a high level parallel Genetic Algorithm that utilizes this method, and make comparisons with other methods. On a network of workstations, our algorithm solves within minutes instances of the problem that would require one billion binary variables in a Quadratic Assignment formulation.

1 Introduction

1.1 Problem Formulation

The Minimum Perimeter Equi-partition problem (MPE) is a geometric problem with applications in scientific computing ([DTR91]), engineering and image processing ([Sch89]). In its most general form it can be stated as follows: *Given a Grid $\mathcal{G}$ of unit cells and a number of processors P, find an assignment of the grid cells to the processors so that the perimeter of the partition is minimized while the loads of the processors are as balanced as possible.* The perimeter of a partition is the sum of the lengths of the boundaries of

---

*This research was partially supported by the Air Force Office of Scientific Research under grant F49620-94-1-0036, and by the NSF under grants CDA-9024618 and CCR-9306807.

[†]Center for Parallel Optimization, Computer Sciences Department, University of Wisconsin, Madison, Wisconsin 53706.

the regions that each processor occupies , while the load of each processor is the area of the region it occupies. The problem is a special case of the (NP-complete) Graph Partitioning problem, and as such, it can be formulated as a Quadratic Assignment problem ([PRW93]), with $|\mathcal{G}|\mathcal{P}$ binary variables and $|\mathcal{G}| + \mathcal{P}$ constraints. Letting $\mathcal{I}$ denote the set of pairs of adjacent cells, and a_i the area for processor i, the QAP formulation is as follows:

$$\min . \sum_{i,j \in \mathcal{G}} \sum_{\substack{p,p'=1 \\ p \neq p'}}^{P} c_{ij} x_i^p x_j^{p'}$$

$$s.t. \begin{cases} \sum_{i \in \mathcal{G}} x_i^p = a_p & p = 1 \dots P \\ \sum_{p=1}^{P} x_i^p = 1 & i \in \mathcal{G} \\ x_i^p \in \mathbf{B} = \{\mathbf{0}, \mathbf{1}\} \end{cases}$$

$$\text{where } c_{ij} = \begin{cases} 1 & \text{if } (i,j) \in \mathcal{I} \\ 0 & \text{else} \end{cases}$$

The objective of this optimization problem is a sum of quadratic terms of binary variables, while the constraints are network constraints. An illustration of the network assignment nature of the problem is in figure 1 where each processor represents a supply node of supply a_i and each grid cell is a demand node of demand 1. The goal is to find a feasible assignment that minimizes the total perimeter.

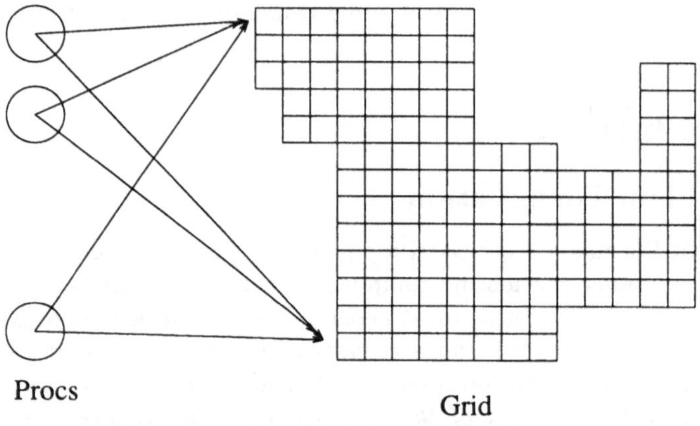

Figure 1: Network Assignment Formulation of MPE

Such problems arise naturally in the numerical solution of Partial Differential Equations (PDEs) over a given domain using finite difference schemes in parallel or distributed computing environments; in such cases, the domain is discretized thus giving rise to a grid, which then must be decomposed among the number of available processors, subject to the constraint that the load (areas) (a_i) of the processors are as balanced as possible. In the 5-point grid scheme, each cell updates its value using the values of its North, East, West and South neighboring cells ([DTR91]). Interprocessor communication occurs when a cell needs the value of another cell that does not belong to the same processor. This means that communication will occur exactly at the boundaries of the regions between the processors, and therefore, since the original boundary is a constant, minimizing the total perimeter of the partition corresponds to minimizing the overall interprocessor communication. The Minimum Perimeter problem arises also in the context of image processing ([Sch89]) and low-level computer vision, where for edge detection an image (a rectangular grid) has to be split among a number of processors in a parallel machine subject to the same load balancing constraints. Many edge detection algorithms use the 5-point grid computation scheme, so the objective of minimizing communication in the parallel machine again reduces to minimizing the perimeter of the partition. As the current trend in parallel computing is towards networks of workstations where the latency of communication between processors can be high, it is important that good solutions to the Minimum Perimeter problem can be found. In the case of networks of workstations in particular, the number of workstations connected together can be any number. Our method differs from most of the popular graph partitioning methods in that it can partition a domain into any number of processors while most methods require the number of processors to be a power of two.

Under the assumption that each processor has the same computing power, the load balancing constraints become ***Equi-partitioning*** constraints: the area a_i of each processor must differ by no more than 1 from the area of any other processor. In the rest of this paper we shall consider the Minimum Perimeter Equi-partitioning problem. We shall further restrict ourselves to the case where the grid $\mathcal{G}$ is a uniform rectangular grid of M rows and N columns. We shall refer to the problem of minimizing the perimeter of a partition of this grid into P processors as MPE(M, N, P).

The area of each processor then, if P divides exactly MN is the same for all processors and is simply $a_i = \frac{MN}{P}$. If, however, P does not divide MN, then we assume that $P \leq MN$, and the first $P_1 > 0$ processors will be assigned a load of A_1 and the remaining P_2 processors will get a load $A_2 = A_1 + 1$. Thus we have,

$$P_1 + P_2 = P$$
$$A_2 = A_1 + 1$$
$$A_1 P_1 + A_2 P_2 = MN$$

from which we get

$$P_1 A_1 + (P - P_1)(A_1 + 1) = MN \implies PA_1 + P_2 = MN.$$

Therefore,

$$\text{for } i = 1 \ldots P_1 \;\; a_i = A_1 = MN \div P = \left\lfloor \frac{MN}{P} \right\rfloor$$

$$P_1 = P - MN \mod P$$

$$\text{and for } i = P_1 + 1 \ldots P \;\; a_i = A_2 = MN \div P + 1 = \left\lceil \frac{MN}{P} \right\rceil$$

$$P_2 = MN \mod P$$

1.2 Related Work

There is a great deal of literature dealing with domain decomposition. By considering the graph of the grid, where for each grid cell, there is a vertex associated with it, and for any two neighboring cells there is an edge joining the associated vertices, one can apply graph partitioning techniques for decomposing the domain. Kernighan and Lin's heuristic ([KL70]) for partitioning a graph into two components is a very well known technique that is still used in many modern codes as a subroutine but has the disadvantage of requiring a relatively good initial partition upon which it attempts to improve. It is a standard local refinement routine incorporated in the Chaco package ([HL95a]). Pothen et. al. ([PSL90]) developed the spectral method in the context of general graph partitioning; discussion of improved spectral partitioning algorithms including spectral quadrisection or octasection can be found in [HL95b]. Laguna et. al. ([LFE94]) also developed a GRASP heuristic for partitioning a general graph into two pieces; and in [CQ95] Crandall and Quinn presented a heuristic for decomposing non-uniform rectangular grids among a number of heterogeneous processors. Also, Miller et. al. ([MTTV93]) have designed a domain decomposer for meshes based on geometric ideas.

The spectral method and its variations have received considerable attention as they are general methods for splitting a graph into two equally sized pieces while minimizing the sum of weights of the arcs with endpoints in both sub-graphs. However, extending the spectral method to decompose a graph among an arbitrary number of components is non-trivial for any number that is not a power of two. The same holds true for the geometric partitioner by Miller, et al.

Finally, Genetic Algorithm approaches to the graph partitioning problem have been proposed ([vL91]) where the length of each individual in the population is at least as big as the size of the graph. Our GA, in contrast, uses the theory of optimal shapes ([YM92a]) in a high-level approach that reduces the length of the individual to P, the number of processors. Performance comparisons of our GA with many of the above approaches is given in section 5.

2 Shapes of Optimal Regions

In [YM92a], Yackel & Meyer showed that for any given area A a processor must occupy, there exists a non-empty collection of configurations of A cells (shapes) with the property that all of the shapes in the collection have minimum perimeter $\Pi^*(A) = 2\left\lceil 2\sqrt{A}\right\rceil$, i.e. there is no configuration of A cells with less perimeter than the perimeter of the shapes in the collection. It turns out that all optimal shapes have a property called slice-convexity (which is the same as convexity in the "polyomino" literature), a consequence of which is the fact that the perimeter of any optimal shape is twice its semi-perimeter, where the semi-perimeter of any shape is the sum of the height and width of the smallest rectangle containing the shape.

It follows immediately that if P shapes from such a collection completely cover the grid, then this partition constitutes an optimal solution to the Minimum Perimeter problem and this optimal solution has a perimeter equal to $P\Pi^*(A)$. If P does not divide MN, an analogous optimal solution will have perimeter $P_1\Pi^*(A_1) + P_2\Pi^*(A_2)$.

By similar arguments, these values give lower bounds on the objective value of the MPE(M, N, P), which are tight in many cases but not always. Specifically, when the dimensions of the grid are not big enough to accommodate the relatively square optimal shapes, the lower bound fails to be tight. In particular, we have the following lemma:

Lemma 1 *Assume that $M < N$ and that the following problem $(\mathcal{P})$ is feasible:*

$$\min_{h,w} \; h + w$$

$$s.t. \begin{cases} hw \geq A \\ h \leq M \\ w \leq N \\ h, w \in \mathbb{N}. \end{cases}$$

Let $(\mathcal{P}_{\nabla\rceil\ddagger})$ denote the relaxed problem

$$\min_{h,w} \; h + w$$

$$s.t. \begin{cases} hw \geq A \\ h, w \in \mathbb{N}. \end{cases}$$

Assume that all optimal solutions of $(\mathcal{P}_{\nabla\rceil\ddagger})$ violate at least one of the constraints of $(\mathcal{P})$. Then, an optimal solution of $(\mathcal{P})$ is $(h^*, w^*) = (M, \lceil \frac{A}{M} \rceil)$.

Proof: For each $h = 1 \ldots M$ that corresponds to a feasible point of $(\mathcal{P})$ (i. e. satisfies $\lceil \frac{A}{h} \rceil \leq N$), the w in the range $\lceil \frac{A}{h} \rceil \ldots N$ that yields the best objective is the value $\lceil \frac{A}{h} \rceil$. Thus, we need only find the number h in the range $1 \dots M$ that corresponds to a feasible solution and minimizes the function $f(h) = h + \lceil \frac{A}{h} \rceil$. But the function $f(h)$ in the range $1 \dots \lfloor \sqrt{A} \rfloor$ is non-increasing. To see this assume $i < \lfloor \sqrt{A} \rfloor$; we are going to show that $i + \lceil \frac{A}{i} \rceil \geq i + 1 + \lceil \frac{A}{i+1} \rceil$, or equivalently that

$$\left\lceil \frac{A}{i} \right\rceil - \left\lceil \frac{A}{i+1} \right\rceil \geq 1.$$

But the number $d := \frac{A}{i} - \frac{A}{i+1} = \frac{A}{i(i+1)} > 1$ as $i(i+1) < \lfloor \sqrt{A} \rfloor^2 \leq A$. Therefore, the ceilings of the two numbers $\frac{A}{i}$, $\frac{A}{i+1}$ are at a distance greater than, or equal to one, so $\lceil \frac{A}{i} \rceil - \lceil \frac{A}{i+1} \rceil \geq 1$.

As $M < \lfloor \sqrt{A} \rfloor$ (otherwise, there exists an optimal solution of $(\mathcal{P}_{\nabla\rceil\ddagger})$ that does not violate the extra constraints of $(\mathcal{P})$ because it is shown in [YM92a] that there always exist an optimal solution of $(\mathcal{P}_{\nabla\rceil\ddagger})$ that has $h^* = \lfloor \sqrt{A} \rfloor$.) an optimal solution of $(\mathcal{P})$ is $(h^*, w^*) = (M, \lceil \frac{A}{M} \rceil)$ and the optimal objective value of $(\mathcal{P})$ is $M + \lceil \frac{A}{M} \rceil$. ∎

The above lemma (1) implies that when the domain is a sufficiently narrow horizontal band (M being small enough) so that no optimal shape from the collection of optimal shapes fits in the domain, then the optimal perimeter is $2(M + \lceil \frac{A}{M} \rceil)$.

Motivated by the above theory of optimal shapes, we may convert the partitioning problem into a tiling problem: find a set of shapes (from the appropriate collection of optimal shapes) that can be tiled together so as to completely cover the grid with no overlap and with minimum distortion. If such a set can be found that completely covers the grid with no distortion of the shapes, then the resulting partition is *provably optimal.*

Many of the shapes in this collection consist of a rectangle of dimensions $h \times w$ plus a fringe of size f, denoted as the tuple (h, w, f). The number of such near rectangular shapes is shown in [YMC95] to grow with A as $\mathcal{O}(\mathcal{A}^{\infty/\triangle})$. In our method, we restrict the initial choices of optimal shapes to this latter subset of the collection of optimal shapes. Furthermore, it follows from [Yac93], that given an area A, and letting $k = \lfloor \sqrt{A} \rfloor$, if $k^2 = A$ then $(k, k, 0)$ is an optimal shape. Else, if $k^2 < A < k(k+1)$ then the shapes (k, k, f), and $(k+1, k-1, f')$ are optimal (with $f < k$ and $f' = A - k^2 + 1 < k + 1$). And finally, if $k(k+1) \leq A$ then $(k, k+1, f)$ and $(k+1, k, f)$ are both optimal shapes (with $f \leq k$). These shapes play a key role in proving the existence of a stripe-form solution of the MPE(M, N, M) where $M \geq N$.

Note that in the combinatorics literature ([Lin91, Mel94]), much research has been published on the generating function approach for developing expressions for the exact number of "convex polyominoes" with various properties.

However, our method (described below) is based on a library comprised of near-rectangular minimum perimeter configurations for a given area, so that the full collection does not have to be counted or generated.

3 Optimal Tilings

In this section we observe that the lower bound described above for MPE may be attained if the number of processors is large relative to the area of the grid. In particular, in the case where the number of processors is such that $\frac{MN}{P} \leq 3$, it is easy to construct an optimal solution that achieves the lower bound (because any connected configuration of 1, 2, or 3 cells is an optimal shape for the respective area size). In the case that $3 < \frac{MN}{P} \leq 4$, some processors will occupy an area of three cells and some will occupy an area of four. Then, a *necessary and sufficient condition* for the existence of an optimal solution that achieves the lower bound is the existence of a subrectangle of the grid that can accommodate all the optimal shapes for the processors having area four (there is only one optimal shape of area four, namely the 2×2 square). Necessity follows immediately from the fact that the grid is a subrectangle of itself, therefore if the squares cannot fit in the original grid, the lower bound cannot be attained. The condition is also sufficient because then, we can obtain an optimal partition that achieves the lower bound by tiling all the square shapes having an area of four in the north-western corner of the grid and filling the remaining area of the grid, row by row, with shapes of area three (any connected configuration of three cells is an optimal shape of area three).

If there exists no subrectangle that can accommodate 2×2 optimal shapes for all of the regions of area 4, then an optimal solution may be obtained by filling the largest (if any) northwest subrectangle of even dimensions with 2×2 shapes, then completing the assignment of the remaining cells (which will all lie in a single row and/or column) by assigning the remaining processor indices connectively left to right then bottom to top along the unassigned border. It is easily seen that this induces the minimum possible perimeter increase in the minimum possible number of shapes of area four whose shapes must be non-optimal and assigns optimal shapes of area three.

For these optimal solutions, we require $P \geq \frac{MN}{4}$, i.e. P is at least $\frac{1}{4}$ of the area of the grid. In the next section we show how to construct asymptotically optimal solutions if P dominates the individual dimensions M, N.

4 Asymptotically Optimal Solutions via Stripe Decomposition

In [CM95], we proved the following theorem:

Theorem 2 *The MPE(M,N,P) with $P = M \geq N$ has a feasible solution whose total perimeter possesses a relative distance δ from the lower bound that satisfies*

$$\delta < \frac{1}{\lceil 2\sqrt{N} \rceil}. \tag{1}$$

The proof of this theorem is an illustration of the stripe-decomposition technique. For any integer $k \geq 0$, if the number of rows of the grid M is at least $k(k-1)$, we can always find two natural numbers a, b such that

$$M = ak + b(k+1). \tag{2}$$

Letting now $k = \lfloor \sqrt{N} \rfloor$, where N is the area of each processor for the problem MPE(M, N, M), we can decompose the rows of the grid into stripes of height k or $k+1$. Each stripe can be filled with optimal and near optimal shapes, using a stripe-filling process ([CM95]) and the perimeter of each non-optimal shape will be at most two more than the optimal. A provably optimal partition of a 200×200 grid among 200 processors that is in stripe-form is shown in figure 2.

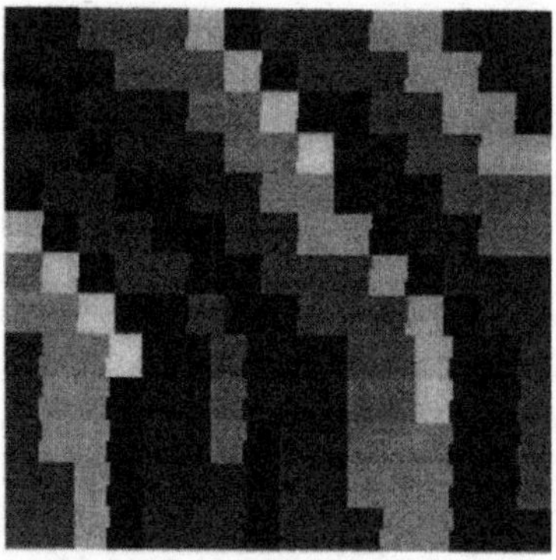

Figure 2: Optimal Partition in Stripe-Form for MPE(200,200,200)

The following theorem establishes an error bound for stripe-decomposition that improves on a related result in [CM95]:

Theorem 3 *Assuming P divides MN and that $P \geq max(M, N)$ the perimeter minimization problem MPE(M,N,P) has a feasible solution whose relative distance δ from the lower bound satisfies*

$$\delta < \frac{1}{\sqrt{A}} + \frac{1}{A}. \tag{3}$$

Thus the error bound δ converges to zero as A (the area of each processor) tends to infinity.

Proof: The grid is shown in figure 3. Note that $A \leq \min\{M, N\}$

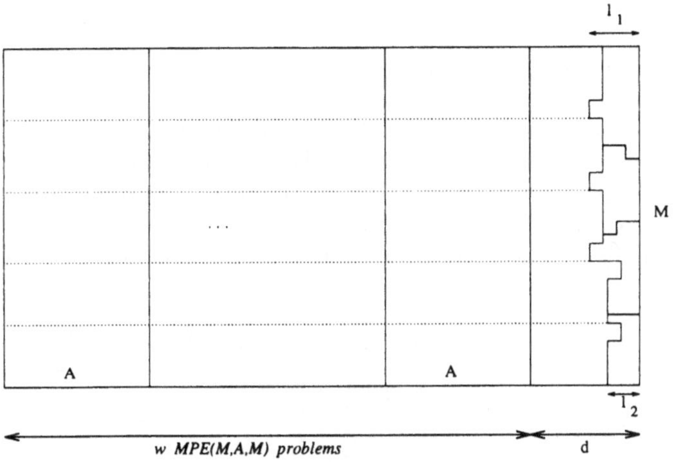

Figure 3: MPE(M, N, P), $P \geq max(M, N)$

and write $N = wA + d$ for some naturals $w \geq 1$ and $d < A$. Define $k = \left\lfloor \sqrt{A} \right\rfloor$. Observe that the problem can be decomposed into w MPE(M, A, M) problems, and a MPE$(M, d, Md/A)$. In each of the w problems MPE(M, A, M), use the stripe decomposition method of theorem 2 to get a total absolute perimeter error $e < 2wM$. This striping technique (which partitions the rows of the grid into $r \leq M/k$ stripes of height $h_1, \ldots, h_r$) is continued over the last d columns in each stripe until no additional shape can be placed in the stripe. Let p denote the number of processors that have not been assigned. The stripe decomposition in the last d columns thus placed $\frac{Md}{A} - p$ processors, each of which may have an error in perimeter of no more than two.

The stripe decomposition for MPE(M, A, M) uses at most two different shapes. Arrange the stripes of the grid so that all stripes that use the first shape are used in the top rows of the grid which we will refer to as area 1, and all the stripes that use the second shape are in the (remaining) bottom rows which we will refer to as area 2. Let l_i $i = 1, 2$ denote the maximum number of columns in area i that contain unassigned grid cells, and without any loss of generality, assume that $l_1 \geq l_2 \geq 0$.

We place the last p processors in the remaining area using the following "orthogonal stripe filling" algorithm that approximates the optimal shapes established in lemma 1: starting from the top row of the grid, keep assigning the unassigned cells row-wise (interchanging left to right and then right to left) until the processor has A cells.

To compute the error bound in perimeter of the last p processors that were placed in the grid using this "orthogonal stripe filling" algorithm we compute the length of the boundary enclosing the region they occupy, plus the length of the border between processors, then subtract the lower bound. Thus, the maximum error in this region is

$$e < (2M + l_1 + l_2) + (2r - 1 + l_1 - l_2) + 2\left[(p-1)l_1 + (p-1)\right] - 2(p\left\lceil 2\sqrt{A}\right\rceil)$$

. The first six terms in the RHS of the inequality account for the left, right, top and bottom borders of this region, the next two terms account for the inner borders, and the last term is the lower bound. Note that the perimeter of the left border includes four terms $(M + (2r-1) + l_1 - l_2)$ because it is not a straight line. Thus the total relative distance of the perimeter of the solution from the lower bound satisfies:

$$\delta < \frac{2\left[wM + \frac{Md}{A} - p + (p-1)l_1 + p - 1 - p\left\lceil 2\sqrt{A}\right\rceil\right]}{2M\frac{wA+d}{A}\left\lceil 2\sqrt{A}\right\rceil} + \frac{2M + l_1 + l_2 + (2r-1) + l_1 - l_2}{2M\frac{wA+d}{A}\left\lceil 2\sqrt{A}\right\rceil}$$

or

$$\delta < \frac{wM + \frac{Md}{A} + (p-1)l_1 - 1 - p\left\lceil 2\sqrt{A}\right\rceil + M + l_1 + r}{M\frac{wA+d}{A}\left\lceil 2\sqrt{A}\right\rceil}.$$

But for all $A \geq 2$, $l_1 \leq \left\lfloor\sqrt{A}\right\rfloor + 2 \leq \left\lceil 2\sqrt{A}\right\rceil$ which implies that $(p-1)l_1 + l_1 = pl_1 \leq p\left\lceil 2\sqrt{A}\right\rceil$, and since $r \leq \frac{M}{\lfloor\sqrt{A}\rfloor}$ we get

$$\delta < \frac{M(wA+d) + MA + \frac{MA}{\lfloor\sqrt{A}\rfloor}}{M(wA+d)\left\lceil 2\sqrt{A}\right\rceil} = \frac{1}{\left\lceil 2\sqrt{A}\right\rceil} + \frac{A}{(wA+d)\left\lceil 2\sqrt{A}\right\rceil} + \frac{A}{(wA+d)\left\lceil 2\sqrt{A}\right\rceil\left\lfloor\sqrt{A}\right\rfloor}.$$

It's easy to show that $\forall x \geq 1$ $x^2 \leq \lceil 2x\rceil\lfloor x\rfloor$ so (since $A \leq M$, $A \leq N$)

$$\delta < \frac{2}{\left\lceil 2\sqrt{A}\right\rceil} + \frac{1}{N} < \frac{1}{\sqrt{A}} + \frac{1}{A}.$$

∎

The following theorem extends the previous discussion to the case in which P does not divide MN.

Theorem 4 *Assume P does not divide MN and that $P \geq max(M, N)$; the perimeter minimization problem MPE(M,N,P) has a feasible solution whose relative distance δ from the lower bound satisfies*

$$\delta < \frac{1}{\sqrt{A_1}} + \frac{1}{\sqrt{A_2}} + \frac{1}{A_1} \tag{4}$$

where $A_1 = \lfloor MN/P \rfloor$ and $A_2 = \lceil MN/P \rceil$. Thus the error bound δ converges to zero as A_1, A_2 (the areas of the processors) tend to infinity.

Proof: Decompose the grid among the P processors using

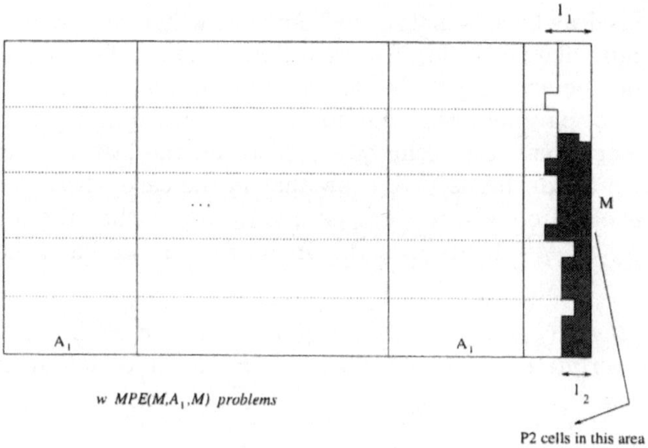

Figure 4: MPE(M, N, P), $P \geq max(M, N)$, $MN \mod P \neq 0$

the stripe decomposition and orthogonal stripe filling techniques discussed in the proof of theorem 3, initially assigning an area $A_1 = MN \div P$ to *all* P processors. Note that N can be written as $wA_1 + d$ for some naturals $w > 0$ and $0 \leq d < A_1$. The w MPE(M, A_1, M) problems cover the first wA_1 columns of the grid. The stripe filling is continued over the last d columns in the same manner as in the previous theorem. Orthogonal stripe filling is used for the remaining processors (if there are any). This leaves $P_2 = MN \mod P$ grid cells unassigned near the bottom right corner of the grid (the gray area in figure 4). Assign each of these cells to the last P_2 processors; the perimeter of these cells is at most $4P_2$. Recall that the lower bound in perimeter is a non-decreasing function of the area of each processor, and therefore, the absolute perimeter error of the last P_2 processors (having area $A_2 = A_1 + 1$) can be no larger than the absolute error computed in the stripe-decomposition of their first A_1 cells plus 4 (for the extra cell). Thus, the relative distance of the perimeter of the partition from the lower bound is

$$\delta < \frac{1}{\sqrt{A_1}} + \frac{1}{A_1} + \frac{4P_2}{2P_2 \lceil 2\sqrt{A_2} \rceil}$$

(since the lower bound is $2(P_1\lceil 2\sqrt{A_1}\rceil + P_2\lceil 2\sqrt{A_2}\rceil) \geq 2P_2\lceil 2\sqrt{A_2}\rceil$) and therefore

$$\delta < \frac{1}{\sqrt{A_1}} + \frac{1}{\sqrt{A_2}} + \frac{1}{A_1}.$$

∎

Essentially, the preceding theorems guarantee the existence of good quality solutions to the Minimum Perimeter problem as long as the number of processors is bigger than the dimensions of the grid. But the proof of the theorems also provide a method (stripe decomposition) for constructing such solutions. Furthermore, the theorems ensure the quality of the theoretical lower bounds.

The obvious drawback is that in environments where communication latencies can be high –like networks of workstations– the number of available processors may not be as large as the domain (which is assumed big enough to require the use of parallel processing for the efficient solution of the problem in hand). Nevertheless, the technique used to fill the last d columns of the grid in theorem 3, can be used to show that in the case where $N \leq P < M$ there exists a partition whose total perimeter approaches the lower bound (asymptotically) if $\frac{M}{PN}$ tends to zero. In particular, we have the following theorem:

Theorem 5 *If M, N and P satisfy $N \leq P < M$ and P divides MN, then the minimum perimeter problem MPE(M, N, P) has a feasible solution whose relative distance δ from the lower bound satisfies $\delta < \frac{1}{\lceil 2\sqrt{A}\rceil} + \frac{1}{N} + \frac{A}{N\lceil 2\sqrt{A}\rceil}$.*

Proof: It is shown in [CM95] that if $M \geq k(k-1)$, then there exist two natural numbers a, b such that $M = ak + b(k+1)$. Now, let k (as before) be $\lfloor\sqrt{A}\rfloor$. From the hypothesis, $P \geq N$, and since $A = \frac{MN}{P}$ we get $M \geq A \geq k^2 \geq k(k-1)$, so M can be written as $ak + b(k+1)$. This means that the grid can be decomposed exactly like the grid of figure 3 except of the first wA columns which do not exist. Following exactly the same arguments for computing the perimeter of the solution in the last d columns of figure 3 then, gives us

$$\delta < \frac{2\left[(P-p) + (p-1)l_1 + (p-1)\right]}{2\frac{MN}{A}\left\lceil 2\sqrt{A}\right\rceil} + \frac{(2M + l_1 + l_2) + (2r-1) + (l_1 - l_2) - 2p\left\lceil 2\sqrt{A}\right\rceil}{2\frac{MN}{A}\left\lceil 2\sqrt{A}\right\rceil}$$

from which, after substitution ($l_1 \leq \left\lceil 2\sqrt{A}\right\rceil$ and $r \leq \frac{M}{\lfloor\sqrt{A}\rfloor}$) we get

$$\delta < \frac{1}{\left\lceil 2\sqrt{A}\right\rceil} + \frac{1}{N} + \frac{A}{N\left\lceil 2\sqrt{A}\right\rceil}.$$

∎

It is easy to check that if M, N and P grow large in such a manner so that $\frac{M}{PN} \to 0$ then $\delta \to 0$ too. For example, by letting $P = N = A^{\frac{1}{2}+\epsilon}$, and $M = A$, then as A tends to infinity, $\delta \to 0$.

In the case where P does not divide MN, we have $A_1 = \lfloor \frac{MN}{P} \rfloor$ and because $\frac{N}{P} \in (0,1)$ we have $M > A_1 \geq k^2 > k(k-1)$ and thus the grid can be decomposed as shown in figure 4, except the first wA_1 columns which do not exist. Using the stripe decomposition method as described in theorem 4, we initially assign an area $A_1 = MN \div P$ to *all* processors, and fill the last bottom-right P_2 unassigned cells with the remaining cell of each of the P_2 processors. The error in perimeter of these last P_2 processors can be no larger than the absolute error computed in the stripe decomposition of their first A_1 cells plus four (for the extra cell). Using theorem 5, the relative distance of the perimeter of the partition from the lower bound is less than $\frac{1}{\lceil 2\sqrt{A_1} \rceil} + \frac{1}{N} + \frac{A_1}{N\lceil 2\sqrt{A_1} \rceil} + \frac{4P_2}{2P_2\lceil 2\sqrt{A_2} \rceil}$ (since the lower bound on perimeter is greater than $2P_2 \lceil 2\sqrt{A_2} \rceil$) and so we have established the following

Theorem 6 *If M, N and P satisfy $N \leq P < M$ and P does not divide MN, then the minimum perimeter problem $MPE(M, N, P)$ has a feasible solution whose relative distance δ from the lower bound satisfies $\delta < \frac{1}{\lceil 2\sqrt{A_1} \rceil} + \frac{1}{N} + \frac{A_1}{N\lceil 2\sqrt{A_1} \rceil} + \frac{1}{\sqrt{A_2}}$ where $A_1 = \lfloor MN/P \rfloor$ and $A_2 = \lceil MN/P \rceil$.*

Finally, note that there are instances of problems for which the best solution that can be found via the stripe-decomposition method are not optimal. For example, the MPE(17, 17, 17) has a provably optimal solution that achieves the lower bound (see figure 5) yet the best solution found by stripe-decomposition is at distance 0.65% from the lower bound. The optimal solution was found after a swapping heuristic was applied to a solution produced by PERIX-GA (to be discussed below).

5 Computational Results

Based on the observation that the partitioning problem can be viewed also as a tiling problem when restricted to the class of uniform 5-point grids, we developed PERIX, an algorithm that given a set of P optimal shapes from the appropriate library of optimal shapes attempts to tile the grid. To achieve this goal, PERIX maintains a list of maximal free rectangles of the grid (a structure used successfully by Yackel in [YM92b] for another tiling problem), into which it attempts to place the next optimal shape, one at a time. The optimal shapes in our library are blocks accompanied by a fringe. PERIX attempts to place the block part of the optimal shape first, then attempts to place the fringe heuristically next to it with as little modification as possible.

To search the huge search space of input combinations to the PERIX algorithm (in order to find the best partition –not only the best stripe-

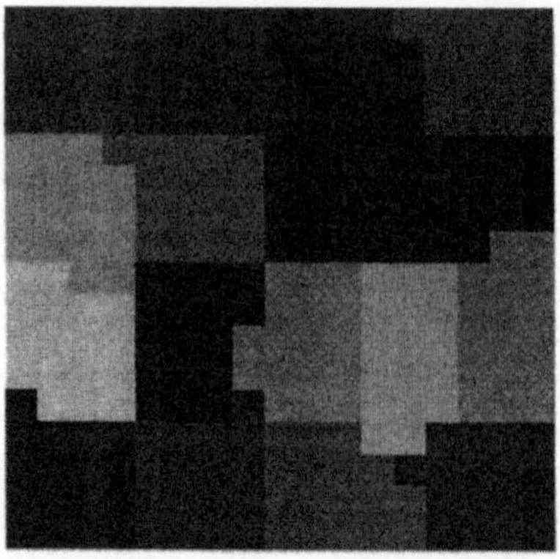

Figure 5: Optimal Solution for MPE(17, 17, 17)

form–) we have developed PERIX-GA, a high level repair Genetic Algorithm ([Hol92, Mic94]). PERIX-GA works with a population of individuals each of which is an array of shape indices to be tiled together by the PERIX algorithm. PERIX-GA breeds a population of such individuals for a certain number of generations using a modified crossover and mutation operator to take advantage of the fact that many good solutions of the problem are in stripe form (and therefore to encourage their appearance). Specifically, each allele in an individual has a tag associated with it, indicating whether the corresponding shape was placed at the beginning of a row or not. Crossover then, occurs at positions that that are marked by both parents as beginning of (possibly) a new stripe. Similarly, the mutation operator attempts to alter all "genes" with the same shape-index between two positions whose shapes are placed in the beginning of a new row. The parents may replace their offspring in the next generation if the offspring's fitness is worse than the worse fitness of the individuals of the parent generation or if the parents are the best individuals found so far in the evolution process (this elitist survival policy ensures the best individual found is not lost in subsequent generations and that the worse fitness value of the population monotonically improves).

We have run our algorithm (PERIX-GA), which is written in C, on a cluster of 33 SUN 20 SPARC-SERVER workstations (COW) using the host-node paradigm (one workstation serving as the host which co-ordinates the selection process, and the other 32 workstations being the nodes; all are connected via Ethernet). Each node maintains 2 individuals, thus the total population size is 64. The communication between workstations used the PVM 3.3.7 message-passing system ([GBD$^+$94]) (before that, we had run earlier versions of our algorithm on a CM-5 with 32 nodes using the CMMD message-passing library ([Thi93]), and we reported the results in [CM95]). We run our algorithm for 20 generations except for the last case (1000×1000 grid) which we run for only 10 generations due to time limitations.

A metric of the size of the test problems is shown in table 1. The column "LIP" indicates the size of the problem formulated as a Mixed Linear Integer Program (MNP of the total variables are binary, the rest are continuous). The QAP dimension of each problem is MN, and the GA dimension is the length of the individuals, i.e. the number of processors to be assigned to the domain.

In table 2 we have compared our algorithm with a GRASP heuristic for the QAP ([LPR94]), run on one node of the cluster of workstations. This GRASP code is a state-of-the-art code, but as the dimension of the problem grows as the product of the dimensions of the grid, it has difficulties dealing with the larger problems in our test-suite.*

PROBLEM			QAP	LIP		GA	Str bd
M	N	P	DIM	VARS	CONSTR	VARS	(%)
7	7	7	49	427	3584	7	16.66
13	13	13	169	2509	48854	13	12.50
17	17	17	289	5457	148274	17	11.11
32	30	64	960	63298	7492480	64	32.48
32	31	8	992	9857	108576	8	-
32	31	256	992	255873	1.25E+08	256	141.06
100	100	8	10000	99800	1118808	8	-
101	101	101	10201	1050501	2.04E+08	101	4.76
128	128	128	16384	2129664	5.28E+08	128	4.34
200	200	200	40000	8079600	3.16E+09	200	3.44
256	256	256	65536	1.690E07	8.52E+09	256	3.12
512	512	512	262144	1.347E08	1.36E+11	512	2.17
1000	1000	1000	1.0E06	1.001E09	1.99E+12	1000	1.56

Table 1: Test Problem Sizes under Various Formulations

We also compared our algorithm against two popular graph-partitioning methods, namely the (recursive) spectral bisection method ([PSL90]) with a Kernighan - Lin local refinement procedure applied, and the geometric mesh partitioning method ([MTTV93]). We obtained an implementation of the geometric mesh partitioner (in MATLAB) as described in [GMT95], and we used the Chaco package version 2.0 ([HL95a]) for the spectral bisection (Chaco is entirely written in ANSI C). We ran our experiments on these two graph partitioning methods on a SUN-20 workstation. Both of these methods require the number of processors to be a power of two, and we tabulated the results of the comparison from another test-suite in table 3. The column labeled "BTm" indicates the time it took PERIX-GA to find the best solution. An asterisk in table 3 indicates the fact that the partition found was not balanced (i.e. there were components that had at least two more nodes

*in the QAP literature, problems of dimension more than 30 are considered large, difficult problems.

PROBLEM			GRASP (COW: 1)		PERIX-GA (COW: 33)		
M	N	P	Err %	Time	Err %	Gens	Time
7	7	7	0.00	153.5	0.0	1	196.1
13	13	13	25.96	10327.6	0.0	15	227.8
17	17	17	-	-	0.0	9	268.6
32	31	8	-	-	2.17	2	201.0
32	31	256	-	-	0.0	5	230.2
101	101	101	-	-	0.04	15	219.1
200	200	200	-	-	0.0	7	261.0
512	512	512	-	-	0.66	5	402.3
1000	1000	1000	-	-	0.45	5	1660.5

Table 2: Computational Results: PERIX-GA and GRASP-QAP

than other components). Also, note that for the last problem, both the geometric and the spectral method ran out of memory when trying to construct the adjacency matrix of the graph. The times are all in seconds. For this

PROBLEM			SPECTRAL		GEOMETRIC		PERIX-GA		
M	N	P	Time	Err%	Time	Err%	Time	BTm	Err%
32	31	8	1.8	6.52	43.6	5.43	84.0	20.9	2.71
32	31	256	4.3	6.73	152.3	-2.73*	80.4	4.1	0.00
32	30	64	3.0	6.25	90.4	6.25	50.9	43.2	0.00
100	100	8	9.0	9.33	111.0	7.39	81.9	12.3	2.64
128	128	128	85.5	14.13	539.9	7.13	67.6	16.9	1.65
256	256	256	227.8	13.25	3304.2	4.15	105.1	4.1	0.00
512	512	512	-	-	-	-	279.0	111.6	1.63

Table 3: PERIX-GA on 9 procs. vs Spectral and Geometric Partitioning

comparison, we ran PERIX-GA on an 9-node partition of the COW (that is, 8 workstations of the cluster were serving as nodes, and one workstation was the host). The results show that the quality of the partition of our method is significantly better when the assumptions of theorem 3 are satisfied; but importantly, even when the assumption on the number of available processors is violated, our method still produces better partitions than spectral or geometric partitioning. The geometric mesh partitioning method proved to be rather fast on smaller problems, as its serial version gives times comparable to our algorithm running on 9 SUN-20 SPARC SERVER workstations except on the three largest problems. The (recursive) spectral bisection was the fastest method on many of the smaller problems. However, the quality of the resulting solutions is not as good as the quality of the partitions given by the other methods in the comparison test. It is also interesting to note that both the spectral bisection and the geometric partitioner fail to find the (provably optimal) partition of the 256×256 grid partitioned among 256 processors which is simply 256 squares of size 16×16 tiled together. This particular test-problem is the only one in our suite for which an optimal solution was known to exist a-priori.

6 Conclusions and Future Directions

Stripe decomposition is a fast and efficient method for constructing very good quality partitions of rectangular grids among processors as long as the number of processors is bigger than both dimensions of the grid. We have developed PERIX, an algorithm that tiles together an input set of near rectangular optimal shapes while seeking to minimize modification of the input shapes. (For suitable inputs, a stripe decomposition is produced). PERIX-GA is a parallel Genetic Algorithm that runs on a network of workstations and searches the space of inputs to PERIX with genetic operators that encourage the occurrence of stripe-form solutions. The computational results indicate that problems that are intractable for many methods because of their size, can be solved by PERIX-GA with good accuracy within minutes on a network of workstations. The quality of the partitions produced by our code is superior to the quality of the solutions provided by other popular codes.

Recently, we extended PERIX to work on arbitrary uniform grids and one immediate goal is to evaluate the performance of our algorithm in general grids. First results on elliptical domains look very promising. Improving the quality of the lower bounds is another goal. Finally, we would like to find ways to relax the assumption on the number of available processors in the theorems on stripe decomposition. This relaxation should be possible as indicated by the computational results for problems with small number of processors.

7 Acknowledgments

We would like to thank P.M. Pardalos and M. Resende for providing us with a version of their high-quality GRASP heuristic for the QAP. We obtained the MATLAB code for the geometric mesh partitioner from the anonymous ftp site indicated in Gilbert et al. ([GMT95]) to whom we are thankful for the easy access to their code. Finally, our thanks to B. Hendrickson and R. Leland for providing us with version 2.0 of the Chaco package.

References

[CM95] I. T. Christou and R. R. Meyer. Optimal equi-partition of rectangular domains for parallel computation. Technical Report MPTR 95-04, University of Wisconsin - Madison, February 1995. To appear in the *Journal of Global Optimization.*

[CQ95] P. Crandall and M. Quinn. Non-uniform 2-d grid partitioning for heterogeneous parallel architectures. In *Proceedings of the 9th International Symposium on Parallel Processing*, pages 428–435, 1995.

[DTR91] R. DeLeone and M. A. Tork-Roth. Massively parallel solution of quadratic programs via succsessive overrelaxation. Technical Report 1041, University of Wisconsin - Madison, August 1991.

[GBD+94] A. Geist, A. Beguelin, J. Dongarra, W. Jiang, R. Manchek, and V. Sunderam. *PVM 3 User's Guide and Reference Manual.* Oak Ridge National Laboratory, 1994.

[GMT95] J. R. Gilbert, G. L. Miller, and S. H. Teng. Geometric mesh partitioning: Implementation and experiments. In *Proceedings of the 9th International Symposium on Parallel Processing*, pages 418–427, 1995.

[HL95a] B. Hendrickson and R. Leland. *The Chaco User's Guide Version 2.0.* Sandia National Laboratories, July 1995.

[HL95b] B. Hendrickson and R. Leland. An improved spectral graph partitioning algorithm for mapping parallel computations. *SIAM J. on Sci. Comput.*, 16:452–469, 1995.

[Hol92] John Holland. *Adaptation in Natural and Artificial Systems.* MIT Press, 1992.

[KL70] B. W. Kernighan and S. Lin. An effective heuristic procedure for partitioning graphs. *Bell Systems Tech. Journal*, pages 291–308, February 1970.

[LFE94] M. Laguna, T. A. Feo, and H. C. Elrod. A greedy randomized adaptive search procedure for the two - partition problem. *Operations Research*, July - August 1994.

[Lin91] K. Y. Lin. Exact solution of the convex polygon perimeter and area generating function. *J. Phys. A. Math Gen.*, 24:2411–2417, 1991.

[LPR94] Y. Li, P. M. Pardalos, and M. G. C. Resende. A grasp for the qap. In P. M. Pardalos and H. Wolkowicz, editors, *Quadratic Assignment and Related Problems.* DIMACS Series Vol. 16, American Mathematical Society, 1994.

[Mel94] M. Bousquet Melou. Codage des polyominos convexes et equation pour l'enumeration suivant l'aire. *Discrete Applied Mathematics*, 48:21–43, 1994.

[Mic94] Zbigniew Michalewicz. *Genetic Algorithms + Data Structures = Evolution Programs.* Springer-Verlag, 1994.

[MTTV93] G. L. Miller, S. H. Teng, W. Thurston, and S. A. Vavasis. Automatic mesh partitioning. In A. George, J. R. Gilbert, and J. W. H. Liu, editors, *Graph Theory and Sparse Matrix Computation.* Springer-Verlag, 1993.

[PRW93] P. M. Pardalos, F. Rendl, and H. Wolkowicz. The quadratic assignment problem: A survey and recent developments. In P. M. Pardalos and H. Wolkowicz, editors, *Quadratic Assignment and Related Problems.* American Mathematical Society, 1993.

[PSL90] A. Pothen, H. D. Simon, and K. P. Liu. Partitioning sparse matrices with eigenvectors of graphs. *SIAM Journal on Matrix Analysis and Applications*, 11:430–452, 1990.

[Sch89] R. J. Schalkoff. *Digital Image Processing and Computer Vision.* John Wiley & Sons, 1989.

[Thi93] Thinking Machines Corporation. *CMMD Reference Manual*, May 1993.

[vL91] von Laszewski. Intelligent structural operators for the k-way graph partitioning problem. In R. Belew and L. Booker, editors, *Proceedings of the Fourth Intl. Conference on Genetic Algorithms*, pages 45–52. Morgan Kaufmann Publishers, Los Altos, CA, 1991.

[Yac93] J. Yackel. *Minimum Perimeter Tiling in Parallel Computation.* PhD thesis, University of Wisconsin - Madison, August 1993.

[YM92a] J. Yackel and R. R. Meyer. Minimum perimeter decomposition. Technical Report 1078, University of Wisconsin - Madison, February 1992.

[YM92b] J. Yackel and R. R. Meyer. Optimal tilings for parallel database design. In P. M. Pardalos, editor, *Advances in Optimization and Parallel Computing*, pages 293–309. North - Holland, 1992.

[YMC95] J. Yackel, R. R. Meyer, and I. T. Christou. Minimum-perimeter domain assignment, February 1995. Submitted to the Math. Programming Journal.

TRUST–REGION INTERIOR–POINT ALGORITHMS FOR MINIMIZATION PROBLEMS WITH SIMPLE BOUNDS *

J. E. DENNIS † AND LUÍS N. VICENTE †

Abstract. Two trust–region interior–point algorithms for the solution of minimization problems with simple bounds are analyzed and tested. The algorithms scale the local model in a way similar to Coleman and Li [1]. The first algorithm is more usual in that the trust region and the local quadratic model are consistently scaled. The second algorithm proposed here uses an unscaled trust region. A global convergence result for these algorithms is given and dogleg and conjugate–gradient algorithms to compute trial steps are introduced. Some numerical examples that show the advantages of the second algorithm are presented.

Keywords. trust–region methods, interior–point algorithms, Dikin–Karmarkar ellipsoid, Coleman and Li affine scaling, simple bounds.

AMS subject classification. 49M37, 90C20, 90C30

1. Introduction.

In this note we consider the box–constrained minimization problem

$$\begin{array}{ll} \text{minimize} & f(x) \\ \text{subject to} & a \le x \le b, \end{array} \tag{1}$$

where $x \in \mathbb{R}^n$, $a \in (\mathbb{R} \cup \{-\infty\})^n$, $b \in (\mathbb{R} \cup \{+\infty\})^n$ and f maps $\mathbb{R}^n$ into $\mathbb{R}$. We assume that f is continuously differentiable in the box

$$\mathcal{B} = \{x \in \mathbb{R}^n : \ a \le x \le b\}.$$

Coleman and Li [1] give an elegant diagonal affine scaling for this problem. It has the flavor of the Dikin–Karmarkar affine scaling, but it has a direct connection to the dual information of the first–order necessary optimality conditions. A diagonal element corresponding to what appears to be a constraining bound is the same as in the Dikin–Karmarkar affine scaling. In their algorithms, they allow the elliptical trust region thus defined to have trust radius greater than one, so that some infeasible points are inside the trust region. As is usual, in their algorithms the trust region and the quadratic model are consistently scaled.

We adopt a version of the Coleman and Li scaling for the local quadratic model in both our algorithms. The first algorithm that we propose here uses the same scaling for the trust region, and so it is similar to the Coleman and Li algorithms. However, the trial steps are computed completely differently. The second algorithm that we suggest maintains the trust region in the unscaled variables. In both algorithms, the trial step computation is very simple

* Support of this work has been provided by INVOTAN, CCLA and FLAD (Portugal) and by DOE DE–FG03–93ER25178, CRPC CCR–9120008 and AFOSR–F49620–9310212.

† Department of Computational and Applied Mathematics, Rice University, Houston, Texas 77005–1892, USA.

and convenient with respect to staying strictly inside $\mathcal{B}$. We present some numerical examples to illustrate the advantages of the second algorithm.

There are three points of novelty here:

1. An improved interior–point algorithm for the solution of (1).
2. A trust–region convergence analysis for a fraction of Cauchy decrease condition in which the scaling of the direction that defines the Cauchy step does not come from the ellipsoidal norm that defines the trust region.
3. A new way of extending dogleg and conjugate–gradient algorithms for the solution of trust–region subproblems that arise in unconstrained optimization, to trust–region subproblems that appear when the simple bounds are included.

A point x_* in $\mathcal{B}$ is said to be a first–order Karush–Kuhn–Tucker (KKT) point if

$$\begin{array}{lcl} a_i < (x_*)_i < b_i & \Longrightarrow & (\nabla f(x_*))_i = 0, \\ (x_*)_i = a_i & \Longrightarrow & (\nabla f(x_*))_i \geq 0, \quad \text{and} \\ (x_*)_i = b_i & \Longrightarrow & (\nabla f(x_*))_i \leq 0, \end{array}$$

or equivalently if,

$$D(x_*)^p \nabla f(x_*) = 0,$$

where $p > 0$ is arbitrary and $D(x)$ is a diagonal matrix whose diagonal elements are given by

$$(D(x))_{ii} = \begin{cases} (b-x)_i & \text{if} \quad (\nabla f(x))_i < 0 \text{ and } b_i < +\infty, \\ 1 & \text{if} \quad (\nabla f(x))_i < 0 \text{ and } b_i = +\infty, \\ (x-a)_i & \text{if} \quad (\nabla f(x))_i \geq 0 \text{ and } a_i > -\infty, \\ 1 & \text{if} \quad (\nabla f(x))_i \geq 0 \text{ and } a_i = -\infty, \end{cases}$$

for $i = 1, \ldots, n$.

We show that a sequence $\{x_k\}$ generated by either of our algorithms satisfies

$$\lim_k \|D(x_k)^p \nabla f(x_k)\| = 0,$$

for any $p \geq \frac{1}{2}$. It is important to note that these results are obtained under very mild assumptions on the trial steps, and that the sequence of approximations to the Hessian matrix of f is assumed only to be bounded.

This paper is organized as follows. In Section 2 we introduce the conditions that we need to impose on the trial steps for the algorithms mentioned before. These are described in Section 3. In Section 4 we give a unified global convergence result for the trust–region interior–point algorithms. In Section 5 we describe dogleg and conjugate–gradient algorithms to compute the trial steps. Finally, in Section 6 we present some numerical examples and some final conclusions. In this paper $\|\cdot\|$ represents the ℓ_2 norm, and I_l the identity matrix of order l.

2. Trust–region subproblems and trial steps. In this section we motivate the computation of the trial steps and present the conditions that these trial steps have to satisfy. The algorithms generate a sequence of iterates $\{x_k\}$ where x_k is strictly feasible, in other words where $a < x_k < b$. Given x_k we compute a trial step s_k, and decide whether to accept it or not. If s_k is accepted then $x_{k+1} = x_k + s_k$, otherwise $x_{k+1} = x_k$.

2.1. Motivation. Our approach begins like sequential quadratic programming (SQP). Before we think about globalization of SQP, we look at the local quadratic programming (QP) subproblem

$$\begin{array}{ll} \text{minimize} & \Psi_k(s) \\ \text{subject to} & \sigma_k(a - x_k) \leq s \leq \sigma_k(b - x_k), \end{array}$$

gotten by building a quadratic model

$$\Psi_k(s) = f(x_k) + g_k^T s + \tfrac{1}{2} s^T H_k s$$

of $f(x_k + s)$ about x_k, where $g_k = \nabla f(x_k)$, and H_k is an approximation to the Hessian matrix $\nabla^2 f(x_k)$ of f evaluated at x_k. Here $\sigma_k \in [\sigma, 1)$ and $\sigma \in (0,1)$ is fixed for all k. Enforcing these bounds at every iteration ensures that the solutions to the subproblems remain strictly feasible for the original problem.

For this quadratic problem, we like the idea of the affine scaling algorithm, i.e., we rewrite the quadratic problem in a basis $\hat{s} = D_k^{-1} s$, for which the distances to the constraints of the current iterate x_k are the same in order that all directions be equally free for use in decreasing the objective function from x_k. Here we follow the concept of Coleman and Li [1] and choose D_k as $D(x_k)$. They actually use $D_k = D(x_k)^{\frac{1}{2}}$. However, either choice has the effect of only rescaling those components that appear, from the sign of the gradient, to threaten to restrict the step. This gives the local QP subproblem:

$$\begin{array}{ll} \text{minimize} & \Psi_k(D_k \hat{s}) \\ \text{subject to} & \sigma_k D_k^{-1}(a - x_k) \leq \hat{s} \leq \sigma_k D_k^{-1}(b - x_k), \end{array}$$

gotten by building a quadratic model

$$\Psi_k(D_k \hat{s}) = f(x_k) + (D_k g_k)^T \hat{s} + \tfrac{1}{2} \hat{s}^T D_k H_k D_k \hat{s}\,.$$

In this subproblem there is an *explicit* scaling given by D_k in the $\hat{s}$ basis. For instance, the $\hat{s}$ steepest–descent direction in the ℓ_2 norm is given by $-D_k g_k$.

Thus, we would like to minimize this quadratic function over a trust region with the requirement that $x_k + s_k = D_k(\hat{x}_k + \hat{s}_k)$ has to be strictly feasible. Although we do this in the original basis s so that we can work always with the same variables, we bring the scaling that is used in the basis $\hat{s}$. The reference trust–region subproblem that we consider, written in the original basis, is the following:

(2) minimize $\Psi_k(s)$

(3) subject to $\|S_k^{-1}s\| \leq \delta_k,$

(4) $\sigma_k(a - x_k) \leq s \leq \sigma_k(b - x_k),$

where δ_k is the trust radius, and S_k is a $n \times n$ nonsingular matrix. This subproblem is *implicitly* scaled by D_k^2 so that the direction $-D_k g_k$ defined in the $\hat{s}$ variables is now given by $-D_k(D_k g_k) = -D_k^2 g_k$ in the s variables.

The two algorithms suggested here differ mainly in their choice of S_k. We discuss that issue now.

If we continue to follow the affine scaling idea, then we use the ellipse defined at each iterate in the original s coordinates by the ℓ_2 norm on these new coordinates $\hat{s}$ to help to enforce the bounds. In other words, we would choose $S_k = D_k$, and the shape of the trust region (3) would be ellipsoidal in the original basis.

This substitution of one ellipsoidal constraints for all the bound constraints was a prime motivation for interior–point methods. However from the beginning of the computational study of interior–point methods, it was found to be important to allow steps past the boundary of this ellipsoid, as long as they still satisfy the subproblem bound constraints. This translates here to saying that if the trust region is to have the ellipsoidal shape, then the trust radius should be allowed to exceed one, and so the trust region really is not used to enforce the bound constraints. Sometimes, the subproblem is further biased away from the bounds by adding a barrier term to the model (2).

The motivation for the second algorithm is that there is no reason to use the ellipsoid to define the shape of the trust region if it is not useful for enforcing the bounds. In fact, there are even more good reasons not to use it here than in the linear programming problem. One of the most important is that for nonlinear programs x_* may lie strictly inside $\mathcal{B}$. This happens in problems where the bounds are really to define the region of interest. If x_k is near a bound which is not active at x_*, then many iterations may be required to move off that bound. Hence we choose S_k to be the identity in the second algorithm.

It is important to point out that Coleman and Li [1] present a different motivation to their algorithms. They see their algorithms as applying Newton's method to the system of nonlinear equations $D(x)\nabla f(x) = 0$. The vector function $D(x)\nabla f(x)$ is continuous, but nondifferentiable if $(\nabla f(x))_i = 0$. As result of applying Newton's method, they add to H_k a diagonal matrix C_k of the form

$$C_k = D_k^{-\frac{1}{2}} diag(g_k) J_k D_k^{-\frac{1}{2}}, \tag{5}$$

where $diag(g_k)$ is a diagonal matrix whose diagonal elements are given by $(diag(g_k))_{ii} = (g_k)_i$, $i = 1, \ldots, n$, and where J_k is defined as the "Jacobian" of $D(x)$ at $x = x_k$. We note that the choice made by Coleman and Li [1] is $S_k = D_k^{\frac{1}{2}}$. See their paper for more details.

The global convergence result given in Section 4 holds for any D_k of the form D_k^p with $p \geq \frac{1}{2}$. We motivated the algorithms by using $p = 1$ but for the rest of this paper we assume that p is any number greater or equal than $\frac{1}{2}$.

2.2. What to impose on the trial steps. We need to define the Cauchy step associated with the trust–region subproblem (2)–(4). This Cauchy step c_k is defined as the solution of

$$\begin{array}{ll} \text{minimize} & \Psi_k(s) \\ \text{subject to} & \|S_k^{-1} s\| \leq \delta_k, \quad s \in span\{-D_k^{2p} g_k\}, \\ & \sigma_k(a - x_k) \leq s \leq \sigma_k(b - x_k). \end{array} \tag{6}$$

As in many trust–region algorithms, s_k is required to give a decrease on $\Psi_k(s)$ smaller than a uniform fraction of the decrease given by c_k for the same function $\Psi_k(s)$. This condition is often called fraction of Cauchy decrease, and in this case is

$$\Psi_k(0) - \Psi_k(s_k) \geq \beta \left(\Psi_k(0) - \Psi_k(c_k)\right), \tag{7}$$

where β is positive and fixed across all the iterations.

Coleman and Li [1] define the fraction of Cauchy decrease condition in a different way by using $\sigma_k = 1$ in (6) although they suggest $\sigma_k \in (0, 1)$ in the computation of the trial step s_k. Our definition leads naturally to the condition (7) holding for any trial step generated by the algorithms that we propose in Section 5.

3. TRIP algorithms. To decide whether to accept or reject a trial step s_k, how ever it is computed, we need to evaluate the ratio $r_k = \frac{ared(s_k)}{pred(s_k)}$, where $ared(s_k) = f(x_k) - f(x_{k+1})$ is the actual decrease and $pred(s_k) = \Psi_k(0) - \Psi_k(s_k)$ is the predicted decrease. We describe next the steps of the algorithms leaving the computation of the trial steps for Section 5.

ALGORITHM 3.1 (Trust–region interior–point (TRIP) algorithms).

1. Choose $\delta_0 > 0$, x_0 such that $a < x_0 < b$, and p, ϵ, σ, α, and η such that $p \geq \frac{1}{2}$, $\epsilon > 0$ and $0 < \sigma, \alpha, \eta < 1$.
2. For $k = 0, 1, 2, \ldots$ do
   - 2.1. If $\|D_k^p g_k\| \leq \epsilon$, stop and return x_k as a solution for (1).
   - 2.2. Compute a trial step s_k satisfying (7), $\|S_k^{-1} s_k\| \leq \delta_k$ and $\sigma_k(a - x_k) \leq s_k \leq \sigma_k(b - x_k)$, where $\sigma_k \in [\sigma, 1)$.

2.3. If $r_k < \eta$ reject s_k, set $\delta_{k+1} = \alpha\|s_k\|$, $x_{k+1} = x_k$.
If $r_k \geq \eta$ accept s_k, choose $\delta_{k+1} \geq \delta_k$, set $x_{k+1} = x_k + s_k$.

Of course the rules to update the trust radius can be much more involved but the above suffices to prove convergence results and to understand the trust–region mechanism.

The choices $S_k = D_k^p$ and $S_k = I_n$ correspond respectively to the first and second algorithms.

4. Global convergence result. To analyze the convergence properties of the TRIP algorithms, all we need to do is to express (7) in a more useful form. We do this in the following technical lemma, and it is not a surprise to see that the proof follows the proof given by Powell (see [5, Theorem 4] and [3, Lemma 4.8]) for the unconstrained minimization case.

LEMMA 4.1. *If s_k satisfies (7) then*

$$\Psi_k(0) - \Psi_k(s_k) \geq \frac{1}{2}\beta\|\hat{g}_k\| \min\left\{\frac{\|\hat{g}_k\|}{\|\hat{H}_k\|}, \min\left\{\frac{\|\hat{g}_k\|}{\|S_k^{-1}D_k^p\hat{g}_k\|}\delta_k, \sigma\frac{\|\hat{g}_k\|}{\|h_k\|_\infty}\right\}\right\},$$

where $\hat{g}_k = D_k^p g_k$, $\hat{H}_k = D_k^p H_k D_k^p$ and h_k is a vector in $\mathbb{R}^n$ defined by

$$(h_k)_i = \frac{|(g_k)_i|}{\min\{(x_k - a)_i{}^{(1-2p)}, (b - x_k)_i{}^{(1-2p)}\}},$$

for $i = 1, \ldots, n$.

Proof. Define $\psi : \mathbb{R}^+ \longrightarrow \mathbb{R}$ as $\psi(t) = \Psi_k(-tD_k^p\frac{\hat{g}_k}{\|\hat{g}_k\|}) - \Psi_k(0)$. Then $\psi(t) = -\|\hat{g}_k\|t + \frac{v_k}{2}t^2$ where $v_k = \frac{\hat{g}_k^T\hat{H}_k\hat{g}_k}{\|\hat{g}_k\|^2}$. Now we need to minimize ψ in $[0, T_k]$ where T_k is given by

$$\begin{aligned} T_k \quad = \quad & \min\left\{\tfrac{\|\hat{g}_k\|}{\|S_k^{-1}D_k^p\hat{g}_k\|}\delta_k, \quad \sigma_k\|\hat{g}_k\|\min\left\{\tfrac{(x_k-a)_i{}^{(1-2p)}}{(g_k)_i} : (g_k)_i > 0\right\},\right. \\ & \left.\sigma_k\|\hat{g}_k\|\min\left\{-\tfrac{(b-x_k)_i{}^{(1-2p)}}{(g_k)_i} : (g_k)_i < 0\right\}\right\}. \end{aligned}$$

Let t_k^* be the minimizer of ψ in $[0, T_k]$. If $t_k^* \in (0, T_k)$ then

$$\psi(t_k^*) = -\frac{1}{2}\frac{\|\hat{g}_k\|^2}{v_k} \leq -\frac{1}{2}\frac{\|\hat{g}_k\|^2}{\|\hat{H}_k\|}. \tag{8}$$

If $t_k^* = T_k$ then either $v_k > 0$ in which case $\frac{\|\hat{g}_k\|}{v_k} \geq T_k$ or $v_k \leq 0$ in which case $v_kT_k \leq \|\hat{g}_k\|$. In either event,

$$\psi(t_k^*) = \psi(T_k) = -T_k\|\hat{g}_k\| + \frac{v_k}{2}T_k^2 \leq -\frac{T_k}{2}\|\hat{g}_k\|. \tag{9}$$

Now we can combine (8) and (9) and get

$$\begin{aligned}\Psi_k(0)-\Psi_k(s_k) &\geq \beta\left(\Psi_k(0)-\Psi_k(c_k)\right)=-\beta\psi(t_k^*)\\ &\geq \tfrac{1}{2}\beta\|\hat{g}_k\|\min\left\{\tfrac{\|\hat{g}_k\|}{\|\hat{H}_k\|},T_k\right\}\\ &\geq \tfrac{1}{2}\beta\|\hat{g}_k\|\min\left\{\tfrac{\|\hat{g}_k\|}{\|\hat{H}_k\|},\min\left\{\tfrac{\|\hat{g}_k\|}{\|S_k^{-1}D_k^p\hat{g}_k\|}\delta_k,\sigma\tfrac{\|\hat{g}_k\|}{\|h_k\|_\infty}\right\}\right\}.\end{aligned}$$

□

Now the following theorem is a consequence of Lemma 4.1. We can see that under its assumptions, the sequences $\{\|\hat{H}_k\|\}$, $\{\|S_k^{-1}D_k^p\hat{g}_k\|\}$, and $\{\|h_k\|_\infty\}$ are bounded. Thus, Lemma 4.1 implies

$$\Psi_k(0)-\Psi_k(s_k)\geq\kappa_1\|\hat{g}_k\|\min\left\{\kappa_2\|\hat{g}_k\|,\kappa_3\delta_k\right\},$$

where the constants κ_1, κ_2, and κ_3 are positive and independent of k. Hence the proof of Theorem 4.1 follows the same steps as the proof of Theorem 3.5 of Coleman and Li [1].

THEOREM 4.1. *Let f be continuously differentiable and bounded below on $\mathcal{L}(x_0)=\{x\in\mathcal{B}:\ f(x)\leq f(x_0)\}$, where $\{x_k\}$ is a sequence generated by the TRIP algorithms. If H_k and $S_k^{-1}D_k^p$ are uniformly bounded and $\mathcal{L}(x_0)$ is compact then*

$$\lim_k\|D_k^pg_k\|=0.$$

The choices $S_k=D_k^p$ and $S_k=I_n$ produce bounded sequences $\{S_k^{-1}D_k^p\}$, under the assumption that $\mathcal{L}(x_0)$ is compact. They correspond to the algorithms that we propose.

5. Algorithms to compute trial steps. As in unconstrained minimization we have dogleg and conjugate–gradient algorithms to compute a trial step s_k that satisfies the fraction of Cauchy decrease condition (7).

The conjugate–gradient algorithm to compute a trial step s_k is very similar to the conjugate–gradient algorithm proposed by Steihaug [6] and Toint [7] for unconstrained minimization. The only difference is caused by the fact that x_k+s_k has to be strictly feasible.

ALGORITHM 5.1 (Conjugate–gradient algorithm for the computation of s_k).

1 Set $s^0=0$, $r_0=-g_k$, $q_0=D_k^{2p}r_0$, and $d_0=q_0$. Choose a small positive tolerance ϵ_1.
2 For $i=0,1,2,\ldots$ do
  2.1 Compute $\gamma_i=\frac{r_i^Tq_i}{d_i^TH_kd_i}$.
  2.2 Compute
  $$\begin{aligned}\tau=\max\{\tau>0\ :\ &\|S_k^{-1}(s^i+\tau d_i)\|\leq\delta_k,\\ &\sigma_k(a-x_k)\leq s^i+\tau d_i\leq\sigma_k(b-x_k)\}.\end{aligned}$$

2.3 If $\gamma_i \leq 0$, or if $\gamma_i > \tau$, then set $s_k = s^i + \tau d_i$, where τ is given as in 2.2 and stop; otherwise set $s^{i+1} = s^i + \gamma_i d_i$.
2.4 Update the residuals $r_{i+1} = r_i - \gamma_i H_k d_i$ and $q_{i+1} = D_k^{2p} r_{i+1}$.
2.5 Check truncation criteria: If $\sqrt{\frac{r_{i+1}^T q_{i+1}}{r_0^T q_0}} \leq \epsilon_1$ then stop and set $s_k = s^{i+1}$.
2.6 Compute $\alpha_i = \frac{r_{i+1}^T q_{i+1}}{r_i^T q_i}$ and update the direction $d_{i+1} = q_{i+1} + \alpha_i d_i$.

As before we have the choices $S_k = D_k^p$ and $S_k = I_n$.

A dogleg algorithm to compute a trial step s_k similar to the dogleg algorithm proposed by Powell [4] for unconstrained minimization can also be applied. Since both dogleg and the conjugate–gradient algorithms start by minimizing the quadratic function $\Psi_k(s)$ along the direction $-D_k^{2p} g_k$, it is a simple matter to see that any trial step s_k computed by using these algorithms satisfies the fraction of Cauchy decrease (7) with $\beta = 1$.

Now we briefly describe how Coleman and Li [1] compute the trial steps. They define p_k as the solution of the trust–region subproblem

$$\begin{array}{ll} \text{minimize} & \Psi_k(s) + \frac{1}{2} s^T C_k s \\ \text{subject to} & \|D_k^{-\frac{1}{2}} s\| \leq \delta_k, \end{array}$$

where C_k is given by (5), and compute the Cauchy point c_k for some $\sigma_k \in (0,1)$. They propose two algorithms. In the first, called double trust–region method, they scale p_k into the interior of $\mathcal{B}$ and accept or reject s_k based on a fraction of Cauchy decrease and on the following ratio between actual and predicted decreases:

$$\frac{f(x_k) - f(x_{k+1}) - \frac{1}{2} s_k^T C_k s_k}{\Psi_k(0) - \Psi_k(s_k) - \frac{1}{2} s_k^T C_k s_k}. \tag{10}$$

In the second, called practical trust–region algorithm, they choose s_k to be either c_k or the scaled p_k according to a fraction of Cauchy decrease condition. Then they accept or reject s_k based on the ratio (10). Their algorithms are, under appropriate assumptions, globally convergent to a nondegenerate point satisfying the second–order KKT conditions with a q–quadratic local rate of convergence.

6. Numerical examples and conclusions. We have implemented the TRIP algorithms using MATLAB 4.2a in a Sun (Sparc) workstation. We have used $\delta_0 = 1$, $p = 1$, $\sigma_k = \sigma = 0.99995$ for all k, $\epsilon_1 = 10^{-4}$, and $\epsilon = 10^{-5}$. We have tested the algorithms in a set of problems given in Conn, Gould and Toint [2]. This set of problems is divided in two groups, labeled by U and C (see Table 1). In problems U, the solution lies in the interior of $\mathcal{B}$ and therefore these problems correspond to the situation described in Section 2.1.

		$S_k = D_k$		$S_k = I_n$	
Problem	n	geval	feval	geval	feval
GENROSE U	8	24	35	28	36
GENROSE C	8	9	9	8	8
CHAINROSE U	25	15	17	16	20
CHAINROSE C	25	22	29	22	26
DEGENROSE U	25	31	39	28	29
DEGENROSE C	25	33	42	27	32
GENSING U	20	25	25	25	25
GENSING C	20	17	17	17	17
CHAINSING U	20	25	25	25	25
CHAINSING C	20	28	28	29	29
DEGENSING U	20	33	33	34	34
DEGENSING C	20	33	33	32	32
GENWOOD U	8	38	58	39	51
GENWOOD C	8	9	9	8	8
CHAINWOOD U	8	38	57	37	50
CHAINWOOD C	8	9	9	10	10
HOSC45 U	10	11	11	12	12
HOSC45 C	10	11	11	13	13
BROYDEN1A U	30	14	14	14	14
BROYDEN1A C	30	21	21	22	22
BROYDEN1B U	30	7	7	7	7
BROYDEN1B C	30	20	20	17	17
BROYDEN2A U	30	16	24	15	15
BROYDEN2A C	30	22	22	23	23
BROYDEN2B U	30	9	9	8	8
BROYDEN2B C	30	24	24	23	23
TOINTBROY U	30	8	8	*9	*37
TOINTBROY C	30	21	21	21	21
TRIG U	10	11	22	10	15
TRIG C	10	13	13	12	12
TOINTTRIG U	10	*6	*27	6	6
TOINTTRIG C	10	22	28	12	12
CRAGGLEVY U	8	25	25	25	25
CRAGGLEVY C	8	22	27	21	21
PENALTY U	15	23	23	24	25
PENALTY C	15	27	27	30	30
AUGMLAGN U	15	19	23	20	24
AUGMLAGN C	15	26	27	27	29
BROWN1 U	10	18	18	18	18
BROWN1 C	10	28	28	28	28
BROWN3 U	10	8	8	8	8
BROWN3 C	10	10	10	9	9
BVP U	10	9	9	9	9
BVP C	10	9	9	10	10
VAR U	20	9	9	9	9
VAR C	20	8	8	8	8

TABLE 1

Numerical solution of small test problems. *n – number of variables, geval – number of gradient evaluations, feval – number of function evaluations.*

	$S_k = D_k$		$S_k = I_n$	
	geval	feval	geval	feval
Problems U	422	526	426	502
Problems C	444	472	429	440
Total	866	998	855	942

TABLE 2
Comparation of the two algorithms. *geval – number of gradient evaluations, feval – number of function evaluations.*

In the cases where the initial point given in [2] is not strictly feasible, we scale it back into the interior of $\mathcal{B}$ according to the rules used in [1]. The scheme 2.3 (see Section 3) used to update the trust radius is the following:

- If $r_k < 10^{-4}$, reject s_k and set $\delta_{k+1} = 0.5\|S_k^{-1}s_k\|$.
- If $10^{-4} \leq r_k < 0.1$, reject s_k and set $\delta_{k+1} = 0.5\|S_k^{-1}s_k\|$.
- If $0.1 \leq r_k < 0.75$, accept s_k and set $\delta_{k+1} = \delta_k$.
- If $r_k \geq 0.75$, accept s_k and set $\delta_{k+1} = 2\delta_k$.

We also stopped the algorithms when the trust radius was reduced below 10^{-16}. These failures are indicated by $*$ in Table 1. Our stopping criteria is different from the stopping criteria used by Coleman and Li. We stop if either $\|D_k \nabla f(x_k)\| \leq 10^{-5}$ or the trust radius is reduced below 10^{-16}. They stop when $D_k^{\frac{1}{2}} H_k D_k^{\frac{1}{2}}$ is positive definite and $\Psi_k(s_k) - \Psi_k(0) + \frac{1}{2}s_k^T C_k s_k < 0.5 * 10^{-12}$. Our stopping criteria is of the type given in [2].

The results are given in Table 1. In Table 2 we list the total number of function and gradient evaluations taken by both approaches to solve problems U and C. From this table we observe that the second algorithm ($S_k = I_n$) performed better. We believe that this will be more clearly the case in larger problems.

Acknowledgments. We are grateful to Matthias Heinkenschloss, Virginia Polytechnic Institute and State University, for his many suggestions that have improved the presentation of this paper.

REFERENCES

[1] T. F. COLEMAN AND Y. LI, *An interior trust region approach for nonlinear minimization subject to bounds*, Tech. Report TR93–1342, Department of Computer Science, Cornell University, 1993. To appear in SIAM J. Optim.

[2] A. R. CONN, N. I. M. GOULD, AND P. L. TOINT, *Testing a class of methods for solving minimization problems with simple bounds on the variables*, Math. Comp., 50 (1988), pp. 399–340.

[3] J. J. MORÉ, *Recent developments in algorithms and software for trust regions methods*, in Mathematical programming. The state of art, A. Bachem, M. Grotschel, and B. Korte, eds., Springer Verlag, New York, 1983, pp. 258–287.

[4] M. J. D. POWELL, *A new algorithm for unconstrained optimization*, in Nonlinear Programming, J. B. Rosen, O. L. Mangasarian, and K. Ritter, eds., Academic Press, New York, 1970.

[5] ———, *Convergence properties of a class of minimization algorithms*, in Nonlinear Programming 2, O. L. Mangasarian, R. R. Meyer, and S. M. Robinson, eds., Academic Press, New York, 1975, pp. 1–27.

[6] T. STEIHAUG, *The conjugate gradient method and trust regions in large scale optimization*, SIAM J. Numer. Anal., 20 (1983), pp. 626–637.

[7] P. L. TOINT, *Towards an efficient sparsity exploiting Newton method for minimization*, in Sparse Matrices and Their Uses, I. S. Duff, ed., Academic Press, New York, 1981, pp. 57–87.

Adaptive Kernel Estimation of a Cusp-shaped Mode

W. Ehm
Technische Universität München

1. Introduction. Consider the nonparametric regression model

$$y_i = f(x_i) + \varepsilon_i, \qquad 1 \le i \le n,$$

with unknown regression function f, deterministic design points $x_1 < \ldots < x_n$, and independent error random variables ε_i with mean zero and variance $\sigma_i^2 = \sigma^2(x_i)$. The problem is to estimate the mode τ (i.e., the location of the maximum) of f in the case where locally at τ, (the graph of) f looks like the minimum of two straight lines. Situations inviting such modelling often occur in applications; cf. Müller (1992).

A straightforward approach to the problem consists in estimating f by means of some kernel estimator and taking the mode of the latter as an estimate of τ. While this works well in the smooth case, when f'' is continuous with $f''(\tau) < 0$ (Eddy, 1980), it implies a suboptimal rate of convergence in the case where the first derivative of f is discontinuous at τ. The rate then cannot be better than $O_p(n^{-1/3})$ in general, even if f is a straight line on each side of τ. However, a cusp-shaped mode should be estimable as well as f itself, that is at the rate $O_p(n^{-p/(2p+1)})$ for f of class C^p except at $t = \tau$.

Our goal in this paper is: to clarify the reason for the breakdown of the simple method described above; and, based on this analysis, to introduce a step by step improvement procedure which yields the proper rate of convergence.

This rate may also be obtained by other approaches. Müller (1992) studied the more general problem of estimating the location (and size) of a discontinuity in some derivative of f. He estimates the left- and right-hand derivatives of f using one-sided kernels and takes the maximizer of the difference as the estimate of τ. Eubank and Speckman (1994) decomposed f into $f(t) = f_s(t) + \beta\,(t - \tau)_+$, with a smooth non-parametric part f_s. Their semiparametric estimator is based on a comparison of raw and smoothed residuals with respect to the second part. As shown in these papers, both estimators of τ are asymptotically normal and achieve the precision indicated above. (Eubank and Speckman consider only the case $p = 2$. They assume $f_s \in C^2$ except at $t = \tau$, and $f_s \in C^1$ everywhere. Müller's assumptions are slightly stronger, requiring $f_s \in C^3$ everywhere when $p = 2$.)

In section 2 we collect some facts concerning the local behaviour of Gasser-Müller kernel estimators near τ. Section 3 outlines the idea of the stepwise improvement procedure. Section 4 fills in the details. The main result, and discussion of a related superefficiency phenomenon may be found in section 5.

2. Preliminaries. Concerning f we make the following

Assumption (A). For every $\eta > 0$, $\sup_{|t-\tau|\geq\eta} f(t) < f(\tau)$.

$f \in C^p$ on $(-\infty, \tau)$ and on $(\tau, +\infty)$, respectively, for some $p \geq 2$. The j-th derivative $f^{(j)}$ has a continuous extension to each of the closed intervals $(-\infty, \tau]$ and $[\tau, +\infty)$, for $j \leq p$. $f_\pm^{(j)} := f^{(j)}(\tau\pm)$ denote the one-sided derivatives at τ.

$f'_- > 0 > f'_+$.

The Gasser-Müller (1979) nonparametric regression estimator with bandwidth b and kernel K satisfying $\int K = 1$ and $K(x) = 0$ for $|x| > 1$ is

$$\hat{f}_b(t) = \sum_{i=1}^{n} \int_{\xi_{i-1}}^{\xi_i} K(\tfrac{t-u}{b}) \tfrac{du}{b}\, y_i, \text{ where the } \xi_i\text{'s are such that } \xi_{i-1} < x_i < \xi_i,$$

for every i. It may be decomposed as

$$(1) \qquad \hat{f}_b(t) = \sum_{i=1}^{n} \int_{\xi_{i-1}}^{\xi_i} K(\tfrac{t-u}{b}) \tfrac{du}{b} f(x_i) + \sum_{i=1}^{n} \int_{\xi_{i-1}}^{\xi_i} K(\tfrac{t-u}{b}) \tfrac{du}{b} \varepsilon_i = f_b(t) + Z_b(t).$$

Under Assumption (A), the deterministic part f_b in (1) can be approximated by the convolution of f with the rescaled kernel $K_b(u) = K(u/b)/b$,

$$(2) \qquad | f*K_b(t) - f_b(t) | = \Big| \sum_{i=1}^{n} \int_{\xi_{i-1}}^{\xi_i} K(\tfrac{t-u}{b}) (f(u) - f(x_i)) \tfrac{du}{b} \Big| \leq \delta L \int_{-1}^{1} |K(s)|\, ds,$$

where $\delta = \max_i (\xi_i - \xi_{i-1})$, $L = \sup \{ |(f(x)-f(y))/(x-y)| \mid |x-y| \leq \delta, x \neq y \}$.

To study the local behaviour of $f*K_b$ near τ, note first that

$$f*K_b(\tau + bs) - f*K_b(\tau) = \int_{-\infty}^{\infty} (f(\tau - bu) - f(\tau)) (K(u+s) - K(u))\, du.$$

Since K(x) vanishes for $|x| \geq 1$, two one-sided Taylor expansions then show that for $b \to 0$, every $q = 1,\ldots,p$, and any $T > 0$ we have uniformly in $|s| \leq T$,

$$(3) \qquad \Big| f*K_b(\tau + bs) - f*K_b(\tau) - \sum_{j=1}^{q} \frac{(-b)^j}{j!} \phi_j(s) \Big| \leq \Omega_q(bT) \frac{b^q}{q!} R_q(s),$$

where

$$\phi_j(s) = f_+^{(j)} \phi_j^-(s) + f_-^{(j)} \phi_j^+(s), \qquad \text{with}$$

$$\phi_j^+(s) = \int_0^\infty u^j\,(K(s+u) - K(u))du, \quad \phi_j^-(s) = \int_{-\infty}^0 u^j\,(K(s+u) - K(u))du,$$
$$R_q(s) = \int_{-\infty}^{\infty} |u|^q\,|K(u+s) - K(u)|\,du, \qquad \text{and}$$
$$\Omega_q(w) = \max\{\Omega_q^+(w), \Omega_q^-(w)\} \qquad \text{where}$$
$$\Omega_q^\pm(w) = \sup\{|f^{(q)}(\tau \pm v) - f^{(q)}(\tau\pm)|: 0 \le v \le w\} \qquad (w > 0).$$

Next we reexpress the random part as a stochastic integral with respect to a Brownian motion like process W_b defined as follows. Let S denote the piecewise constant random function

$$S(u) = \varepsilon_i \text{ for } \xi_{i-1} \le u < \xi_i, \quad S(u) = 0 \text{ for u outside } (\xi_0, \xi_n]\,.$$

Put $\omega(u) = \int_{\xi_0}^u S(v)\,dv$ and introduce the process

$$W_b(u) = V_b^{-1/2}\,(\omega(\tau + bu) - \omega(\tau)) \qquad (u \in \mathbf{R}),$$

where $V_b = \mathrm{var}\,(\omega(\tau + b) - \omega(\tau))$. A change of variables then gives

$$\text{(4)} \qquad Z_b(t) = b^{-1}\,V_b^{1/2} \int_{-\infty}^{\infty} K(\tfrac{t-\tau}{b} - u)\,dW_b(u)\,.$$

Combining (1) to (4) and writing

$$G_b(s) = \int_{-\infty}^{\infty} (K(s-u) - K(-u))\,dW_b(u),$$
$$\psi_b(s) = b^{-1} \sum_{j=1}^{q} \frac{(-b)^j}{j!}\,\phi_j(s) \qquad (s \in \mathbf{R}),$$

we get, for $b \to 0$, $1 \le q \le p$, uniformly in $|s| \le T$,

$$\text{(5)} \qquad \Delta_b(s) := b^{-1}(\hat{f}_b(\tau + bs) - \hat{f}_b(\tau))$$
$$= \psi_b(s) + b^{-2}\,V_b^{1/2}\,G_b(s) + O(\delta/b) + o(1)\,b^{q-1}\,O(s).$$

In the sequel everything (except for f) will depend on n, but this is not made explicit notationally. In particular, $G_b(s)$, $\Delta_b(s)$ etc. are in fact sequences of stochastic processes indexed by n (thus may be "tight" or "converge weakly to a Gaussian process").

If the ξ_i's are approximately equidistant locally around τ, and if the variance function $\sigma^2(x)$ is continuous at $x = \tau$, then V_b behaves like a constant times b/n, and the process W_b should converge weakly to standard two-sided Brownian motion (because $\omega(u)$ is the partial sum process $\sum_{\xi_i \le u} \varepsilon_i(\xi_i - \xi_{i-1})$ up to linear interpolation). To simplify matters we make this an assumption.

Assumption (B). $\delta = \max_i (\xi_i - \xi_{i-1}) = o(n^{-(2p-1)/(2p+1)})$; $V_b = b/n$; and for every $T > 0$, the process $W_b(u)$, $|u| \le T$, converges weakly to the standard (two-sided) Wiener process on $[-T, T]$.

Assumptions (A) and (B) are supposed to hold throughout the remainder of this paper.

Lemma 1. If K' exists a. e. and is of bounded variation, then for every $T > 0$ the process of first derivatives of G_b, $\{G_b'(s) : |s| \le T\}$, is tight. Moreover, $G_b'(0)$ is asymptotically normally distributed, $G_b'(0) \to_{\mathcal{D}} N(0, \int K'(-u)^2\, du)$.

Proof. Consider the process $H(s) = \int_{-\infty}^{\infty} K'(s-u)\, dW_b(u)$, $s \in \mathbf{R}$. H(s) is well-defined as a Lebesgue-Stieltjes integral since ω, and hence W_b, is absolutely continuous and K' is of bounded variation. Partial integration gives

$$H(s) \;=\; -\int_{-\infty}^{\infty} W_b(u)\, d_u K'(s-u) \;=\; -\int_{-\infty}^{\infty} W_b(s-v)\, d_v K'(v).$$

Since K' considered as a signed measure has no mass outside [−1, 1] we get

$$|H(s)-H(t)| \;\le\; \sup_{-1 \le v \le 1} |W_b(s-v)-W_b(t-v)| \times TV(K'),$$

TV(K') denoting the total variation norm of K'. Therefore the tightness properties of W_b which follow from Assumption (B) carry over to H. This in turn implies that $H = G_b'$, the derivative (in the usual sense) of G_b. Again by (B), the second claim is now obvious.

3. Outline of basic argument. For simplicity we will assume that there already exists a consistent preliminary estimator of τ of a certain precision.

Assumption (C). There exists an estimator $\tilde{\tau}$ such that $\tilde{\tau} = \tau + O_p(n^{-3/10})$.

This could be proved using well-known results about convergence rates of nonparametric estimators of Lipschitz-continuous regression functions. In fact, one can show that τ is estimable at the rate $O(n^{-1/3})$ if the design points are not too irregularly spaced, but the above rate suffices for our purposes.

Let $b_q = b_{q,n} = n^{-1/(2q+1)}$ ($q \in \mathbf{N}$). We shall show that the following holds for every $q = 2,\dots,p$:

Suppose there exists an estimator $\tilde{\tau}$ of τ such that $\tilde{\tau} = \tau + O_p(b_q^{q-1/2})$.

Then $\tilde{\tau}$ can be used to construct an improved estimator $\hat{\tau}$ satisfying

$$\hat{\tau} = \tau + O_p(b_q^q).$$

For $q = 2$, $b_q^{q-1/2} = n^{-3/10}$, and the premise follows from Assumption (C). Now $b_2^2 = o(b_3^{5/2})$, so the improved estimator $\hat{\tau}$ can be used as the preliminary estimator $\tilde{\tau}$ for $q = 3$. Since $b_{q-1}^{q-1} = o(b_q^{q-1/2})$ for every $q \ge 2$, one may iterate this argument which finally yields an estimator having the desired precision $O_p(b_p^p) = O_p(n^{-p/(2p+1)})$.

To explain the construction we need to discuss the local behaviour of the kernel estimators of section 2 in more detail. Let $\varepsilon = \varepsilon_n$ be a sequence of reals tending to 0 such that $b^{q-1} = o(\varepsilon)$, and put $b = b_q$, to ease the notation. Consider a kernel estimator $\hat{f}_b$ as above and let

$$(6)\qquad t^* = \inf\{t \in J(\tau, \varepsilon b) : \hat{f}_b(t) \geq \hat{f}_b(x) \text{ for every } x \in J(\tau, \varepsilon b)\}$$

denote the location of the first maximum of $\hat{f}_b$ in the interval $J(\tau, \varepsilon b) = [\tau - \varepsilon b, \tau + \varepsilon b]$. Observe that $b^{-2} V_b^{1/2} = (nb^3)^{-1/2} = n^{-(q-1)/(2q+1)} = b^{q-1}$, and $\delta/b = o(b^{2(q-1)})$, by (B). Therefore we get from (5)

$$(7)\qquad \Delta_b(s) = \psi_b(s) + b^{q-1}\{G_b(s) + o(b^{q-1}) + o(s)\}$$

uniformly in $|s| \leq \varepsilon$. Two cases are of particular interest.

CASE 1. $\liminf |\psi_b'(0)| > 0$.

In this case, $\max_{|s| \leq \varepsilon} |\Delta_b(s) - s\psi_b'(0)| = o(\varepsilon)$ (at least if $\psi_b''(s) = O(1)$ for $|s| \leq \varepsilon$, which is true here; see below), whence it follows by "unscaling" (7) that

$$t^* = \tau + b\varepsilon\,\mathrm{sign}(\psi_b'(0)) + o_p(b\varepsilon).$$

Thus, since ε may tend to 0 arbitrarily slowly, t^* cannot tend to τ at a rate faster than $O(b)$ in this case.

CASE 2. $\psi_b'(0) = O(b^{q-1})$.

Let $Y_b = b^{1-q}\psi_b'(0)$. Then $\Delta_b(s)$ can be approximated by the parabola $s\,b^{q-1}(Y_b + G_b'(0)) + \frac{s^2}{2}\psi_b''(0)$ (for details see below), and we may expect that

$$t^* \approx \tau + b^q (Y_b + G_b'(0)) / (-\psi_b''(0)).$$

Thus t^* should tend to τ at the rate $O_p(b^q)$ in this case.

These considerations suggest that also in all cases between the two extremes the convergence rate of t^* is given by b times the order of magnitude of $\psi_b'(0)$.

In order to evaluate the derivatives of ψ_b, we use partial integration with the functions $\phi_j^{\pm}$ entering the definition of ψ_b. Uniformly in $|s| \leq \varepsilon$,

$$(8)\qquad \psi_b(s) = b^{-1} \sum_{j=1}^{q} \frac{(-b)^j}{j!}\{s\,\phi_j'(0) + \frac{s^2}{2}\phi_j''(0) + o(s^2)\}, \quad \text{where}$$

$$\phi_j'(0) = -j\,(f_+^{(j)} K_{j-1}^- + f_-^{(j)} K_{j-1}^+) \qquad (1 \leq j \leq q),$$

$$\phi_j''(0) = j\,(j-1)\,(f_+^{(j)} K_{j-2}^- + f_-^{(j)} K_{j-2}^+) \qquad (2 \leq j \leq q),$$

$$\phi_1''(s) = K(s)\,(f_-' - f_+'), \qquad \text{with}$$

$$K_j^+ = \int_0^1 u^j K(u)du, \quad K_j^- = \int_{-1}^0 u^j K(u)du \qquad (j \geq 0).$$

Now

(9) $$\lim \psi_b'(0) = -\phi_1'(0) = f_+' K_0^- + f_-' K_0^+,$$

which will not vanish in general unless we make K depend on the unknown regression function f:

$$\phi_1'(0) = 0 \text{ if and only if } K_0^+ = -f_+' / (f_-' - f_+').$$

Thus, for a fixed kernel K CASE 1 applies, and t* cannot tend to τ at a rate faster than $O_p(b)$. The optimal bandwidth then is $b = n^{-1/3}$ (q = 1), because for smaller bandwidths the drift term ψ_b of Δ_b would be dominated by the noise component $(nb^3)^{-1/2} G_b$.

This shows that if one estimates τ by taking the mode of a kernel estimator of f with fixed kernel K and bandwidth b not smaller than $O(n^{-1/3})$, then this estimator will have a bias of the order of the bandwidth. The bias can be explained geometrically by the smoothing effect of the kernel estimator. For a symmetric kernel the mode of the smoothed version f_b of f clearly is shifted to that side of τ where f has smaller slope. Of course this may also be seen by inspecting the analytic expression for $-\phi_1'(0)$ obtained above.

To reduce the bias one thus may proceed as follows: estimate $-f_+' / (f_-' - f_+') = \pi$ by $\tilde{\pi}$, say, and select the kernel $\tilde{K}$ such that $\tilde{K}_0^+ = \tilde{\pi}$. Now, if $\tilde{\pi}$ is consistent we will have $\psi_b'(0) = o_p(1)$ – ψ_b is now random! –, so the rates of convergence of argmax Δ_b and of $\psi_b'(0)$ to 0 should be identical and equal to the rate at which $\tilde{\pi}$ approaches π. The details of this construction are given in the next section.

4. Details.

4.1. *Construction of "adaptive" kernels.*

For $q \in \mathbf{N}$ let $\mathcal{K}(q)$ denote the family of all kernels K satisfying the conditions

$$\int K = 1,\ K(x) = 0 \text{ for } |x| > 1,\ K_j^+ = 0 \text{ and } K_j^- = 0 \text{ for } 1 \le j < q.$$

Our adaptive method requires kernels $K \in \mathcal{K}(q)$ which can be tuned such that for given $\pi \in [0,1]$, $K_0^+ = \pi$. These can be constructed as piecewise continuous polynomials. For $x \in [0, 1]$ we make the ansatz

$$K(x) = (1-x)\,Q(x), \text{ with } Q(x) = \sum_{j=0}^{q} a_j x^j.$$

Then

$$K_i^+ = \int_0^1 x^i (1-x)\,Q(x)\,dx = \sum_{j=0}^{q} \left(\frac{1}{i+j} - \frac{1}{i+j+1}\right) a_j,$$

so if we add the further condition $K(0) = m$, or $a_0 = m$ (for some $m > 0$), the moment conditions become $Ma = c$, where M is the matrix with elements $M_{ij} = 1/(i+j) - 1/(i+j+1)$ $(1 \le i,j \le q)$, $a = (a_1,\ldots,a_q)^T$, and $c_i = \delta_{i1}\,\pi - m\,(i^{-1}-(i+1)^{-1})$ $(1 \le i \le q)$.

M is positive definite.To see this, let $A(x) = \sum_{j=1}^{q} a_j x^j$. Then $a^T M a = \int_0^1 u^{-1}(1-u)\, A(u)^2\,du$, and this integral equals zero if and only if A vanishes identically, i.e., if and only if $a = 0$.

Let $Q_{q,\pi,m}$ denote the unique polynomial of degree q obtained in this manner. We have obtained the following result.

Lemma 2. Given q $(2 \le q \le p)$, $\pi \in [0,1]$, and $m > 0$, let kernel $K = K_{q,\pi,m}$ be defined by

$$K(x) = (1-x)\,Q_{q,\pi,m}(x) \quad \text{for } 0 \le x \le 1,$$
$$K(x) = (1+x)\,Q_{q,1-\pi,m}(-x) \quad \text{for } -1 \le x \le 0,$$
$$K(x) = 0 \quad \text{for } |x| > 1.$$

Then $K \in \mathcal{K}(q)$, $K_0^+ = \pi$, $K(0) = m$, and $\sup_{0\le\pi\le1} TV(K'_{q,\pi,m}) < \infty$.

To simplify the discussion we put $m = 1$ and restrict attention to kernels K belonging to the families $\mathcal{P}(q) = \{K_{q,\pi,1} : 0 \le \pi \le 1\}$.

4.2. *Stochastic expansion of* t^*. For the following it may be helpful to recall our convention to write $b = b_q = n^{-1/(2q+1)}$.

Lemma 3. For every $q = 2,\ldots,p$ we have uniformly in all $K \in \mathcal{P}(q)$ for which $Y_b := -\phi_1'(0)\,b^{1-q} = O(1)$

$$t^* \;=\; \tau \;+\; b^q\,\frac{Y_b + G_b'(0)}{f_-' - f_+'} \;+\; o_p(b^q).$$

Proof. By (5), (8) and our assumptions, the following expansion holds uniformly in $|s| \le \varepsilon$ and $K \in \mathcal{P}(q)$,

$$\Delta_b(s) \;=\; -s\,\phi_1'(0) \;-\; \frac{s^2}{2}\,(\phi_1''(0) - b\phi_2''(0) + o(1)) \;+\; b^{q-1}\,G_b(s) \;+\; O(\delta/b) \;+\; o(b^{q-1}\,|s|)$$
$$= b^{q-1}\,s\,(Y_b + G_b'(0) + o_p(1)) \;-\; \frac{s^2}{2}\,(\phi_1''(0) + o(1)) \;+\; o(b^{2(q-1)}). \tag{10}$$

Observe that indeed $G_b(s) - s\,G_b'(0) = o_p(|s|)$ uniformly in $|s| \le \varepsilon$ and $K \in \mathcal{P}(q)$, which is clear from Lemma 2 and the proof of Lemma 1. Also, the Taylor expansion of the functions ϕ_j clearly is valid uniformly in those pairs (s, K).

To maximize the approximate parabola (10) one zooms in, once again, by putting

$s = u\eta$, $\eta = b^{q-1}$, and considering the process $Q(u) = \eta^{-2}\Delta_b(u\eta)$, $|u| \le \varepsilon/\eta$. The mode u^* of Q is easily shown to satisfy $u^* = (Y_b + G_b'(0))/\phi_1''(0) + o_p(1)$, and since $\phi_1''(0) = K(0)(f_-' - f_+')$, with $K(0) = 1$ for $K \in \mathcal{P}(q)$, unscaling twice, from u to s and then from s to $t = \tau + bs$, completes the proof.

Notice that for the validity of the expansion it is decisive that $K(0)(f_-' - f_+')$ stay bounded away from zero. This is why we fixed the value of K(0) in the preceding subsection, additionally to the other conditions.

4.3. *Estimating* $f_\pm'$. Again, let q be fixed, $2 \le q \le p$. $f_\pm'$ will be estimated by extrapolating (via a Taylor expansion) the derivative of a kernel estimate of f from "safe" regions (where f is regular) to the critical region near τ. We denote the kernel used in this step by $\overline{K}$ and the corresponding Gasser-Müller estimate by $\hat{\bar{f}}_b$, in order to distinguish them from the adaptively chosen kernels $\widetilde{K} \in \mathcal{P}(q)$ to be plugged in into the Gasser-Müller estimate of f, which is then denoted as $\tilde{f}_b$; cf. section 4.4. Concerning $\overline{K}$ we shall assume the following.

Assumption (D). $\overline{K} \in C^p$ on **R**. $\overline{K}(x) = 0$ for $|x| \ge 1$.

$$\int \overline{K}(x)\, x^j\, dx = \delta_{j0} \quad (0 \le j \le p).$$

Given a preliminary estimator $\tilde{\tau}$, f_-' will be estimated as follows. Fix some number $d_0 \le -3$. Then let

$$\hat{f}_-' = \sum_{j=0}^{q-1} \frac{(-b\, d_0)^j}{j!} \hat{\bar{f}}_b^{(j+1)}(\tilde{\tau} + b d_0). \tag{11}$$

Lemma 4. Suppose the preliminary estimator $\tilde{\tau}$ satisfies $\tilde{\tau} = \tau + o_p(b^{q-1})$. Then

$$\hat{f}_-' = f_-' + b^{q-1}\,\xi_b(d_0, d_0) + o_p(b^{q-1}), \qquad \text{where}$$

$$\xi_b(r,s) = \int_{-\infty}^{\infty} \Big\{ \sum_{j=0}^{q-1} \frac{(-r)^j}{j!} \overline{K}^{(j+1)}(s-u) \Big\}\, dW_b(u) \qquad (d \in \mathbf{R}).$$

Proof. If t is such that $|t - \tau| \ge b$ then for every $j \le q$

$$\begin{aligned}(f * \overline{K}_b)^{(j)}(t) - f^{(j)}(t) &= \int \{ f^{(j)}(t - ub) - f^{(j)}(t) \}\, \overline{K}(u)\, du \\ &= \sum_{k=1}^{q-j} \frac{f^{(j+k)}(t)}{k!} \int (-ub)^k\, \overline{K}(u)\, du + o(b^{q-j}) \\ &= o(b^{q-j}).\end{aligned}$$

Therefore, if $s \leq -1$, $s = O(1)$, then we get by a Taylor expansion around $\tau + bs$, letting $\bar{f}_b$ denote the analogue of f_b in (1) when K is replaced by $\overline{K}$,

$$
\begin{aligned}
(12) \qquad f'_- &= \sum_{j=0}^{q-1} \frac{(-bs)^j}{j!} f^{(j+1)}(\tau + bs) + o(b^{q-1}) \\
&= \sum_{j=0}^{q-1} \frac{(-bs)^j}{j!} \{ (f * \overline{K}_b)^{(j+1)}(\tau + bs) + o(b^{q-j-1}) \} + o(b^{q-1}) \\
&= \sum_{j=0}^{q-1} \frac{(-bs)^j}{j!} \{ \bar{f}_b^{(j+1)}(\tau + bs) + O(b^{-j-1}\delta) \} + o(b^{q-1}) \\
&= \sum_{j=0}^{q-1} \frac{(-bs)^j}{j!} \bar{f}_b^{(j+1)}(\tau + bs) + o(b^{q-1}).
\end{aligned}
$$

Now, putting $\tilde{d}_0 = d_0 + (\tilde{\tau} - \tau)/b$ we have $\tilde{\tau} + bd_0 = \tau + b\tilde{d}_0 \leq \tau - b$ with probability tending to 1, so by (12), (11), (1) and (4)

$$
\begin{aligned}
\widehat{f'_-} &= \sum_{j=0}^{q-1} \frac{(-b\,d_0)^j}{j!} \left[\bar{f}_b^{(j+1)}(\tau + b\tilde{d}_0) + b^{-j-2} V_b^{1/2} \int_{-\infty}^{\infty} \overline{K}^{(j+1)}(\tilde{d}_0 - u)\, dW_b(u) \right] \\
&= \sum_{j=0}^{q-1} \frac{(-b\,\tilde{d}_0)^j}{j!} \bar{f}_b^{(j+1)}(\tau + b\tilde{d}_0) + \sum_{j=0}^{q-1} \frac{(-b\,d_0)^j - (-b\tilde{d}_0)^j}{j!} \bar{f}_b^{(j+1)}(\tau + b\tilde{d}_0) \\
&\quad + b^{q-1} \xi_b(d_0, \tilde{d}_0) \\
&= f'_- + o(b^{q-1}) + O(|bd_0 - b\tilde{d}_0|) + b^{q-1} \{ \xi_b(d_0, d_0) + o_p(1) \} \\
&= f'_- + b^{q-1} \xi_b(d_0, d_0) + o_p(b^{q-1}),
\end{aligned}
$$

on using $bd_0 - b\tilde{d}_0 = \tilde{\tau} - \tau = o_p(b^{q-1})$ and tightness of W_b similarly as in the proof of Lemma 1. This proves the lemma.

An analogous result holds for the estimation of f'_+ when d_0 is replaced by some $d_1 \geq 3$.

4.4. ***Improving upon the preliminary estimator*** $\tilde{\tau}$.

Fix q, $2 \leq q \leq p$, and let $b = b_q$, as above.

Construction.

(i) Suppose there exists an estimator $\tilde{\tau}$ of τ such that $\tilde{\tau} = \tau + O_p(b^{q-1/2})$.

(ii) Given $\tilde{\tau}$, compute the slope estimates $\widehat{f'_-}$, $\widehat{f'_+}$ as in subsection 4.3 and select the kernel $\widetilde{K} \in \mathcal{P}(q)$ such that

$$\widetilde{K}_0^+ = \min\{\max\{-\widehat{f'_+} / (\widehat{f'_-} - \widehat{f'_+}), 0\}, 1\}.$$

(iii) Define $\hat{\tau}$ to be the first maximum of $\tilde{f}_b(t)$ over the interval

$$S = [\tilde{\tau} - b^{q-2/3}, \tilde{\tau} + b^{q-2/3}] = J(\tilde{\tau}, b^{q-2/3}), \quad \text{where}$$

(13) $$\tilde{f}_b(t) = \sum_{i=1}^{n} \int_{\xi_{i-1}}^{\xi_i} \tilde{K}\left(\frac{t-u}{b}\right)\frac{du}{b}\, y_i \,.$$

Proceeding to the analysis of this estimator, we note first that certainly $\tilde{\tau} = \tau + o_p(b^{q-1})$, by (i). After a straightforward expansion we get from Lemma 4

(14) $$f'_+ \tilde{K}_0^- + f'_- \tilde{K}_0^+ = -b^{q-1} U_b + o_p(b^{q-1}), \qquad \text{with}$$

$$U_b = \pi\, \xi_b(d_0,d_0) + (1-\pi)\, \xi_b(d_1,d_1), \quad \pi = -\, f'_+ / (f'_- - f'_+).$$

Let $\tilde{t}^*$ be defined as t^* in (6) but with $\tilde{f}_b$ replacing $\hat{f}_b$ and with ε tending to 0 so that $b^{1/3} = o(\varepsilon)$. Suppose Lemma 3 were applicable (see below). Then certainly

$$\tilde{t}^* \in J(\tau, b^{q-1/3}) \text{ w. pr. t. } 1$$

– this being short for "with probability tending to 1" – and hence

$$\hat{\tau} = \tilde{t}^* \text{ w. pr. t. } 1\,,$$

since clearly $J(\tau, b^{q-1/3}) \subset S \subset J(\tau, \varepsilon b)$ w. pr. t. 1. In particular, $\hat{\tau}$ admits the same stochastic approximation as $\tilde{t}^*$, which in view of (14), (9) and Lemma 3 becomes

(15) $$\hat{\tau} = \tau + b^q \frac{G_b'(0) - U_b}{f'_- - f'_+} + o_p(b^q).$$

The difficulty with the application of Lemma 3 comes from $\tilde{K}$ being data-dependent. The construction avoids the resulting complications, however.

In order to see this, note first that since $\tilde{K}$ has support [–1,+1], the weight of y_i in (13) can be positive only if the intervals (ξ_{i-1}, ξ_i) and $(t - b, t + b)$ have some point in common, hence only if $|x_i - t| \le b + \delta$. Let B denote the event $[|\tilde{\tau} - \tau| \le b^{q-2/3}]$. On B, the search interval S over which $\tilde{f}_b(t)$ is to be maximized is contained in $[\tau - 2b^{q-2/3}, \tau + 2b^{q-2/3}]$. But for $t \in [\tau - 2b^{q-2/3}, \tau + 2b^{q-2/3}]$ the sum in (13) may be restricted to those indices i for which $|x_i - \tau| \le b + 2b^{q-2/3} + \delta$, by the first observation. On the other hand, the estimate of f'_- depends only on those y_i's for which $|x_i - (\tilde{\tau} + bd_0)| \le b + \delta$, and similarly for f'_+. Therefore, *on the event B*, the slope estimates and hence the kernel $\tilde{K}$ depend only on those y_i's for which $|x_i - \tau| \ge 3b - b^{q-2/3} - b - \delta$. Now $b + 2b^{q-2/3} + \delta < 3b - b^{q-2/3} - b - \delta$ if n is large enough. Thus, on the event B and for $t \in S$, the nonzero weights in (13) depend on a set of y_i's disjoint from those y_i's which are weighted.

Furthermore, since the probability of B tends to 1, the conditional joint distribution of all y_i's (conditional on B) approximates their unconditional distribution in total variation norm.

Putting all this together we find that *we may proceed as if* $\tilde{K}$ *were chosen by an independent chance mechanism.* The above argument also shows that $G_b'(0)$ and the random variables $\xi_b(d_0,d_0)$, $\xi_b(d_1,d_1)$ entering the definition of U_b are (exactly) independent for n large enough.

5. Statement of main result. A superefficiency phenomenon.

Collecting the results of the preceding sections we get the following theorem, in which $\hat{\tau}$ is the estimator constructed above, for $q = p$.

Theorem. Under Assumptions (A) to (C),

$$n^{p/(2p+1)}(\hat{\tau}-\tau) \to_{\mathcal{D}} N(0, S),$$

with

$$S = \int \left(\pi^2 g(u;d_0,d_0)^2 + (1-\pi)^2 g(u;d_1,d_1)^2 + K'(-u)^2\right) du \,/\, (f_-' - f_+')^2,$$

where $\pi = -f_+'/(f_-' - f_+')$, $K = K_{p,\pi,1}$ (cp. Lemma 2), $d_0 \le -3$, $d_1 \ge 3$, and, for some kernel $\overline{K}$ satisfying Assumption (D),

$$g(u;r,s) = \sum_{j=0}^{p-1} \frac{(-r)^j}{j!} \overline{K}^{(j+1)}(s-u).$$

The bandwidth b in this result simply was defined as $b = n^{-1/(2p+1)}$ (in the last improvement step). A closer look at the effect of the bandwidth choice reveals a phenomenon which may appear rather surprising at first sight.

If $d > 0$ denotes the local density of the points ξ_i near τ (so that $\xi_i - \xi_{i-1} = (dn)^{-1}(1 + o(1))$ as $\xi_i \to \tau$) and if $\sigma^2(\xi) \to \sigma^2 > 0$ as $\xi \to \tau$ then $V_b = (dn)^{-1}\sigma^2 b(1 + o(1))$; cp. Assumption (B). Going through the above arguments one finds that without our simplifying assumptions the stochastic approximation for $\hat{\tau}$ becomes

$$\hat{\tau} = \tau + b^{-1} V_b^{1/2} (G_b'(0) - U_b) / (f_-' - f_+') + o_p(b^p),$$

provided the condition $b^{-1}V_b^{1/2} = O(b^p)$ is satisfied.

Suppose that, given $a > 0$, the bandwidth b is selected such that $b^{-1}V_b^{1/2} = ab^p$, or $b = (\sigma^2/a^2dn)^{1/(2p+1)}$. Then $b^{-1}V_b^{1/2} = a^{1/(2p+1)}(\sigma^2/dn)^{p/(2p+1)}$, and the theorem may be stated in the following form,

$$n^{p/(2p+1)}(\hat{\tau}-\tau) \rightarrow_{\mathcal{D}} N(0, (a(\sigma^2/d)^p)^{2/(2p+1)} S).$$

Thus, since $a > 0$ was arbitrary, the asymptotic variance can be made arbitrarily small.

This superefficiency phenomenon is a consequence of our assumption that $f^{(p)}$ is *continuous* on both sides of τ, which for fixed f means slightly more than just "regularity of order p". It is well-known that if $f^{(p)}$ satisfies a Hölder condition of order $\alpha \in (0,1)$ then (for $t \neq \tau$) f(t) is estimable at the rate $O(n^{-s/(2s+1)})$, where $s = p + \alpha$, and the same should be true for the estimation of a cusp-shaped mode. Thus $n^{-p/(2p+1)}$ need not necessarily be the optimal rate of convergence.

Another important aspect in this context is that superefficiency does not hold uniformly in f, even if $f'_- - f'_+$ is assumed to stay bounded away from zero within the class of f's considered. This is because in general the continuity modulus Ω_p (see around (3)) is not equi-continuous at zero (uniformly in f), and then the remainder in the approximation of the deterministic term will be no more of the order $o(b^p)$.

Acknowledgment. I am grateful to Hans-Georg Müller for stimulating discussions on the subject of this paper.

References.

Eddy, W.F. (1980). Optimum kernel estimators of the mode. *Ann. Statist.* **8** 870-882.

Eddy, W.F. (1982). The asymptotic distributions of kernel estimators of the mode. *Z. Wahrsch. verw. Geb.* **59** 279-290.

Eubank, R.L. and Speckman, P.L. (1994). Nonparametric estimation of functions with jump dicontinuities. In: *Change-point Problems.* IMS Lecture Notes - Monograph Series Vol. 23, 130-143.

Gasser, T. and Müller, H.-G. (1979). Kernel estimation of regression functions. In *Lecture Notes in Mathematics*, T. Gasser and M. Rosenblatt (eds), Vol. 757, 23-68.

Müller, H.-G. (1985). Kernel estimators of zeros and of location and size of extrema of regression functions. *Scand. J. Statist.* **12** 221-232.

Müller, H.-G. (1992). Change-points in nonparametric regression analysis. *Ann. Statist.* **20** 737-761.

Automatic Differentiation: The Key Idea and an Illustrative Example

Herbert Fischer

Abstract. For many problems of Numerical Analysis the derivatives of some functions need to be computed. In practice, programs for computing derivatives may be (1) hand-coded, (2) generated by symbolic manipulation of formulas, (3) set up via function calls and divided differences, or (4) obtained using automatic differentiation. Numerically, the divided differences approach (3) is still the standard technique. But in many cases, derivatives can be obtained cheaper and more accurately by automatic differentiation. In this paper we are concerned mainly with demonstrating the key idea of automatic differentiation.

1 Introduction

The computational solution of many mathematical problems involves derivatives. As is well known, derivatives of functions can be computed (1) by hand-coded programs, (2) using symbolic manipulation of formulas e.g. with MAPLE, or (3) numerically with divided differences. None of these three processes is considered to be "automatic differentiation". What exactly this paper illustrates are some interesting points of automatic differentiation, which is a modern computer method with many benefits.

In 1959, researchers at the USSR Academy of Science (L.M. Beda, L.N. Korolev, N.V. Sukkikh, and T.S. Frolova) wrote a report [5] "Programs for automatic differentiation for the machine BESM", which is apparently the earliest paper on this subject. Since then, roughly 200 relevant papers have been published. In 1981, Rall's book [38] appeared, which is now a standard reference. An overall presentation of the state of the art as of 1991 can be found in the book [23] edited by

Griewank and Corliss. The proceedings of the Second International Workshop on Computational Differentiation, Santa Fe, New Mexico, in February 1996, shall provide the latest information on our topic.

Automatic differentiation is a method in which a program for evaluating a function f is transformed into another program that evaluates both the function f and its derivative f'. The key idea is the repeated, hierarchical use of the chain–rule for composing the derivative of that function f from derivatives of "parts" of f. Available software [26] shows that automatic differentation (where it can be applied) is superior to divided differences. At present automatic differentiation research is relatively active.

2 The forward mode

Let $f : D \subseteq \mathbb{R}^n \to \mathbb{R}$ be a differentiable function, and let f' denote the derivative of f. Recall that

$$f'(x) = \left[\frac{\partial f(x)}{\partial x_1}, \frac{\partial f(x)}{\partial x_2}, \ldots, \frac{\partial f(x)}{\partial x_n}\right] \in \mathbb{R}^{1\times n},$$

that means $f'(x)$ is a matrix with 1 row and n columns.
Define a black box FD, which accepts the function f and the point $x \in D$, and that produces the pair $f(x), f'(x)$.

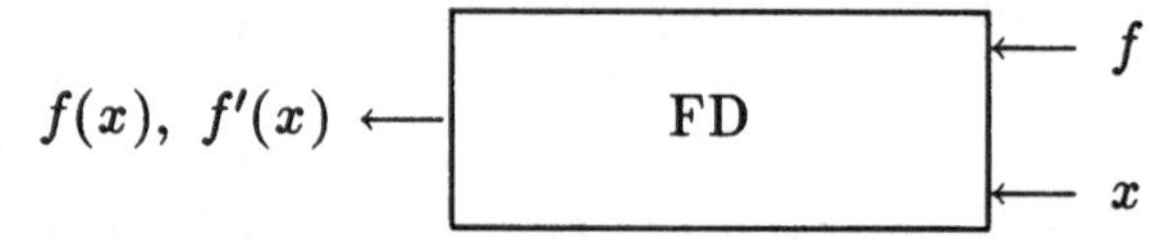

(FD stands for Function value and Derivative value.) Our aim is to implement the black box FD for fairly general f and arbitrary $x \in D$.

To begin with, we state trivial cases. Let $r : D \subseteq \mathbb{R}^n \to \mathbb{R}$ be a projection, that means $r(x) = x_k = k$-th component of x. Then

$$r'(x) = [0 \ldots 0\ 1\ 0 \ldots 0] \text{ with 1 in } k\text{-th position.}$$

Let $r : D \subseteq \mathbb{R}^n \to \mathbb{R}$ be a constant function. Then

$$r'(x) = [0 \ldots\ldots 0] = \text{ zero-matrix.}$$

Hence, the black box FD can be implemented for projections and constant functions. Though this is trivial, we need it as basis for the sequel.

Now we consider two familiar ways for building new functions from old ones, first the rational composition, and second the library composition.

2.1 Rational composition

Consider two functions

$$a : D \subseteq \mathbb{R}^n \to \mathbb{R} \quad \text{and} \quad b : D \subseteq \mathbb{R}^n \to \mathbb{R}.$$

Let r be one of the functions $a+b$, $a-b$, $a \cdot b$, a/b with the restriction $b(x) \neq 0$ for all $x \in D$ in the case $r = a/b$. Assume that the functions a and b are differentiable. Then the function r is differentiable too. Table 1 shows formulas for the derivative r'.

type	function	derivative
A	$r = a + b$	$r' = a' + b'$
S	$r = a - b$	$r' = a' - b'$
M	$r = a \cdot b$	$r' = b \cdot a' + a \cdot b'$
D	$r = a/b$	$r' = (a' - r \cdot b')/b$

Table 1: Derivative of rational composition

We have to distinguish strictly between functions and function values! So it should be clear that Table 1 shows equations of functions. Applying any of the functions r, r' to some $x \in D$ we get equations of function values. For instance in case of multiplication we get

$$r'(x) = b(x) \cdot a'(x) + a(x) \cdot b'(x).$$

From the formulas in Table 1 we conclude:

The pair $r(x), r'(x)$ can be computed from the pairs $a(x), a'(x)$ and $b(x), b'(x)$.

Note that the pair $r(x), r'(x)$ is not a pair of formulas, nor is it a pair of functions; instead it is an element of $\mathbb{R} \times \mathbb{R}^{1 \times n}$. The mechanism to compute the pair $r(x), r'(x)$ from the pairs $a(x), a'(x)$ and $b(x), b'(x)$ does not depend on x, nor does it depend on the special forms of the functions a and b; it merely depends on the type of r. This observation allows definition of a black box RAT, which accepts the type T $\in$

{A,S,M,D} of r, and the pairs $a(x), a'(x)$ and $b(x), b'(x)$, and that produces the pair $r(x), r'(x)$.

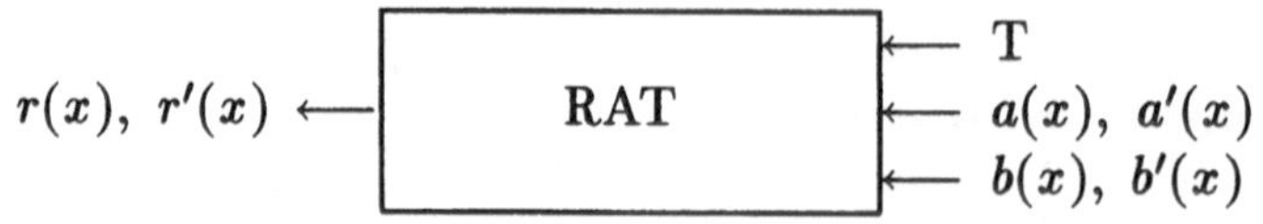

It is obvious that RAT can be implemented easily as a procedure in PASCAL, as a subroutine in FORTRAN, or as a function in a more flexible programming language. Of course, the black box RAT can be replaced by four black boxes ADD, SUB, MUL, DIV, each of these designed for one particular type T. Using a programming language with operator–overloading, the pair $r(x), r'(x)$ can be computed from the pairs $a(x), a'(x)$ and $b(x), b'(x)$ with properly defined operators, see for instance [39].

A very short example shall show the use of RAT. Consider the function f of three variables

$$f : D \subseteq \mathbb{R}^3 \to \mathbb{R} \text{ with } f(x) = \frac{x_1 \cdot x_2}{x_3},$$

where $D = \{x | x = (x_1, x_2, x_3) \in \mathbb{R}^3, x_3 \neq 0\}$.
Define the functions $f_1,\ f_2,\ f_3,\ f_4,\ f_5 : D \to \mathbb{R}$ by

$$\begin{aligned} f_1(x) &= x_1 \\ f_2(x) &= x_2 \\ f_3(x) &= x_3 \\ f_4(x) &= f_1(x) \cdot f_2(x) \\ f_5(x) &= f_4(x)/f_3(x) \end{aligned}$$

Of course, $f_5 = f$. For given $x \in D$ we compute

$$\begin{aligned} Y_1 &\leftarrow (x_1, [1,0,0]) \\ Y_2 &\leftarrow (x_2, [0,1,0]) \\ Y_3 &\leftarrow (x_3, [0,0,1]) \\ Y_4 &\leftarrow RAT(M, Y_1, Y_2) \\ Y_5 &\leftarrow RAT(D, Y_4, Y_3) \end{aligned}$$

So we obtain the pair $Y_5 = (f_5(x),\ f_5'(x)) = (f(x),\ f'(x))$.
This example indicates that the black box FD can be implemented for *any* explicitly given rational function.

2.2 Library composition

Let Λ be a collection of real functions of one real variable. For brevity these functions are called *library functions*. One may choose sin, ln, sqrt, ... and the like as library functions.

Consider some library function

$$\lambda : E \subseteq \mathbb{R} \to \mathbb{R}$$

and some function

$$a : D \subseteq \mathbb{R}^n \to \mathbb{R}.$$

Under the restriction $a(D) \subseteq E$ we define the function

$$r : D \subseteq \mathbb{R}^n \to \mathbb{R} \text{ with } r(x) = \lambda(a(x)).$$

Assume that the functions λ and a are differentiable. Then the function r is differentiable too, and by the chain–rule we have

$$r'(x) = \lambda'(a(x)) \cdot a'(x).$$

From the formulas for $r(x)$ and $r'(x)$ we conclude:

> *The pair $r(x), r'(x)$ can be computed from the pair $a(x), a'(x)$ using λ, λ'.*

The mechanism to compute the pair $r(x), r'(x)$ from the pair $a(x), a'(x)$ does not depend on x nor does it depend on the special form of the function a; it is merely a matter of the library function λ. We assume we are able to evaluate λ and λ' at any suitable value of the argument. This is no problem as long as λ is one of the commonly used library functions sin, ln, sqrt and the like. Hence, we can define a black box LIB, which accepts the name N of λ, and the pair $a(x), a'(x)$, and that produces the pair $r(x), r'(x)$.

It is obvious that LIB can be implemented easily as a procedure in PASCAL, as a subroutine in FORTRAN, or as a function in a more flexible programming language. Of course, the black box LIB can be replaced by a collection of black boxes, each of these designed for one particular library function.

A second very short example now illustrates the use of RAT and LIB. Consider the function f of three variables

$$f : D \subseteq \mathbb{R}^3 \to \mathbb{R} \text{ with } f(x) = \frac{(x_1 - 7) \cdot \sin(x_1 + x_2)}{x_3}.$$

For given $x \in D$ the function value $f(x)$ can be computed step by step as shown in the first column of Table 2. Here $y_9 = f(x)$.

$y_1 \leftarrow x_1$	$f_1(x) = x_1$	$Y_1 \leftarrow (x_1, [1,0,0])$
$y_2 \leftarrow x_2$	$f_2(x) = x_2$	$Y_2 \leftarrow (x_2, [0,1,0])$
$y_3 \leftarrow x_3$	$f_3(x) = x_3$	$Y_3 \leftarrow (x_3, [0,0,1])$
$y_4 \leftarrow 7$	$f_4(x) = 7$	$Y_4 \leftarrow (7, [0,0,0])$
$y_5 \leftarrow y_1 - y_4$	$f_5(x) = f_1(x) - f_4(x)$	$Y_5 \leftarrow RAT(S, Y_1, Y_4)$
$y_6 \leftarrow y_1 + y_2$	$f_6(x) = f_1(x) + f_2(x)$	$Y_6 \leftarrow RAT(A, Y_1, Y_2)$
$y_7 \leftarrow \sin(y_6)$	$f_7(x) = \sin(f_6(x))$	$Y_7 \leftarrow LIB(SIN, Y_6)$
$y_8 \leftarrow y_5 \cdot y_7$	$f_8(x) = f_5(x) \cdot f_7(x)$	$Y_8 \leftarrow RAT(M, Y_5, Y_7)$
$y_9 \leftarrow y_8 / y_3$	$f_9(x) = f_8(x) / f_3(x)$	$Y_9 \leftarrow RAT(D, Y_8, Y_3)$

Table 2: Computation of $Y_9 = (f(x),\ f'(x))$.

The setup for the intermediate values $y_1, \ldots, y_9$ gives rise to the definition of intermediate functions $f_1, \ldots, f_9$ as shown in the second column of Table 2. Of course $f_9 = f$.

Now we introduce pairs $Y_k = (f_k(x), f_k'(x))$ for $k = 1, \ldots, 9$, consisting of a function value and the corresponding derivative value. For $k = 1, 2, 3, 4$ the pair Y_k is obvious. And for $k = 5, 6, 7, 8, 9$ in this order the pair Y_k can be computed from previous pairs using RAT and LIB as shown in the third column of Table 2. The final pair is

$$Y_9 = (f_9(x),\ f_9'(x)) = (f(x),\ f'(x)).$$

In Table 2 the first column represents a *characterizing sequence* or *code list* [38] for the function f, the second column only serves to show that a code list defines an iterative sequence of functions. And the third column can be considered as an informal program for computing $f(x)$ and $f'(x)$.

3 More general setting

Let $f : D \subseteq \mathbb{R}^n \to \mathbb{R}$ be a differentiable function. Assume that for $x \in D$ the function value $f(x)$ can be computed using the algorithm A, see Table 3.

(1)	for $k = 1, \ldots, n$ $y_k \leftarrow x_k = k$-th component of x
(2)	for $k = n+1, \ldots, n+t$ $y_k \leftarrow \lambda_k(y_1, \ldots, y_{k-1})$
(3)	$f(x) \leftarrow y_{n+t}$

Table 3: Algorithm A for computing $f(x)$.

Hereby we assume that the λ's used in block (2) are differentiable, and that the respective derivatives can be evaluated for any argument of interest. For instance λ_k may be a constant function, or λ_k may represent a rational operation

$$\lambda_k(y_1, \ldots, y_{k-1}) = y_i * y_j \text{ with } * \in \{+, -, \cdot, /\},$$

or λ_k may represent a library function e.g.

$$\lambda_k(y_1, \ldots, y_{k-1}) = \sin(y_i).$$

So the two examples of section 2 fit into the schedule of Algorithm A. Define for $k = n+1, \ldots, n+t$

$$E_k : D_{k-1} \subseteq \mathbb{R}^{k-1} \to \mathbb{R}^k \text{ with } E_k(z) := \begin{bmatrix} z \\ \lambda_k(z) \end{bmatrix},$$

where $D_n = D$. The function E_k corresponds to the k-th loop of Algorithm A. E_k copies all values of its argument and appends one "operation". Furthermore define

$$L : \mathbb{R}^{n+t} \to \mathbb{R} \text{ with } L(z) := \text{ last component of } z.$$

Then we get a representation of f in the form

$$f(x) = L(E_{n+t}(E_{n+t-1}(\dots E_{n+1}(x)\dots))).$$

For shorter notation, we set for $k = n+1, \dots, n+t$

$$z_k = E_k(E_{k-1}(\dots E_{n+1}(x)\dots)).$$

Using the chain–rule, we obtain

$$f'(x) = L'(z_{n+t}) \cdot E'_{n+t}(z_{n+t-1}) \cdot \dots \cdot E'_{n+2}(z_{n+1}) \cdot E'_{n+1}(x).$$

Now $f'(x)$ is expressed as a product of very sparse matrices. This product can be evaluated in two modes:

mode 1	*mode 2*
right to left	left to right
bottom up	top down
forward	backward
	reverse

Of course, in multiplying the matrices, one takes advantage of their sparsity. A closer look at *mode 1* reveals that it is just the forward method described in section 2. With an economic organization of *mode 2* one can obtain $f(x)$ and $f'(x)$ surprisingly cheaply, see [19], [21], [25], [28], [33], [43].

4 Use of interpreter or converter

Let P be a program for evaluating $f(x)$.

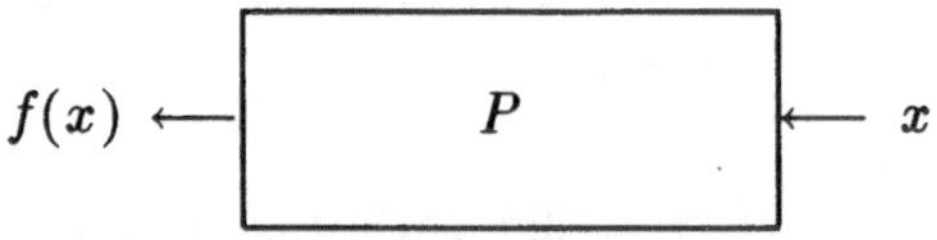

Assume that P consists only of assignments of the form

y_k	$\leftarrow$	k-th component of x
y_k	$\leftarrow$	some real constant
y_k	$\leftarrow$	$y_i * y_j$ with $* \in \{+, -, \cdot, /\}$
y_k	$\leftarrow$	$\lambda(y_i)$ where λ is a library function

Then we can think of a program INT (=interpreter) that scans P, accepts x, and produces the values $f(x)$ and $f'(x)$.

A typical action of a forward mode interpreter would be

If in program P the assignment k reads $y_k \leftarrow y_i/y_j$
then compute $Y_k \leftarrow RAT(D, Y_i, Y_j)$.

A reverse mode interpreter is much more complicated.

If $f(x)$ and $f'(x)$ are needed for several x's, then INT scans P for every x. Now we can think of a program CON (=converter) that scans P just once and writes down the appropriate actions in a new program P'.

Then the program P' can be used to compute $f(x)$ and $f'(x)$ for several x's without reference to the original program P.

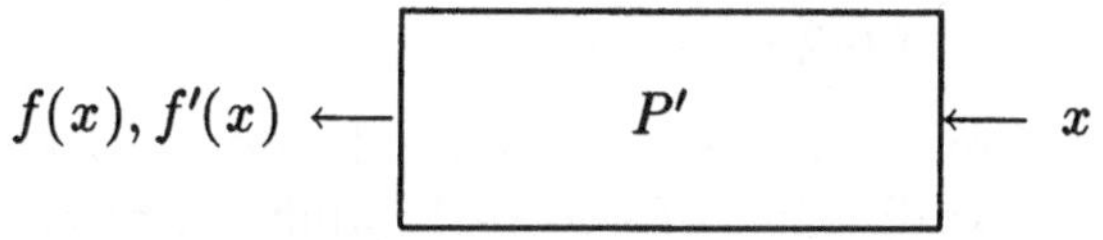

A typical action of a forward mode converter would be

If in program P the assignment k reads $y_k \leftarrow y_i/y_j$
then write in program P' the assignment $Y_k \leftarrow RAT(D, Y_i, Y_j)$.

Our example in Table 2 contains the program P in the first column and the program P' in the third column. The similarity between these columns shows that the task of a forward mode converter is a simple matter of replacements line by line. A reverse mode converter is much more complicated.

5 An illustrative example

We consider several methods for computing the derivative of a function, and we compare the respective computational costs, see also [18]. For this comparison we use a specific function which drastically exhibits the differences between the methods considered. We choose the function

$$f : D \subseteq \mathbb{R}^n \to \mathbb{R} \quad \text{with} \quad f(x) := x_1 \cdot x_2 \cdot \ldots \cdot x_n,$$

where $D = \{x | x \in \mathbb{R}^n,\ x_\nu \neq 0 \text{ for } \nu = 1, 2, \ldots, n\}$ and $n \geq 3$. For any given $x \in D$ the function value $f(x)$ can be computed by means of the algorithm *FUN* shown in Table 4.

(1)	for $k = 1, 2, \ldots, n$ $y_k = x_k = k\text{–th component of } x$
(2)	$y_{n+1} = y_1 \cdot y_2$ $y_{n+2} = y_{n+1} \cdot y_3$ $y_{n+3} = y_{n+2} \cdot y_4$ $\vdots$ $y_{2n-1} = y_{2n-2} \cdot y_n$
(3)	$f(x) = y_{2n-1}$

Table 4: Algorithm *FUN* for $f(x) = x_1 \cdot x_2 \cdot \ldots \cdot x_n$

Let us agree on the following assumptions:

a) By a rational operation we understand addition, subtraction, multiplication, or division of two real numbers, or changing the sign of a real number. (This point of view agrees with [2], Theorem 2.)

b) $\#(f, FUN)$:=the number of rational operations for computing $f(x)$ using *FUN*. Note that this definition refers to an algorithm for computing $f(x)$. Without such a reference "the number of rational operations" would be meaningless.

c) If A is an algorithm for computing $f(x)$ and $f'(x)$, then $\#(f, f', A) :=$ the number of rational operations for computing $f(x)$ and $f'(x)$ using A.

For the function we introduced above we obviously have

$$\#(f, FUN) = n - 1.$$

Now we employ five methods for computing $f(x)$ and $f'(x)$: the forward mode *FM*, the forward mode in a sparse version *FMS*, the reverse mode *RM*, numerical differentiation *NUM*, and symbolic differentiation *SYM*.

Forward mode *FM*

A straightforward implementation of the forward mode gives

$$\#(f, f', FM) = (3n + 1) \cdot \#(f, FUN).$$

Forward mode sparse *FMS*

This modification of *FM* avoids rational operations with system-zeros. As shown in [18], [19], we have

$$\#(f, f', FMS) = (\frac{1}{2}n + 2) \cdot \#(f, FUN).$$

Reverse mode *RM*

With a thorough implementation of the reverse mode we obtain

$$\#(f, f', RM) = 3 \cdot \#(f, FUN).$$

Numerical differentiation *NUM*

Using forward-differences we get the well-known algorithm *NUM* for computing an approximation a for $f'(x)$.

(1)	compute $f(x)$ using FUN
(2)	for $\nu = 1, 2, \ldots, n$ choose step-size $h_\nu > 0$ compute $\bar{x}_\nu = x_\nu + h_\nu$ for $\mu = 1, 2, \ldots, n$ and $\mu \neq \nu$ set $\bar{x}_\mu = x_\nu$ compute $f(\bar{x})$ using FUN compute $a_\nu = \dfrac{f(\bar{x}) - f(x)}{h_\nu}$
(3)	$a = [a_1, a_2, \ldots, a_n]$

Table 5: Algorithm *NUM*

It is easy to verify the relation

$$\#(f, f', NUM) = (n + 4 + \frac{3}{n-1}) \cdot \#(f, FUN).$$

Symbolic differentiation *SYM*

For the present purpose it suffices to consider a formula-manipulator *FO-MA* that accepts a formula for $f(x)$ and produces formulas for the partial derivatives $f_{(1)}(x), f_{(2)}(x), \ldots$

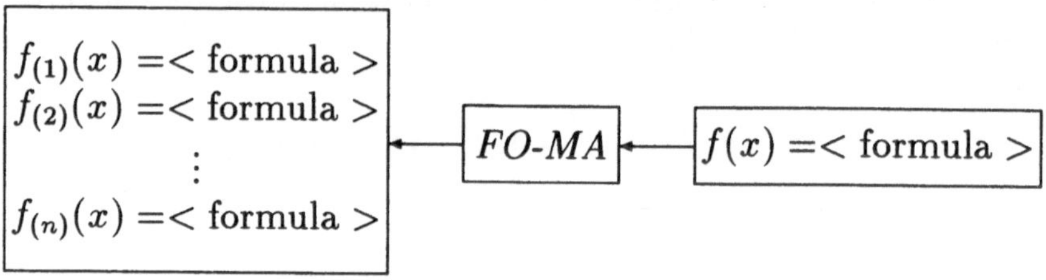

In Table 6 we state the algorithm *SYM* for computing $f(x)$ and $f'(x)$.

(1)	compute $f(x)$ using *FUN*
(2)	employ *FO-MA* to produce formulas for $f_{(1)}(x), f_{(2)}(x), \ldots, f_{(n)}(x)$
(3)	for $\nu = 1, 2, \ldots, n$ compute $f_{(\nu)}(x)$ using the formula produced in (2)
(4)	$f'(x) = [f_{(1)}(x), f_{(2)}(x), \ldots, f_{(n)}(x)]$

Table 6: Algorithm *SYM*

What does it cost to compute $f(x)$ and $f'(x)$ by means of *SYM* ? To simplify matters, we assume (in favour of *SYM*) that employing the formula-manipulator *FO-MA* is free. We further assume that *FO-MA* produces "nice" formulas for the partial derivative, i.e.

$$\begin{aligned}
f_{(1)}(x) &= x_2 \cdot x_3 \cdot x_4 \cdot \ldots \cdot x_n \\
f_{(2)}(x) &= x_1 \cdot x_3 \cdot x_4 \cdot \ldots \cdot x_n \\
&\vdots \\
f_{(\nu)}(x) &= x_1 \cdot x_2 \cdot x_3 \cdot \ldots \cdot x_{\nu-1} \cdot x_{\nu+1} \cdot \ldots \cdot x_n \\
&\vdots \\
f_{(n)}(x) &= x_1 \cdot x_2 \cdot x_3 \cdot \ldots \cdot x_{n-1}
\end{aligned}$$

The operation count yields

$$\#(f, f', SYM) = (n - \frac{1}{n-1}) \cdot \#(f, FUN).$$

Let us compare the costs. For each method X we established a relation of the form

$$\#(f, f', X) = K(n) \cdot \#(f, FUN).$$

The factor $K(n)$ characterizes the quality of the respective method as far as computational costs are concerned: The smaller the $K(n)$, the better the method. In Table 7 we give an overview.

method	algorithm	$K(n)$
automatic forward mode	*FM*	$3n + 1$
automatic forward sparse mode	*FMS*	$\frac{1}{2}n + 2$
automatic reverse mode	*RM*	3
numerical	*NUM*	$n + 4 + \frac{3}{n-1}$
symbolic	*SYM*	$n - \frac{1}{n-1}$

Table 7: Factor $K(n)$ for $f(x) = x_1 \cdot x_2 \cdot \ldots \cdot x_n$

An immediate impression of the factor $K(n)$ can be gathered from the following figure.

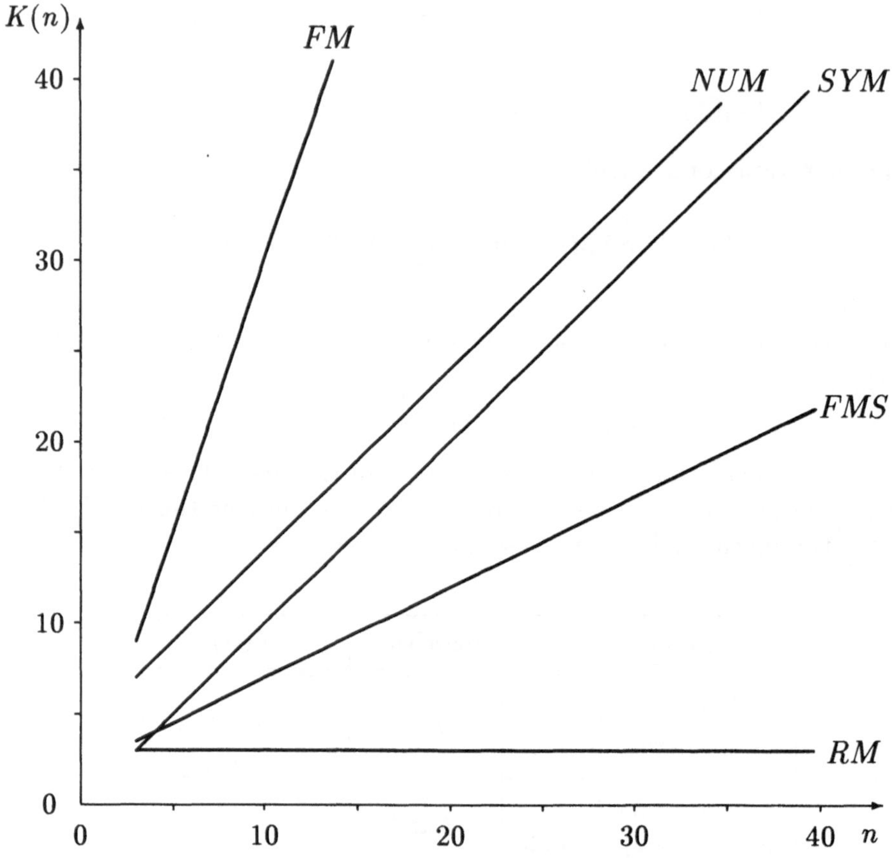

Figure 1: Factor $K(n)$ for $f(x) = x_1 \cdot x_2 \cdot \ldots \cdot x_n$

The methods considered differ drastically in their computational costs. For the forward mode *FM* of automatic differentiation and its sparse version *FMS*, for numerical differentiation *NUM*, and for symbolic differentiation *SYM*, the factor $K(n)$ grows linearly with n. The reverse mode *RM* of automatic differentiation is extremely cheap, here the factor $K(n)$ is 3 for all n.

6 Final remarks

Derivatives are used in many areas of computational mathematics. Here we can only provide references concerning specific fields where automatic differentiation has been used:

1. Nonlinear optimization, see [12], [17], [20], [23], [27], [29], [38].

2. Nonlinear equations, see [3], [8], [23], [27], [38], [39].

3. Sensitivity analysis, see [1], [15], [23], [25], [31], [33], [38].

4. Optimal control, see [13], [23].

5. Differential equations, see [23], [27], [38].

6. Interval mathematics, see [9], [10], [23], [24], [31], [32], [34], [35], [38], [40], [41].

7. Parallel computation, see [6], [7], [11], [12], [14], [20], [23], [30], [36], [42].

A passage in [37] p.461 reads "... the calculation of derivatives requires typically about 60 per cent of the total CPU time for a complete solution of a structural optimization problem in aircraft design". Hence, investigating differentiation techniques may contribute to improve methods in optimization, and in other areas as well.

References

[1] Aberth, O., *Precise Numerical Analysis.* William C. Brown Publishers, Dubuque, Iowa, USA, 1988.

[2] Baur, W., Strassen, V., *The complexity of partial derivatives.* Theoretical Computer Science 22, 1983, 317–330.

[3] Beck, T., *Automatic differentiation of iterative processes.* Journal of Computational and Applied Mathematics 50, 1994, 109–118.

[4] Beck, T., Fischer, H., *The if-problem in automatic differentiation.* Journal of Computational and Applied Mathematics 50, 1994, 119–131.

[5] Beda, L.M., Korolev, L.N., Sukkikh, N.V., Frolova, T.S., *Programs for automatic differentiation for the machine BESM (in Russian).* Technical Report, Institute for Precise Mechanics and Computation Techniques, Academy of Science, USSR, Moscow, 1959.

[6] Bischof, Ch.H., *Issues in parallel automatic differentiation.* In [23], 100–113.

[7] Bischof, Ch., Griewank, A., Juedes, D., *Exploiting parallelism in automatic differentiation.* In: Proceedings of the 1991 International Conference on Supercomputing, edited by E. Houstis and Y. Muraoka, ACM Press, 1991, 146–153.

[8] Carlile, B.R., *Solution of nonlinear systems of equations on the FPS 64-bit family of scientific computers using automatic differentiation.* In: Proceedings of the 1986 Array Conference (Portland, Oregon), 1986, 142–169.

[9] Corliss, G.F., *Applications of differentiation arithmetic.* In: Reliability in Computing – The Role of Interval Methods in Scientific Computing, edited by R.E. Moore, Academic Press, London, 1988, 127–148.

[10] Corliss, G.F., *Overloading point and interval Taylor operators.* In [23], 139–146.

[11] Dixon, L.C.W., *Automatic differentiation and parallel processing in optimisation.* In: Optimization, Parallel Processing and Applications, ed. by A.Kurzhanski, K.Neumann, D.Pallaschke, (Lecture Notes in Economics and Mathematical Systems 304), Springer-Verlag, Berlin, 1988, 86–93.

[12] Dixon, L.C.W., *On the impact of automatic differentiation on the relative performance of parallel truncated Newton and variable metric algorithms.* SIAM Journal on Optimization 1, 1991, 475–486.

[13] Evtushenko, Yu.G., *Automatic differentiation viewed from optimal control theory.* In [23], 25–30.

[14] Fischer, H., *Automatic differentiation: Parallel computation of function, gradient, and Hessian matrix.* Parallel Computing 13, 1990, 101–110.

[15] Fischer, H., *Automatic differentiation of the vector that solves a parametric linear system.* Journal of Computational and Applied Mathematics 35, 1991, 169–184.

[16] Fischer, H., *Special problems in automatic differentiation.* In [23], 43–50.

[17] Fischer, H., *Automatic differentiation: Reduced gradient and reduced Hessian matrix.* Computational Optimization and Applications 1, 1992, 327–344.

[18] Fischer, H., *Automatisches Differenzieren.* In: Wissenschaftliches Rechnen – Eine Einführung in das Scientific Computing, edited by J.Herzberger, Akademie Verlag, Berlin, 1995, 53–104.

[19] Fischer, H., Warsitz, H., *Complexity investigations concerning the derivative of a rational function.* Technical Report 138, Institut für Angewandte Mathematik und Statistik, Technische Universität München, 1995.

[20] Grandinetti, L., Conforti, D., *Numerical comparisons of nonlinear programming algorithms on serial and vector processors using automatic differentiation.* Mathematical Programming 42, 1988, 375–389.

[21] Griewank, A., *On automatic differentiation.* In: Mathematical Programming – Recent Developments and Applications, edited by M. Iri and K. Tanabe, Kluwer Academic Publishers, Dordrecht, Holland, 1989, 83–107.

[22] Griewank, A., *Some bounds on the complexity of gradients, Jacobians and Hessians.* In: Complexity in Numerical Optimization, edited by P.M. Pardalos, World Scientific Publishing Company, Singapore, 1993, 128-162.

[23] Griewank, A., Corliss, G.F. (editors), *Automatic Differentiation of Algorithms: Theory, Implementation, and Application.* SIAM, Philadelphia, 1991.

[24] Hammer, R., Hocks, M., Kulisch, U., Ratz, D., *Numerical Toolbox for Verified Computing I.* Springer-Verlag, Berlin, 1993.

[25] Iri, M., *Simultaneous computation of functions, partial derivatives and estimates of rounding errors, complexity and practicality.* Japan Journal of Applied Mathematics 1, 1984, 223–252.

[26] Juedes, D.W., *A taxonomy of automatic differentiation tools.* In [23], 315–329.

[27] Kagiwada, H., Kalaba, R., Rasakhoo, N., Spingarn, K., *Numerical Derivatives and Nonlinear Analysis.* Plenum Press, New York, 1986.

[28] Kim, K.V., Nesterov, Yu.E., Cherkasskii, B.V., *An estimate of the effort in computing the gradient.* Soviet Math. Dokl. 29, 1984, 384–387.

[29] Kredler, Ch., *PADMOS, ein MS-DOS-Programm für nichtlineare Optimierung mit automatischem Differenzieren.* In: Operations Research Proceedings 1990, edited by W.Bühler, G.Feichtinger, R.F.Hartl, F.J.Radermacher, P.Stähly, Springer, Berlin, 1992, 149–156.

[30] Kubota, K., *PADRE2, A Fortran precompiler yielding error estimates and second derivatives.* In [23], 251-262.

[31] Kubota, K., Iri, M., *Estimates of rounding errors with fast automatic differentiation and interval analysis.* Journal of Information Processing 14, 1991, 508–515.

[32] Layne, J.D., *Applying automatic differentiation and self-validating numerical methods in satellite simulations.* In [23], 211–217.

[33] Linnainmaa, S., *Taylor expansion of the accumulated rounding error.* BIT 16, 1976, 146-160.

[34] Lohner, R.D., *Enclosing the solutions of ordinary initial and boundary value problems.* In: Computer Arithmetic — Scientific Computation and Programming Languages, edited by E.W. Kaucher, U.W. Kulisch, Ch. Ullrich, Wiley-Teubner Series in Computer Science, Stuttgart, 1987, 255–286.

[35] Matijasevich, Yu.V., *A posteriori interval analysis.* In: EUROCAL '85, European Conference on Computer Algebra, Linz, Austria, April 1–3, 1985, Proceedings Volume 2: Research Contributions, edited by Bob F. Caviness, Lecture Notes in Computer Science 204, Springer, Berlin, 1985, 328-334.

[36] Mohseninia, M., *Parallel automatic differentiation in ADA applied to the Navier Stokes equations.* Technical Report 223, Numerical Optimisation Centre, Hatfield Polytechnic, 1989.

[37] Morris, A.J. (editor), *Foundations of Structural Optimization: A Unified Approach.* Wiley, Chichester, 1982.

[38] Rall, L.B., *Automatic Differentiation: Techniques and Applications.* Lecture Notes in Computer Science 120, Springer, Berlin, 1981.

[39] Rall, L.B., *Differentiation in PASCAL-SC: type GRADIENT.* ACM Trans. Math. Software 10, 1984, 161–184.

[40] Rall, L.B., *Improved interval bounds for ranges of functions.* In: Interval Mathematics 1985, edited by K. Nickel, Lecture Notes in Computer Science 212, Springer, Berlin, 1986, 143–155.

[41] Rall, L.B., *Point and interval differentiation arithmetics.* In [23], 17–24.

[42] Shiriaev, D., *Reduction of spatial complexity in reverse automatic differentiation by means of inverted code.* In: Computer Arithmetic and Enclosure Methods, edited by L. Atanassova and J. Herzberger, Elsevier (North-Holland), Amsterdam, 1992, 475–484.

[43] Volin, Yu.M., Ostrovskii, G.M., *Automatic computation of derivatives with the use of the multilevel differentiation technique – 1. Algorithmic basis.* Computers and Mathematics with Applications 11, 1985, 1099–1114.

An Approach to Parallelizing Isotonic Regression

Anthony J. Kearsley* Richard A. Tapia†
Michael W. Trosset‡

Abstract. Isotonic regression is the problem of fitting data to order constraints. We demonstrate that the isotonic regression of a finite set of numbers Y can be obtained by decomposing Y into subsets, performing parallel isotonic regressions on each subset, then performing a trivial isotonic regression on the resulting combined set. Numerical experiments confirm the efficacy of this approach.

1 Introduction

Given a finite set of real numbers, $Y = \{y_1, \ldots, y_n\}$, the problem of isotonic regression with respect to a complete order is the following quadratic programming problem:

$$\begin{array}{ll} \text{minimize} & \displaystyle\sum_{i=1}^{n} w_i(x_i - y_i)^2 \\ \text{subject to} & x_1 \leq \cdots \leq x_n, \end{array}$$

where the w_i are strictly positive weights. Many important problems in statistics and other disciplines can be posed as isotonic regression problems. Comprehensive surveys of this subject were made by Barlow, Bartholomew, Bremner, and Brunk [1] and by Robertson, Wright, and Dykstra [5] in their respective monographs.

The fundamental concern of the present report is revealed by considering a simple example. Suppose that $\{1, 3, 2, 4, 5, 7, 6, 8\}$ is the given set of real

*Department of Mathematics, University of Massachusetts at Dartmouth.

†Department of Computational & Applied Mathematics and Center for Research in Parallel Computation, Rice University. This author was supported in part by NSF Cooperative Agreement No. CCR9120008 and DOE DEFG05-86ER25017.

‡Department of Computational & Applied Mathematics (adjunct), Rice University; Department of Psychology (adjunct), University of Arizona; and Consultant, P. O. Box 40993, Tucson, AZ 85717-0993. This author was supported in part by NSF Cooperative Agreement No. CCR9120008, as a visiting member of the Center for Research in Parallel Computation, Rice University, August 1993 and 1994.

numbers, and that the weights are all identically one. This set is almost isotonic; however, {3,2} and {7,6} violate the requirement that the numbers are nondecreasing. The antidote to this difficulty is very simple: replacing each "block" of "violators" with the average of the numbers in the block produces $\{1, 2.5, 2.5, 4, 5, 6.5, 6.5, 8\}$, which turns out to be the unique solution of the isotonic regression problem. This is an example of the well-known "Pool Adjacent Violators" algorithm.

What is intriguing about this very simple example is that the two computations required to produce the isotonic regression do not depend on each other and could be performed simultaneously. Furthermore, this property appears to be a characteristic of the isotonic regression problem itself, not of the algorithm used to solve it. Whatever the computational algorithm that is employed, it is obvious that the isotonic regressions of the subsets $\{1, 3, 2, 4\}$ and $\{5, 7, 6, 8\}$ are easily combined to produce the isotonic regression of the entire set. It is this observation that motivates the rigorous derivation of a foundation for using parallel computation to solve isotonic regression problems.

2 A Decomposition Theorem

To attain the necessary rigor, we exploit a famous and very elegant characterization of the solution to the isotonic regression problem. Let $W_j = \sum_{i=1}^{j} w_j$, let P_0 denote the point $(0, 0)$, and let P_j denote the point $(W_j, \sum_{i=1}^{j} w_i y_i)$, for $j = 1, \ldots, n$. We interpret $P_0, \ldots, P_n$ as points in the graph of a function, which we extend to the interval $[0,\ W_n]$ by linear interpolation. Both the function and its graph are called the *cumulative sum diagram* (CSD) of the isotonic regression problem.

The *greatest convex minorant* (GCM) of a function f is the convex function defined by

$$GCM[f] \ = \ \sup \{\phi \ : \ \phi \text{ convex}, \ \phi \leq f\}.$$

It is a well-known and beautiful result that the isotonic regression problem is solved by taking x_j^* to be the left derivative of $GCM[CSD]$ at W_j. Thus, theorems about isotonic regressions can be stated and proved as theorems about greatest convex minorants.

The particular theorem on which the ideas in this report are based is quite elementary, yet it has profound implications for parallel computation. Suppose that we decompose the set Y into $Y_1 \oplus Y_2$, where $Y_1 = \{y_1, \ldots, y_k\}$ and $Y_2 = \{y_{k+1}, \ldots, y_n\}$. Analogously, we can decompose a function f with domain $[0, W_n]$ into $f_1 \oplus f_2$, where f_1 is the restriction of f to $[0, W_k]$ and f_2 is the restriction of f to $(W_k, W_n]$. Then the following result is easily demonstrated.

Theorem 1 $GCM[\,GCM[f_1] \oplus GCM[f_2]\,] \ = \ GCM[f]$

Proof: Because this result is of fundamental importance to this report, we provide a detailed proof.

Since $GCM[f_1] \le f_1$ and $GCM[f_2] \le f_2$,

$$GCM[f_1] \oplus GCM[f_2] \le f_1 \oplus f_2 = f.$$

It follows that, if $\phi \le GCM[f_1] \oplus GCM[f_2]$, then $\phi \le f$, and hence that

$$GCM[\, GCM[f_1] \oplus GCM[f_2]\,] \;\le\; GCM[f]. \tag{1}$$

Conversely, suppose that $\phi \le f$ is convex and write $\phi = \phi_1 \oplus \phi_2$. Then $\phi_1 \le f_1$ and $\phi_2 \le f_2$, so $\phi_1 \le GCM[f_1]$ and $\phi_2 \le GCM[f_2]$. It follows that $\phi \le GCM[f_1] \oplus GCM[f_2]$, and hence that

$$GCM[f] \;\le\; GCM[\, GCM[f_1] \oplus GCM[f_2]\,]. \tag{2}$$

Combining inequalities (1) and (2) gives the desired result. □

3 Implications for Parallel Computation

If one takes the function f to be the CSD for the isotonic regression problem, then Theorem 1 states the following: decomposing Y into $Y_1 \oplus Y_2$, performing separate isotonic regressions on Y_1 and Y_2, and then performing a final isotonic regression on the combined result, produces the isotonic regression on Y. Because the separate isotonic regressions on Y_1 and Y_2 can be performed simultaneously, parallel computations of isotonic regressions will be desirable if the final isotonic regression on the combined result is easy to compute. In point of fact, this is the case.

Suppose that Y_1 satisfies $y_1 \le \cdots \le y_k$ and Y_2 satisfies $y_{k+1} \le \cdots \le y_n$. If $y_k \le y_{k+1}$, then Y is isotonic. If Y is not isotonic, then it must be because some of the largest numbers in Y_1 exceed some of the smallest numbers in Y_2. The antidote to this difficulty is to identify this central block of offending numbers and to replace each of these numbers with the weighted average of the block. (This is just the Pool Adjacent Violators algorithm again.) To accomplish this, let

$$\begin{aligned} m &= \min\{i : y_i > y_{k+1}\}, \\ M &= \max\{i : y_i < y_k\}, \end{aligned}$$

and

$$\bar{y} = \sum_{i=m}^{M} w_i y_i \Big/ \sum_{i=m}^{M} w_i.$$

Then, replacing y_i with $\bar{y}$ for $i = m, \ldots, M$ gives the isotonic regression of Y. Thus, if one decomposes the isotonic regression problem and performs two smaller, separate isotonic regressions, it becomes fairly simple to obtain the solution to the original problem.

By now it should be apparent that what is being proposed in this report is *not* a new, parallel algorithm for isotonic regression that will compete with existing algorithms. Rather, it is the isotonic regression problem itself that has been parallelized. (An instructive analogy is the familiar exercise of sorting a list of numbers by subdividing the list, sorting each sublist, then interweaving the sorted sublists.) Because the problem itself has been parallelized, *any* isotonic regression algorithm can be used to compute the separate isotonic regressions assigned to separate processors. The efficiency of various isotonic regression algorithms has been discussed by Best and Chakravarti [2]. A very fast formulation of the Pool Adjacent Violators algorithm was provided by Grotzinger and Witzgall [3].

In light of the preceding arguments, we are virtually assured that a parallel approach to isotonic regression will speed up computation when n is sufficiently large. This phenomenon is demonstrated in Section 4. Notice, however, that we should not expect that the most efficient strategy will necessarily be the one that uses the largest number of processors, since the more that the original problem is decomposed, the more difficult it becomes to obtain the final solution from the separate isotonic regressions. As an extreme example of this limitation, one might decompose Y into n subsets of singleton values, in which case nothing whatsoever has been accomplished. Furthermore, the more that the original problem is decomposed, the greater the communication costs of parallelization. Hence, it is impossible to anticipate the most efficient decomposition strategy.

4 Numerical Experiments

To obtain a suite of isotonic regression problems, we imagined the problem of measuring the viscosity of a fluid at different temperatures. This problem motivates the models that we describe, although ultimately we are more concerned with varying conditions that might affect computational performance than with faithfully modelling physical reality.

Viscosity is a nonincreasing function of temperature; however, due to measurement error, the observed viscosities may not be nonincreasing when ordered by temperature. In this case, one might want to replace the vector of observed viscosities with the nearest vector that is nonincreasing when ordered by temperature. This can be posed as an isotonic regression problem with unit weights.

As have Kearsley [4] and others, we assumed that viscosity (η) is exponentially dependent on temperature (T):

$$\eta = \eta_0 \exp(-\alpha T).$$

For our experiments, we set $\eta_0 = 1$ and $\alpha = 10^{-4}$. Then, in order to obtain an increasing function, we set

$$y = f(t) = 100 - \eta_0 \exp(-\alpha t)$$

and computed $y_k = f(t_k)$ at $n = 10^6$ equally-spaced grid points in the interval $[0, 100]$. The resulting set Y of n increasing numbers was perturbed in various ways to obtain the data sets that we subjected to isotonic regression. Each of ten strategies for perturbing the original set of numbers was replicated $R = 5$ times, resulting in a total of fifty data sets.

Let $\sigma = \log(2)/1.95996$ be fixed. In what follows, whenever we perturb a value y_k, we do so by replacing y_k with $y_k \exp(\sigma z)$, where z is a standard normal deviate. This multiplicative model of measurement error was constructed so that approximately 95 percent of the perturbed values would be at least one half and no more than twice the replaced value.

The following loops describe our perturbation strategies. In each case, the intent was to perturb P values in the form of B blocks of length L.

For $R = 1$ to 5 repetitions:

1. Perturb *each* of the n values in the original data set Y to obtain data set Y1000.R.
2. For $P = .49n, .25n, .01n$ and $L = 1, \sqrt{P}, P$:
   (a) Randomly select $B = P/L$ numbers from $\{1, \ldots, n/L\}$ without replacement. Call these numbers $s_1, \ldots, s_B$.
   (b) Let $\pi = n/P$. For $i = 1, \ldots, B$ and $j = 0, \ldots, L-1$, let $k = \pi s_i + j$ and perturb each original value y_k.
   (c) Denote the resulting data set by Y0ppl.R, where $pp = 100P/n$ and
   $$l = 2\log(L)/\log(P).$$

   For example, the data set produced on the fourth repetition of the case for which $P = .49n$ and $L = \sqrt{P}$ is denoted by Y0491.4.

Thus, we generated five data sets (Y1000) in which all values were perturbed, fifteen data sets (Y049l) in which 49 percent of the values were perturbed, fifteen data sets (Y025l) in which 25 percent of the values were perturbed, and fifteen data sets (Y001l) in which 1 percent of the values were perturbed. Furthermore, in each of the cases that $P = .49n, .25n, .01n$ of the values were perturbed, we generated five data sets (Y0pp0) in which we perturbed P isolated values, five data sets (Y0pp1) in which we perturbed $\sqrt{P}$ blocks of $\sqrt{P}$ consecutive values, and five data sets (Y0pp2) in which we perturbed one block of P consecutive values. This allowed us to investigate the effect of different data structures on the efficacy of parallel computation.

Each of the fifty data sets was submitted to six isotonic regressions on the Intel Touchstone Delta parallel computing system at the California Institute of Technology. These regressions used respectively $A = 1, 2, 4, 8, 16, 32$ of the Delta's processors. For each regression, the data set was decomposed into A subsets of (approximately) equal size. Each subset was simultaneously sent to a separate processor, where its isotonic regression was computed using Grotzinger's and Witzgall's [3] formulation of the Pool Adjacent Violators algorithm. As soon as the isotonic regressions of two consecutive subsets were computed, the combined result was sent to one of the available processors, which then computed the combined isotonic regression by means of the device described in Section 3. This process was continued until the isotonic regression of the entire data set was obtained. The elapsed time from job submission to completion was measured by the Delta's intrinsic timer. The results are summarized in Table 1.

Table 1: Sample means and standard deviations ($\bar{y} \pm s_y$) of elapsed times in milliseconds for five repetitions of ten isotonic regression experiments.

Data	Number of Processors					
Sets	1	2	4	8	16	32
Y1000	2278 ± 93	1376 ± 22	1158 ± 16	930 ± 7	1062 ± 25	1058 ± 15
Y0490	2406 ± 142	1416 ± 5	1182 ± 4	938 ± 8	1062 ± 4	1068 ± 8
Y0491	2376 ± 180	1436 ± 29	1208 ± 50	958 ± 19	1060 ± 14	1080 ± 27
Y0492	2370 ± 171	1378 ± 8	1152 ± 8	922 ± 4	1032 ± 4	1036 ± 15
Y0250	2396 ± 54	1386 ± 9	1144 ± 11	944 ± 48	1040 ± 14	1040 ± 12
Y0251	2298 ± 128	1410 ± 7	1174 ± 5	936 ± 5	1058 ± 4	1052 ± 4
Y0252	2330 ± 95	1406 ± 9	1172 ± 4	942 ± 4	1066 ± 9	1058 ± 4
Y0010	2232 ± 30	1378 ± 4	1150 ± 7	922 ± 4	1034 ± 5	1032 ± 8
Y0011	2368 ± 156	1410 ± 10	1188 ± 24	940 ± 0	1062 ± 4	1062 ± 4
Y0012	2380 ± 152	1418 ± 8	1184 ± 22	944 ± 5	1068 ± 18	1070 ± 12

Table 1 exhibits several striking features. First, the variations in times produced by $R = 5$ replications are extremely small relative to the magnitudes of the times. In retrospect this is not surprising: each data set contains a very large number of independent errors, so that one should expect that most data sets constructed in accordance with a specific perturbation strategy will be quite similar.

Second, there is very little variation in mean timing profiles between the ten perturbation strategies. This suggests that the phenomena described below are not unique to a particular data structure.

As anticipated, it is apparent that some degree of parallelization decreases the time required to perform an isotonic regression. For the 50 data sets that we considered, the time required by $A = 2$ processors divided by the time required by $A = 1$ processor ranged from a minimum of 53.2% to a maximum of 65.9%, with a median of 60.3%. The time required by $A = 4$ processors divided by the time required by $A = 1$ processor ranged from a minimum of 44.9% to a maximum of 54.8%, with a median of 50.1%. The time required by $A = 8$ processors divided by the time required by $A = 1$ processor ranged from a minimum of 35.7% to a maximum of 43.5%, with a median of 40.5%.

Thus, there is compelling evidence that, for $n = 10^6$ and these types of data sets, using $A = 8$ processors is more efficient than using $A = 4, 2, 1$ processors.

For $A = 16, 32$ processors, the communication costs of the parallelization strategy begin to dominate and the times are actually slower than for $A = 8$ processors. This phenomenon was also anticipated. With larger data sets, we know that we can take advantage of additional processors, but the tradeoff between n and the optimal A must be empirically determined for the data structures and parallel computing system of interest.

Finally we note that, although the proportional improvements in efficiency produced by parallel processing are impressive, the absolute times for serial processing are small. At present, it is difficult to forsee applications involving isotonic regressions on data sets so large that the absolute savings in time will warrant parallel computation. Perhaps that day will come; for now, our primary interest in parallelizing isotonic regression is for the pedagogical value of so doing. In our view, isotonic regression is a remarkably simple and elegant example of a problem for which mathematical theory virtually guarantees that parallelization will be beneficial.

Acknowledgements

The authors thank Christoph Witzgall and Andrea Reiff for sharing their thoughts about isotonic regression algorithms. This research was performed in part using the Intel Touchstone Delta System operated by the California Institute of Technology on behalf of the Concurrent Supercomputing Consortium. Access to this facility was provided by the Center for Research on Parallel Computing under NSF Cooperative Agreement No. CCR9120008.

References

[1] R. E. Barlow, J. M. Bartholomew, J. M. Bremner, and H. D. Brunk. *Statistical Inference Under Order Restrictions.* John Wiley & Sons, New York, 1972.

[2] M. J. Best and N. Chakravarti. Active set algorithms for isotonic regression; a unifying framework. *Mathematical Programming*, 47:425–439, 1990.

[3] S. J. Grotzinger and C. Witzgall. Projections onto order simplexes. *Applied Mathematics and Optimization*, 12:247–270, 1984.

[4] A. J. Kearsley. A steady state model of Couette flow with viscous heating. *International Journal of Engineering Science*, 32:179–186, 1994.

[5] T. Robertson, F. T. Wright, and R. L. Dykstra. *Order Restricted Statistical Inference.* John Wiley & Sons, New York, 1988.

Mathematical Programming at Oberwolfach

Bernhard Korte, Bonn

Abstract: These are some personal reminiscences on the history of the mathematical programming meetings at Oberwolfach over the last fifteen years which were organized by Klaus Ritter and the author.

I.

It was in October 1977 that the institute of the author organized a meeting entitled "Optimization and Operations Research" at the Elly-Hölterhoff-Böcking-Stift in Bad Honnef near Bonn (cf. Henn, Korte, Oettli (1978)). In those days mathematical programming was not yet on the list of mathematical subjects which were considered for meetings at Oberwolfach. Actually, at that time applied mathematics in general had only a minor share of Oberwolfach meetings. There were some irregularly scheduled meetings on different aspects of optimization and even of discrete optimization like graph theory organized by the late Lothar Collatz and his school, and a series of meetings on the mathematics of operations research organized by Rudolf Henn, Horst Schubert and Hans Künzi.

At that time the Elly-Hölterhoff-Böcking-Stift, as an endowment of the University of Bonn, was already used as a conference center. Now it is run by the Deutsche Physikalische Gesellschaft as the "Oberwolfach of Physics", as is Schloß Dagstuhl near Saarbrücken, which is considered the "Oberwolfach of Computer Science".

The Elly-Hölterhoff-Böcking-Stift is a nice Jugendstil villa with a wonderful garden with majestic trees. However, its charm cannot compete with that of Oberwolfach. It was after lunch one day at that meeting in October 1977 that Klaus Ritter and I went for a stroll in his garden. We discussed the status and development of mathematical programming which was - at least speaking for Germany - quite a young field. Eventually we encouraged each other to apply for a week at Oberwolfach exclusively devoted to the area of mathematical programming. Since Klaus was in the field of continuous optimization and I in discrete optimization, we felt that this duo would guarantee a strong bond between these

two main areas of mathematical programming. Already then we agreed that it should not just be our goal to organize a meeting at Oberwolfach, but also to care for and develop the whole field. We wanted to establish an Oberwolfach attitude within the scientific community. Probably in those days only a few scholars in mathematical programming knew about Oberwolfach at all.
We were not sure how colleagues from North America would react to an invitation to come for a whole week to the very remote Black Forest. But, as I will report later, they came like a pilgrimage.

II.

Our application was accepted by the advisory body of the Oberwolfach institute and the first meeting was scheduled for

May 6 - 12, 1979.

We asked Heinz König who was at that time a member of the advisory board to join us in organizing the meeting. This was to ensure the usual structure, organization and quality of meetings at Oberwolfach. He was co-organizer of the first three meetings and helped to convince the advisory board that mathematical programming deserved its place at Oberwolfach.
We made up a list of about 60 invitees. This number was agreed by Martin Barner, the director of the Oberwolfach institute at that time. Since many scholars from foreign countries were on the invitation list, we felt that we would get about 40 participants in all which was considered to be optimal.
But what a surprise: 58 scholars from more than 10 countries accepted the invitation. A quarter of them came from overseas. One should remember that in those days participants from North America were a not too frequent event at Oberwolfach. Actually, the organizer could apply to the institute for travel support for participants from overseas. However, we did not do so on purpose. We did not want to pay for participation but to make the meeting so attractive that all peers wanted to participate.
This philosophy worked well, actually too well. This was the only major, and serious, problem we were faced with.
The very first meeting at Oberwolfach was not only a great success but it also bore a great sensation: it was in May 1979 at

Oberwolfach that western mathematicians for the first time became aware of a strange paper by L.G. Khachian [1979] on the now famous ellipsoid method. Rainer Burkard brought the Russian version of this paper from Poland to Oberwolfach. The late Gene Lawler organized special sessions to understand the paper and Laszlo Lovász, a participant, wrote the first article on it explaining details and giving the lacking proofs (cf. P. Gács and L. Lovász [1981]). About six months later the New York Times carried a front-page article with the headline: "A Soviet discovery rocks the world of mathematics". Further details can be obtained from an article by Phil Wolfe in Optima [Wolfe, 1980], also a participant of that meeting and president of the Mathematical Programming Society at the time.
Since this meeting set the corner-stone for mathematical programming at Oberwolfach, we collected some of the keynote papers and published them as volume 14 of Mathematical Programming Studies (cf. König, Korte, Ritter [1981]). Our general aim was not to publish proceedings at all. We deviated from this rule once more at the third meeting in 1983, and published selected papers as a Mathematical Programming Study (cf. Korte, Ritter [1984]).

III.
To date Klaus and I have organized in all eight meetings on Mathematical Programming at Oberwolfach. As a memory aid this is the list:
May 6 - 12, 1979,
January 25 - 31, 1981,
January 9 - 15, 1983,
January 6 - 12, 1985,
January 3 - 9, 1988,
January 7 - 13, 1990,
January 6 - 10, 1992,
January 8 - 14, 1995.
The ninth meeting is already scheduled for

January 5 - 11, 1997.

As one can deduce, we have managed to keep the biennial rhythm of Oberwolfach meetings with two exceptions. However, we sometimes had to fight for it. We wanted to keep the whole field of mathematical programming within one meeting. By this we believed to exert the strongest influence on the field. Especially, we wanted to have the two mainstreams, namely discrete and continuous optimization, together at one meeting. This should help to keep track of the fast development of the whole field. The mutual competition of different areas at Oberwolfach was always a great stimulant. However, there were colleagues who wanted to organize meetings at Oberwolfach on very specific subtopics of mathematical programming. Hence we are grateful to the advisory board that - sometimes after additional discussions - they granted the whole field of mathematical programming a permanent place at Oberwolfach.

Mathematical programming is still a relatively young domain of mathematics. For a very mature area, say algebra, it would make no sense to run an Oberwolfach meeting on such a major topic, whereas it is unanimously agreed in the mathematical programming community that it is optimal to have one meeting for the whole field.

It would go far beyond the scope of these reminiscences to list here all major events of the previous Oberwolfach meetings. But it can be stated without exaggeration that Oberwolfach was an essential factor in the development of the whole field. Every second year a remote place in the Black Forest became the central market place for the community. Many new theorems, papers, and mainstreams of research had their origin at Oberwolfach. As an example, Laszlo Lovász and the author started to think about greedoids on an afternoon hike during the second Oberwolfach meeting and they finished the preface of a monograph on this subject - much later than intended - at the 6th Oberwolfach meeting in January 1990 (cf. Korte, Lovász, Schrader [1991]). Another example are Claude Lemarechal and Jochem Zowe who were introduced to each other's work at Oberwolfach. As a consequence, they wrote a whole series of Oberwolfach papers.

In some sense the Oberwolfach meetings were the key points of the backbone of development of the field on an international as well as a national level. About 15 years ago mathematical programming was very much on the fringes of mathematics, particularly in Germany. Now it can be considered a solid area of applied, or better applicable, mathematics, keeping its place in the very subtle (not necessarily linear and sometimes not even transitive) rank ordering of mathematical subjects. As a proof of this, many new chairs of optimization, combinatorial optimization and discrete mathematics have been established during the last years at mathematics departments of universities in Germany.

IV.

In organizing the meetings at Oberwolfach we had one sole, though serious problem, namely that of size. The community got such a strong Oberwolfach attitude that everyone important in the field had to participate (some people thought this was not only a necessary but also a sufficient condition!).

Klaus and I did not make too many additional friends but many more enemies when organizing the Oberwolfach meeting. Scholars who could not be invited came from overseas, stayed on their own at the Hirschen and attended only the sessions. This was evidently not liked by the administration. One participant (A.C.) wanted to hide for two years in the basement of the institute as the ghost of Oberwolfach in order to be sure to participate at the next meeting. Oberwolfach became a must in the community and this became a pain in the neck for the organizers.

For example, the meetings in 1988 and 1990 had more than 60 participants and roughly 50 (!) talks. About 50% of the participants came from overseas and 70% from abroad. One should keep in mind that it is not a pure pleasure to travel in early January for only one week from North America to Oberwolfach, passing such famous places as Hassloch, Hausach, Wolfach and Walke. The first week of the year was not liked too much by other organizers. We had chosen it on purpose since this is a non-teaching period at many North American universities.

After meetings Klaus and I had to travel quite often to Freiburg in order to explain to Professor Barner the very special reasons for the oversize of the meeting and promise that this would not happen again. Nevertheless, the character and charm of Oberwolfach meetings was not seriously spoilt by their size. One might raise the question whether different branches of mathematics have different ways of communication, even at such a unique place as Oberwolfach.

At the 6th meeting in 1990 we eventually realized that we could not cure oversizing except by very radical treatment. We discussed the problem openly with all participants, namely that even very eminent and active peers should not expect an invitation for every meeting, but say every second one. Immediately after that meeting we wrote a circular letter to everybody in the community explaining the new philosophy. As a result, the 7th and 8th meetings had only about 40 participants, the normal and expected size of an Oberwolfach meeting.

V.

Let me add some comments on personal matters: for the community Oberwolfach was not only an excellent place for scholarly exchange, but also for private communications. Although many colleagues suffered from jet-lag, the Sunday evening was always like a family reunion with chats on private or professional topics far beyond midnight. And next morning at the opening session a distinguished colleague from North America (S.R.) appeared without jet-lag and with dark suit and tie, and gave an excellent lecture. Scholars from different parts of the world became acquainted for the first time at Oberwolfach, some became friends. A social highlight of each meeting was, of course, Wednesday afternoon. When climbing hills, the late Lothar Collatz always wanted to improve the gradient method, but how? The longest and strongest hike was probably the one via Brandenkopf to Glaswald-Lake. When reaching the winning post the higher order derivatives of the movement of some participants (G.N.) were not smooth at all.

Another hike to remember was through snow of more than half a meter. Here one participant (D.S.) insisted on riding a pony. This pony story became famous and got extended by new (and true?) anecdotes every time it was retold.

The poker association met on a regular basis on some evenings.

And even the author learned to handle the cue without attacking the billiard table.
Another highlight of some meetings were the excellent piano recitals by Therese Lemarechal. Even on Friday evenings she had a full house.

VI.
As mentioned above, the 7th and 8th meetings were organized in a new style with an appropriate number of about 40 participants. At the 8th meeting in 1995 we also tried to adopt the policy of the new director Matthias Kreck, namely to have fewer and longer talks. This was somewhat different from the style of our previous meetings. Originally we wanted to have a complete survey of current research, which could be discussed informally with the aim of encouraging new and joint research. And we have to admit that this has worked well so far. However, we will now experience the new style, which allows the detailed development of some selected research lines by extensive talks or by a series of talks. We will see how this style will enhance research activities at the meeting. One might also ask whether a series of meetings over 15 years organized by the same people is in danger of burning out. We are watching this possibility very carefully. It is not our aim just to add another meeting, but to continue the very high standard of the past with a group of participants that changes substantially with time. More than one third of the participants of the last meeting attended an Oberwolfach meeting on mathematical programming for the first time. This is, we feel, a very good indicator for the liveliness of the subject and the Oberwolfach meetings. As long as this is the case, the Oberwolfach meetings are - as always - an important asset to the field.
The new meetings policy also recommends that one organizer should be from abroad. This is on the one hand an easy requirement for us, but on the other a quite difficult one: easy insofar as our meetings were internationally oriented from the very beginning, but difficult since it is not a simple matter to select just one from among so many active peers from abroad which all have a strong Oberwolfach attitude. We have now solved this prob-

lem by an ex-officio decision by asking John Dennis, the current president of the Mathematical Programming Society, to join us in organizing the 1997 meeting.

VII.

A very final sentence: it was always a pleasure to organize the Oberwolfach meetings jointly with Klaus Ritter. Therefore, ad multos annos - in two different respects.

References:

Henn, R., Korte, B. and Oettli, W. (eds.) [1977]: Optimization and Operations Research. Proceedings of a Workshop held at the University of Bonn, October 2 - 8, 1977. Berlin/Heidelberg/New York: Springer 1978.

König, H., Korte, B. and Ritter, K. (eds.) [1981]: Mathematical Programming at Oberwolfach. Mathematical Programming Study 14 (1981).

Korte, B. and Ritter, K. (eds.) [1984]: Mathematical Programming at Oberwolfach II. Mathematical Programming Study 22 (1984).

Gács, P. and Lovász, L. [1981]: Khachiyan's algorithms for linear programming, in König, H., Korte, B. and Ritter, K. (eds) [1981], p. 61-68.

Wolfe, P. [1980]: The Ellipsoid Algorithm. Optima 1 (June 1980), p. 1-5.

Khachian, L.G. [1979]: A polynomial algorithm in linear programming. Doklady Akademiia Nauk USSR 244 (No. 5, February 1979) p. 1093-1096 (translated in Soviet Mathematics Doklady 20, p. 191-194.

Korte, B., Lovász, L. and Schrader, R. [1991]: Greedoids. Algorithms and Combinatorics, Vol. 4. Berlin/Heidelberg/New York: Springer 1991.

A SQP-Method for Linearly Constrained Maximum Likelihood Problems

Christian Kredler

Institut für Angewandte Mathematik und Statistik
Technische Universität München

Abstract. Newton-like methods are standard algorithms for unrestricted parameter estimation in a wide class of nonlinear regression models. The search directions of the here presented algorithm are solutions from a sequence of quadratic (sub-)problems (SQP) with linear constraints. The practically important generalized linear models with natural link functions, e.g. log-linear or logistic regression, lead to strictly convex optimization problems for which this easy to implement extension of Newton's method converges globally with quadratic rate. The numerical results are demonstrated at some ship damage data.

Key Words: Generalized linear models (GLM), natural link function, restricted maximum likelihood (ML-) estimation, SQP-method, global and quadratic convergence, uniform convexity.

1 Restricted ML-estimation in Generalized Linear Models

In a wide and important area of statistical data analysis we are faced with strictly convex optimization problems for which Newton-like optimization methods are more appropriate than the widely used standard algorithms, e.g. BFGS or the gradient based SQP-method of [Schittkowski (1981)]. The here proposed approach is a robust optimization tool, simply substitutable into statistical programs without using external libraries, provided a solver for quadratic programs is available; e.g. [Best & Ritter (1988)]. The convergence theorems illuminating the global and local behaviour of the suggested algorithm are proved using a technique similar to that presented in [Ritter (1982)].

The convergence properties of the investigated algorithm depend on strong convexity assumptions. Especially, these conditions are satisfied in statistical regression problems where the canonical parameters $\boldsymbol{\theta} \in \mathbb{R}^q$ of some exponential families are involved. Consider a random variable $\boldsymbol{Y} \in \mathbb{R}^q$ having a density

$$h(\boldsymbol{y}, \boldsymbol{\theta}) = c(\boldsymbol{y})\, exp\,(\boldsymbol{y}^T \boldsymbol{\theta} - d(\boldsymbol{\theta})) \tag{1}$$

with respect to the Lebesgue or counting measure. The scalar valued c does not depend on $\boldsymbol{\theta}$. Note that normal, multinomial, Poisson and Gamma random variables belong to our exponential class. For the treatment of nuisance parameters like σ^2 in the normal case see e.g. [McCullagh & Nelder (1983)]. The partial derivatives $d_j, d_{jk}, \ldots$ of the analytic function d generate the cumulants of $\boldsymbol{Y}$ with

$$\begin{aligned} \boldsymbol{E}(\boldsymbol{Y}) &= (d_j) \\ \boldsymbol{cov}(\boldsymbol{Y}) = \boldsymbol{cov}(\boldsymbol{Y}_j, \boldsymbol{Y}_k) &= (d_{jk}) \quad etc. \end{aligned} \tag{2}$$

The third order cumulants of Y can be exploited to obtain explicit bounds for variable selection techniques based on certain quadratic approximations, cf. [Kredler (1986)].
The positive definiteness of (d_{jk}) ensures the convexity properties needed for the objective functions to be analyzed below.

To clarify presentation we now restrict ourselves to the univariate case $q = 1$, although all statements apply analogously to multivariate models from which the multinomial seems to be the most important, e.g. for contingency tables and logistic discriminant analysis (see [Fahrmeir & Kredler (1984)]). Generalized linear models (GLM), i.e. regression models for exponential type random variables, have first been proposed by [Nelder & Wedderburn (1972)]. Given independent observations and covariates

$$y_t \in \mathbb{R} \quad \text{and} \quad \boldsymbol{z}_t \in \mathbb{R}^p\,,\ t = 1, \ldots, T\,, \tag{3}$$

where the canonical or natural parameters θ_t vary in a linear subspace

$$\theta_t = \boldsymbol{z}_t^T \boldsymbol{\beta} \tag{4}$$

with an unknown parameter vector $\boldsymbol{\beta} \in \mathbb{R}^p$ to be estimated. Neglecting superfluous constants the log-likelihood function for all T observations is

$$l(\boldsymbol{\beta}) = \sum_{t=1}^{T} \{y_t \cdot \boldsymbol{z}_t^T \boldsymbol{\beta} - d(\boldsymbol{z}_t^T \boldsymbol{\beta})\}\,. \tag{5}$$

According to (2)

$$-\boldsymbol{\nabla}^2 \boldsymbol{l}(\boldsymbol{\beta}) = \sum_{t=1}^{T} d''(\boldsymbol{z}_t^T \boldsymbol{\beta})\, \boldsymbol{z}_t \boldsymbol{z}_t^T \qquad \text{is positive definite} \tag{6}$$

if the matrix $(z_t)_{1 \le t \le T}$ has full rank. Hence $-l$ is strictly convex for GLM's in the natural parameters θ_t. The existence of the ML-estimate $\hat{\beta}$ with $l(\hat{\beta}) \ge l(\beta)$ is not trivial, but can be ensured a priori by certain properties of the data y_t, z_t, cf. [Wedderburn (1976)] and [Kaufmann (1988)]. [Albert & Anderson (1984)] guarantee especially for the binomial and multinomial case the existence of the ML-estimate $\hat{\beta}$ under easy to verify linear separation properties of the data. For instance, $\hat{\beta}$ cannot exist if there is a hyper-plane $H : 0 = z^T \tilde{\beta}$ such that

$$\begin{aligned} z_t^T \tilde{\beta} &< 0 \quad \text{for all} \quad y_t = 0 \\ z_t^T \tilde{\beta} &> 0 \quad \text{for all} \quad y_t = 1 . \end{aligned} \tag{7}$$

In section 3 a log-linear GLM for ship damage data with linear constraints for the parameter vector β is analyzed. This leads to

$$\max_{\beta} \{ l(\beta) \mid \beta \in R \subseteq \mathbb{R}^p \} \tag{8}$$

where

$$R = \{ \beta \mid A\beta \le b \}, \quad A \in \mathbb{R}^{m,p}, \; b \in \mathbb{R}^m . \tag{9}$$

In the next section we discuss the convergence properties of an SQP-algorithm to compute a linearly restricted ML-estimate of a GLM with natural link function and existing $\hat{\beta}$.

2 SQP-Algorithm, Convergence Properties

To go along with the usual notation in the optimization literature let in this section

$$x := \beta \in \mathbb{R}^p .$$

The problem to solve is then

$$\min_{x} \{ F(x) \mid A x \le b \} \tag{10}$$

where $F(x) \equiv -l(\beta)$ is twice continuously differentiable and has additionally the convex properties stated below. A linear equality constraint $a^T x = b$ can be replaced with two linear inequalities

$$a^T x \le b \qquad \text{and} \qquad -a^T x \le -b . \tag{11}$$

Hence, for the sake of a brief notation, the restrictions in (10) cover this case, too. However, the numerical treatment of linear equalities is much easier than that of linear inequalities, and any good numerical algorithm handles linear equality constraints in a separate way; see also formulation of corollary 2.12. Throughout we denote gradient and Hessian of F by

$$g(\boldsymbol{x}) := \boldsymbol{\nabla F}(\boldsymbol{x}) \quad \text{and} \quad G(\boldsymbol{x}) := \boldsymbol{\nabla}^2 \boldsymbol{F}(\boldsymbol{x}) .$$

Def. 2.1 (Uniform Convexity)

$F \in C^2(\mathbb{R}^p)$ is said to be **uniformly convex** *on a convex set $\boldsymbol{D} \subseteq \mathbb{R}^p$, if and only if there are $0 < \mu \leq \eta$ such that*

$$\mu \|\boldsymbol{x}\|^2 \leq \boldsymbol{x}^T G(\boldsymbol{u}) \boldsymbol{x} \leq \eta \|\boldsymbol{x}\|^2 \tag{12}$$

for all $\boldsymbol{x} \in \mathbb{R}^p$ and all $\boldsymbol{u} \in \boldsymbol{D}$.

Uniformly convex functions are strictly convex and attain their unique minimum. $F(\boldsymbol{x}) = \boldsymbol{x}^4$ shows that the reverse is not true.
However, for certain applications like natural link function GLM's the following theorem gives a concrete characterization depending on the existence of a minimizer $\bar{\boldsymbol{x}}$. As mentioned in the previous section this can be done, for instance in practically relevant GLM's, by an a priori inspection of the data involved in the nonlinear function F.

Theorem 2.2

If $F \in C^2(\mathbb{R}^p)$ has a positive definite Hessian for all $\boldsymbol{x} \in \mathbb{R}^p$ and attains its minimum at some $\bar{\boldsymbol{x}}$, then F is **uniformly convex** *on the level sets $\boldsymbol{N}_o := \{\, \boldsymbol{x} \in \mathbb{R}^p \mid F(\boldsymbol{x}) \leq F(\boldsymbol{x}_0) \,\}$ for all $\boldsymbol{x}_0 \in \mathbb{R}^p$.*

Proof:
We first show that the level sets are bounded. The compactness follows, because they are closed. Pick $\omega > 0$, let $\alpha := F(\bar{\boldsymbol{x}}) + \omega$, and consider $\boldsymbol{N}_\alpha := \{\, \boldsymbol{x} \mid F(\boldsymbol{x}) \leq \alpha \,\}$, which is convex since F is strictly convex. If it was not bounded there would exist a $\boldsymbol{s} \neq \boldsymbol{0} \in \mathbb{R}^p$ and $\varphi(\sigma) := F(\bar{\boldsymbol{x}} - \sigma \, \boldsymbol{s}) \leq \alpha$ for all $\sigma \geq 0$. φ is strictly convex, and hence $F(\bar{\boldsymbol{x}}) = \varphi(0) < \varphi(\sigma)$, for all $\sigma \neq 0$. With $\delta := \varphi(1) - \varphi(0) > 0$ we obtain for all $\sigma > 1$

$$\varphi(0) + \delta = \varphi(1) = \varphi\left(\frac{1}{\sigma}\sigma + (1 - \frac{1}{\sigma})0\right) < \frac{1}{\sigma}\varphi(\sigma) + (1 - \frac{1}{\sigma})\varphi(0),$$

which implies $\varphi(0) + \delta < \frac{1}{\sigma}\Big(\varphi(\sigma) - \varphi(0)\Big) + \varphi(0)$, and further

$$0 < \delta < \frac{1}{\sigma}\Big(\varphi(\sigma) - \varphi(0)\Big) \leq \frac{1}{\sigma}\Big(\alpha - \varphi(0)\Big) = \frac{1}{\sigma}\Big(\omega + \varphi(0) - \varphi(0)\Big) = \frac{\omega}{\sigma}$$

for all $\sigma > 1$, which contradicts $\delta = \varphi(1) - \varphi(0) > 0$. Now, the extremal eigenvalues attain their minimum and maximum on the compact set $\boldsymbol{N}_\alpha$. □

The theorem covers most important generalized linear models including binomial, multinomial and Poisson variables. For the Γ−distribution and related cases the theorem can be extended, with some technical caution, to $F \in C^2(\boldsymbol{D})$ where $\boldsymbol{D}$ is an appropriate, convex subset of $\mathbb{R}^p$. In the proofs of the following statements we need only the uniform convexity of F on $\boldsymbol{N}_0 \cap \boldsymbol{R}$, the intersection of the level set and the feasible region $\boldsymbol{R} = \{\boldsymbol{x} \mid \boldsymbol{A}\boldsymbol{x} \leq \boldsymbol{b}\}$. Trivially, this property can always be obtained by restricting F with positive definite Hessian to a box $\boldsymbol{v} \leq \boldsymbol{x} \leq \boldsymbol{w}$.

2.1 A General Descent Method

We consider an iterative algorithm which, starting from $\boldsymbol{x}_0 \in \mathbb{R}^p$, computes for $j = 0, 1, 2, \ldots$ a search direction $\boldsymbol{s}_j$ and a stepsize σ_j defining the new point

$$\boldsymbol{x}_{j+1} := \boldsymbol{x}_j - \sigma_j\, \boldsymbol{s}_j\,,$$

where $\boldsymbol{g}_j^T \boldsymbol{s}_j > 0$ for $\boldsymbol{g}_j := \boldsymbol{g}(\boldsymbol{x}_j)$.

Def. 2.3 (Armijo-Goldstein Rule)

Choose $0 < \delta < \frac{1}{2}$. In $\boldsymbol{x}_j$ find the smallest integer $\nu_j \geq 0$ such that

$$F(\boldsymbol{x}_j) - F(\boldsymbol{x}_j - (\tfrac{1}{2})^{\nu_j} \boldsymbol{s}_j) \geq \delta \cdot (\tfrac{1}{2})^{\nu_j} \cdot \boldsymbol{g}_j^T \boldsymbol{s}_j\,. \tag{13}$$

Finally, let $\sigma_j = (\frac{1}{2})^{\nu_j}$.

Def. 2.4 (Quadratic Approximation)

With $\boldsymbol{g}_j := \boldsymbol{g}(\boldsymbol{x}_j)$, $\boldsymbol{G}_j := \boldsymbol{G}(\boldsymbol{x}_j)$ the quadratic Taylor approximation is $F(\boldsymbol{x}_j - \boldsymbol{s}) \approx F(\boldsymbol{x}_j) + Q_j(\boldsymbol{s})$ where

$$Q_j(\boldsymbol{s}) := -\boldsymbol{g}_j^T \boldsymbol{s} + \tfrac{1}{2}\, \boldsymbol{s}^T \boldsymbol{G}_j\, \boldsymbol{s}\,. \tag{14}$$

Let throughout $\boldsymbol{\lambda} := (\lambda_1, \ldots, \lambda_m)^T \leq \boldsymbol{0}$ and

$$\boldsymbol{A} := \begin{pmatrix} \boldsymbol{a}_1^T \\ \cdots \\ \boldsymbol{a}_m^T \end{pmatrix}, \quad \boldsymbol{b} := \begin{pmatrix} b_1 \\ \cdots \\ b_m \end{pmatrix}.$$

Def. 2.5 (Kuhn-Tucker Conditions, Lagrange Multipliers)

The tupel $(\bar{\boldsymbol{x}}, \bar{\boldsymbol{\lambda}}) \in \mathbb{R}^{p+m}$, $\bar{\boldsymbol{\lambda}} \leq \mathbf{0}$, *is called* **Kuhn-Tucker pair** *of problem (10) if the following conditions hold*

$$\begin{aligned} \boldsymbol{g}(\bar{\boldsymbol{x}}) - \boldsymbol{A}^T \bar{\boldsymbol{\lambda}} &= \mathbf{0}, \quad \bar{\boldsymbol{\lambda}} \leq \mathbf{0}, & (15) \\ \boldsymbol{A}\bar{\boldsymbol{x}} &\leq \boldsymbol{b}, & (16) \\ \bar{\lambda}_k (\boldsymbol{a}_k^T \bar{\boldsymbol{x}} - b_k) &= 0, \quad 1 \leq k \leq m. & (17) \end{aligned}$$

$\bar{\boldsymbol{x}}$ *satisfying the above conditions is called* **stationary point** *and* $\bar{\boldsymbol{\lambda}}$ *is the vector of optimal* **Lagrange multipliers**.

In our case F is convex and differentiable, hence each stationary point $\bar{\boldsymbol{x}}$ is a global optimum of (10).

Algorithm SQP

Step 0: *Initialization*

Choose a feasible $\boldsymbol{x}_0$, i.e. $\boldsymbol{A}\boldsymbol{x}_0 \leq \boldsymbol{b}$, $0 < \delta < \frac{1}{2}$.
Compute $\boldsymbol{g}_0 := \boldsymbol{g}(\boldsymbol{x}_0)$, let $j := 0$, and goto Step 1.

Step 1: *Search Direction, Stopping Criterion*

Compute

$$\boldsymbol{s}_j := arg \min_{\boldsymbol{s}} \{ Q_j(\boldsymbol{s}) \mid \boldsymbol{A}(\boldsymbol{x}_j - \boldsymbol{s}) \leq \boldsymbol{b} \} \tag{18}$$

If $\boldsymbol{s}_j = \mathbf{0}$, then **STOP**, else goto Step 2.

Step 2: *Stepsize*

Choose σ_j according to Armijo-Goldstein.
Goto Step 3.

Step 3: *Update:*

Define

$$\begin{aligned} \boldsymbol{x}_{j+1} &:= \boldsymbol{x}_j - \sigma_j \boldsymbol{s}_j, \\ \boldsymbol{g}_{j+1} &:= \boldsymbol{g}(\boldsymbol{x}_{j+1}); \end{aligned}$$

replace j with $j+1$, and goto Step 1.

Theorem 2.6 *If F is uniformly convex on $N_0 \cap R$, then algorithm SQP is well defined. The generated iterates $\{x_j\}$ ensure $F(x_{j+1}) < F(x_j)$. If $s_j = 0$, then x_j is a stationary point of (10).*

Proof:

The subproblem (18) is strictly convex, hence s_j is unique. We now have to show that $-s_j$ is a descent direction, and that $s_j = 0$ can only occur in a stationary point.

$\nabla Q_j(s) = -g_j + G_j\, s$. The Kuhn-Tucker conditions for (18) imply

$$-g_j + G_j\, s_j + A^T \lambda_j = 0; \quad \lambda_j \leq 0, \tag{19}$$

$$A\,(x_j - s_j) \leq b, \tag{20}$$

$$\lambda_j^T (A\,(x_j - s_j) - b) = 0. \tag{21}$$

(21) follows from the usual complementarity condition. (19) yields

$$s_j = G_j^{-1}(g_j - A^T \lambda_j) \tag{22}$$

which is well defined since G_j is positive definite by assumption. With $\lambda_j \leq 0$ from (19) and $A\,x_j - b \leq 0$ we obtain by the complementarity condition (21)

$$\lambda_j^T A\, s_j \geq 0. \tag{23}$$

Further, (19) multiplied by s_j gives

$$g_j^T s_j = s_j^T G_j\, s_j + \lambda_j^T A\, s_j. \tag{24}$$

(23) and the uniform convexity yield finally

$$g_j^T s_j \geq \mu\, \|s_j\|^2 > 0. \tag{25}$$

This guarantees the descent property of the algorithm. We remark further that x_{j+1} stays within the feasible region $R = \{\, x \mid Ax \leq b\}$ because $x_j \in R$, $x_j - s_j \in R$ and $\sigma_j \in (0, 1]$. In case $s_j = 0$ we conclude from (19) that x_j must be a stationary point of (10), and hence is optimal due to the convexity of F. □

Def. 2.7 (Lagrangian)

$$L(x, \lambda) := F(x) - \lambda^T (A\, x - b), \quad \lambda \leq 0, \tag{26}$$

denotes the **Lagrangian function** *of problem (10).*

Theorem 2.8 (Global Convergence)

Provided F is uniformly convex on $\boldsymbol{N}_o \cap \boldsymbol{R}$, the intersection of level set and feasible region, the SQP-algorithm converges to a stationary point $\bar{\boldsymbol{x}}$ of (10) with

$$\lim_{j\to\infty} \boldsymbol{\nabla L}(\boldsymbol{x}_j) = \boldsymbol{0}. \tag{27}$$

Proof:

We follow the scheme of [Ritter (1982)]. To get a contradiction assume that $\|\boldsymbol{\nabla L}(\boldsymbol{x}_j)\| = \|\boldsymbol{g}_j - \boldsymbol{A}^T\boldsymbol{\lambda}_j\| > \varepsilon > 0$ for all $j \in J$ where J is an infinite subset of $\mathbb{N}$. Hence by (22)

$$\|\boldsymbol{s}_j\| \geq \frac{1}{\eta}\|\boldsymbol{g}_j - \boldsymbol{A}^T\boldsymbol{\lambda}_j\| \geq \frac{\varepsilon}{\eta}, \quad \text{for all } j \in J. \tag{28}$$

Further, by (25)

$$\frac{\boldsymbol{g}_j^T\boldsymbol{s}_j}{\|\boldsymbol{s}_j\|} \geq \mu\|\boldsymbol{s}_j\| \geq \frac{\varepsilon\mu}{\eta}, \quad \text{for all } j \in J. \tag{29}$$

According to Armijo-Goldstein we choose σ_j such that

$$a_j(\sigma_j) := \frac{F(\boldsymbol{x}_j) - F(\boldsymbol{x}_j - \sigma_j\boldsymbol{s}_j)}{\sigma_j\boldsymbol{g}_j^T\boldsymbol{s}_j} \geq \delta. \tag{30}$$

Taylor expansion with $\boldsymbol{u}_j \in [\boldsymbol{x}_j, \boldsymbol{x}_j - \boldsymbol{s}_j]$ gives

$$\begin{aligned} a_j(\sigma) &\geq 1 - \frac{\|\boldsymbol{g}(\boldsymbol{u}_j) - \boldsymbol{g}_j\| \cdot \|\boldsymbol{s}_j\|}{\boldsymbol{g}_j^T\boldsymbol{s}_j} \\ &= 1 - \frac{\|\boldsymbol{g}(\boldsymbol{u}_j) - \boldsymbol{g}_j\|}{\frac{\boldsymbol{g}_j^T\boldsymbol{s}_j}{\|\boldsymbol{s}_j\|}} \geq 1 - \frac{\eta}{\varepsilon\mu}\|\boldsymbol{g}(\boldsymbol{u}_j) - \boldsymbol{g}_j\|. \end{aligned} \tag{31}$$

As $\boldsymbol{g}(\cdot)$ is uniformly continuous on the compact set $\boldsymbol{N}_o$, there is a $\tau > 0$ such that $\|\boldsymbol{g}(\boldsymbol{u}_j) - \boldsymbol{g}_j\| \leq (\varepsilon\mu)/(2\eta)$ for $\|\sigma\boldsymbol{s}_j\| \leq \tau$. Then (31) with $\|\sigma\boldsymbol{s}_j\| \leq \tau$ yields

$$a_j(\sigma) \geq 1 - \frac{1}{2} = \frac{1}{2} > \delta, \quad \text{for all } j \in J,$$

which means that (13) is satisfied. In the Armijo-Goldstein rule we choose the smallest ν_j (i.e. $\sigma_j = \frac{1}{2}^{\nu_j}$ as large as possible). Hence for $j \in J$ with $\sigma_j \leq 1$

$$\|\sigma_j\boldsymbol{s}_j\| = \|(\frac{1}{2})^{\nu_j}\boldsymbol{s}_j\| \geq \min\{\|\boldsymbol{s}_j\|, \frac{\tau}{2}\},$$

and further for all $j \in J$

$$\begin{aligned} F(\boldsymbol{x}_j) - F(\boldsymbol{x}_j - \sigma_j \boldsymbol{s}_j) &\geq \delta \sigma_j \boldsymbol{g}_j^T \boldsymbol{s}_j = \delta \|\sigma_j \boldsymbol{s}_j\| \frac{\boldsymbol{g}_j^T \boldsymbol{s}_j}{\|\boldsymbol{s}_j\|} \geq \delta \min\{\|\boldsymbol{s}_j\|, \frac{\tau}{2}\} \frac{\varepsilon\mu}{\eta} \\ &\geq \delta \min\{\frac{\varepsilon}{\eta}, \frac{\tau}{2}\} \frac{\varepsilon\mu}{\eta} > 0\,. \end{aligned} \tag{32}$$

Note that the bound is independent of j and $\boldsymbol{x}_j$. As J is infinite F cannot be bounded below, which contradicts the compactness of $\boldsymbol{N}_0$. The remaining Kuhn-Tucker conditions are obtained from those holding for subproblem (18).

F is uniformly, hence strictly convex. Therefore, each stationary point $\bar{\boldsymbol{x}}$ is a global minimum of the linearly constrained problem (10). Further, $\bar{\boldsymbol{x}}$ is unique such that the sequence $\{\boldsymbol{x}_j\}$ converges to $\bar{\boldsymbol{x}}$. □

2.2 Convergence Rate

A specific property of quadratically convergent Newton-like algorithms is that the stepsize σ_j converges to 1. This is also the case for our SQP-algorithm.

Lemma 2.9 (Stepsize $\sigma_j = 1$)

If F is uniformly convex on $\boldsymbol{N}_0 \cap \boldsymbol{R}$ and $\{\boldsymbol{s}_j\}, \{\sigma_j\}$ are generated by algorithm SQP there is a $j_0 \in \mathbb{N}$ such that $\sigma_j = 1$ for $j \geq j_0$.

Proof:

According to Taylor there is some $\boldsymbol{u}_j \in [\boldsymbol{x}_j, \boldsymbol{x}_j - \boldsymbol{s}_j]$ such that

$$F(\boldsymbol{x}_j - \boldsymbol{s}_j) = F(\boldsymbol{x}_j) - \boldsymbol{g}_j^T \boldsymbol{s}_j + \tfrac{1}{2} \boldsymbol{s}_j^T G(\boldsymbol{u}_j) \boldsymbol{s}_j\,. \tag{33}$$

We have to show $a_j(1) \geq \delta > 0$ for the Armijo-Goldstein function from (30).

$$\begin{aligned} a_j(1) &= \frac{F(\boldsymbol{x}_j) - F(\boldsymbol{x}_j - 1 \cdot \boldsymbol{s}_j)}{1 \cdot \boldsymbol{g}_j^T \boldsymbol{s}_j} \\ &= \frac{1}{\boldsymbol{g}_j^T \boldsymbol{s}_j} \left(\tfrac{1}{2} \boldsymbol{g}_j^T \boldsymbol{s}_j + \tfrac{1}{2} \boldsymbol{g}_j^T \boldsymbol{s}_j - \tfrac{1}{2} \boldsymbol{s}_j^T (G_j + G(\boldsymbol{u}_j) - G_j) \boldsymbol{s}_j \right). \end{aligned} \tag{34}$$

With $\boldsymbol{\lambda}_j^T \boldsymbol{A} \boldsymbol{s}_j \geq 0$ from (23) we obtain

$$a_j(1) \geq \frac{1}{2} + \frac{\boldsymbol{g}_j^T \boldsymbol{s}_j - \boldsymbol{s}_j^T G_j \boldsymbol{s}_j - \boldsymbol{\lambda}_j^T \boldsymbol{A} \boldsymbol{s}_j}{2\, \boldsymbol{g}_j^T \boldsymbol{s}_j} - \frac{\boldsymbol{s}_j^T (G(\boldsymbol{u}_j) - G_j) \boldsymbol{s}_j}{2\, \boldsymbol{g}_j^T \boldsymbol{s}_j}. \tag{35}$$

According to (19) the second term on the right hand side vanishes. Hence with (25)

$$\begin{aligned} a_j(1) &\geq \frac{1}{2} - \frac{1}{2}\frac{s_j^T(G(u_j) - G_j)s_j}{g_j^T s_j} \\ &\geq \frac{1}{2} - \frac{1}{2}\frac{\|G(u_j) - G_j\| \cdot \|s_j\|^2}{\mu\|s_j\|^2} \\ &= \frac{1}{2} - \frac{1}{2}\frac{\|G(u_j) - G_j\|}{\mu} \end{aligned} \tag{36}$$

which converges to $\frac{1}{2} > \delta$ since $\lim G(u_j) = \lim G_j = G(\bar{x})$. □

So far the only used regularity condition was the uniform convexity of F. For technical reasons we now need assumptions on the linear independence of the gradients a_k corresponding to the active constraints. Denote by $\bar{K} := \{1 \leq k \leq m \mid a_k^T \bar{x} = b_k\}$ the index set of active constraints at $\bar{x}$, and by $\bar{\lambda} := (\bar{\lambda}_1, ..., \bar{\lambda}_m)^T$ the optimal Lagrange multipliers.

Def. 2.10 (Assumption A)

The optimization problem (10) is said to satisfy **Assumption A** *if and only if F is twice continuously differentiable and if $(\bar{x}, \bar{\lambda})$, $\bar{\lambda} \leq 0$, is a Kuhn-Tucker pair of (10) according to Def. 2.5 and*

$$s^T G(\bar{x})\, s > 0 \quad \text{for all} \quad s \neq 0 \quad \text{with} \quad a_k^T s = 0\,, \quad k \in \bar{K}\,, \tag{37}$$

and

$$\bar{\lambda}_k < 0 \quad \text{for} \quad k \in \bar{K}\,, \qquad (\textit{strict complementary slackness}) \tag{38}$$

and if

$$\bar{A} := (a_k)_{k \in \bar{K}}^T \in I\!R^{r,p} \quad \text{has full rank}\,. \tag{39}$$

Clearly, (37) follows for the special models of section 1 directly from the uniform convexity of F on the level set N_0. The gradients of the active constraints in (39) can only be linearly independent if linear equalities are stated separately, and not in the way proposed for theoretical purposes in (11). Without loss of generality the corresponding Lagrange multiplier can be chosen positive; otherwise we redefine the equation by $-a_k^T x = -b_k$.

With stepsize $\sigma_j = 1$ our algorithm SQP is nothing else but the method of [Wilson (1963)] for linear constraints; see also [Gill et al. (1989)]. Among others the quadratic convergence properties of Wilson's method are investigated in [Robinson (1974)]. Here, we only mention those aspects of the proof which can be exploited algorithmically. The subproblems (18) are viewed as perturbations of the original problem for which, under certain conditions, the solution converges to that of (10).

Theorem 2.11 (Active Constraints, Lagrange Multipliers)

If (10) satisfies **Assumption A** *at* $(\bar{x}, \bar{\lambda})$, *and if* $x_j \to \bar{x}$ *as* $j \to \infty$, *then the Lagrange multipliers* λ_j *of the quadratic subproblem (18) converge to the optimal multipliers* $\bar{\lambda}$. *Further, there is an* j_0 *such that*

$$(\lambda_j)_k < 0 \quad \text{if} \quad \bar{\lambda}_k < 0 \qquad \text{and} \tag{40}$$

$$a_k^T x_j < b_j \quad \text{if} \quad a_k^T \bar{x} < b_k \tag{41}$$

for all $j \geq j_0$.

Proof: See theorem 2.1 of [Robinson (1974)], which is an application of the implicit function theorem. □

Hence we can be sure to pick ultimately the correct active constraints. In such a case, however, the solution of the subproblem (18) can be found very efficiently.

Corollary 2.12 (Newton Direction in Subspaces)

Choose $S \in \mathbb{R}^{p-r,p}$ *with orthonormal columns and* $\bar{A}^T S = 0$, *where according to (39),* $r = |\bar{K}| = \text{rank}(\bar{A})$. *Under the assumptions of theorem 2.11 there is an index* j_0 *such that the subproblem (18) can be replaced with*

$$w_j := \arg\min_{w} \{ F(x_j - S\,w) \mid w \in \mathbb{R}^{p-r} \}. \tag{42}$$

$s_j := S\,w_j$ *of (18) can explicitly be computed from*

$$S^T G_j S\, w_j = S^T g_j\,. \tag{43}$$

Theorem 2.13 (Quadratic Convergence)

Under the assumptions of theorem 2.11 the algorithm SQP converges quadratically to the optimal point $\bar{x}$ *if* F *satisfies a Lipschitz condition in a neighbourhood of* $\bar{x}$.

Proof:

Consider $\boldsymbol{x}_{j_0}$ with $\bar{\boldsymbol{A}}\,\boldsymbol{x}_{j_0} = \boldsymbol{b}$ and

$$\bar{F}(\boldsymbol{w}) := F(\boldsymbol{x}_{j_0} - \boldsymbol{S}\,\boldsymbol{w})\,, \tag{44}$$

being uniformly convex with

$$\nabla^2 \bar{F}(\boldsymbol{w}) = \boldsymbol{S}^T G(\boldsymbol{x}_{j_0} - \boldsymbol{S}\,\boldsymbol{w})\,\boldsymbol{S}\,. \tag{45}$$

With F also $\bar{F}$ satisfies a Lipschitz condition. Hence Newton's method converges quadratically. According to theorem 2.11 and corollary 2.12 in a neighbourhood close enough to $\bar{\boldsymbol{x}}$ the sequences generated by Newton's method for $\bar{F}$ and by algorithm SQP coincide. This completes the proof. □

Corollary 2.14 (Unconstrained Case)

All the stated global and local convergence properties remain valid for the case without linear restrictions. This fact is well known and has been exploited since ever to compute unconstrained ML-estimates in generalized linear models (GLM's) efficiently.

Remark:

1. The convergence results can be extended to problems with nonlinear constraints, cf. e.g. [Robinson (1974)]. From the data analysis point of view distribution results are hard to obtain even for linear constraints (see [McDonald & Diamond (1990)], [Piegorsch (1990)] and [Nyquist (1991)]). This may explain, why this paper is restricted to the linearly constrained case, where convergence results and algorithms can be formulated in quite a simple manner. Nevertheless, a review of the proofs shows that the assumptions can be weakened. The identities (19) - (25) play a crucial role throughout, and there may be scenarios where (23) and (25) hold on $\boldsymbol{N}_0 \cap \boldsymbol{R}$ and F is far from being convex.

2. General link functions for GLM's are of considerable practical interest. Like for unconstrained ML-estimates the Hessian should then be replaced with the positive (semi-)definite Fisher scoring matrix $\tilde{\boldsymbol{G}}_j$, cf. e.g. [Fahrmeir & Kredler (1984)], which is of the same approximation type as the Gauß-Newton-Matrix in Nonlinear Regression. The resulting modified SQP-algorithm can be expected to have satisfactory numerical behaviour. Weaker convergence properties than those stated here can be obtained under appropriate conditions, e.g. work with $\tilde{\boldsymbol{G}}_j + \kappa\,\boldsymbol{I}$ instead of $\tilde{\boldsymbol{G}}_j$, if necessary, to obtain good enough descent directions.

3 A GLM with Parameter Restrictions

Ship Damage Data (Lloyd's Register of Shipping)
cf. [McCullagh & Nelder (1983)], p. 137

Ship type	Year of construction	Period of operation	aggregate months service	number of damage incidents
A	1960-64	1960-74	127	0
A	1960-64	1975-79	63	0
A	1965-69	1960-74	1095	3
A	1965-69	1975-79	1095	4
A	1970-74	1960-74	1512	6
A	1970-74	1975-79	3353	18
A	1975-79	1960-74	0	0*
A	1975-79	1975-79	2244	11
B	1960-64	1960-74	44882	39
B	1960-64	1975-79	17176	29
B	1965-69	1960-74	28609	58
B	1965-69	1975-79	20370	53
B	1970-74	1960-74	7064	12
B	1970-74	1975-79	13099	44
B	1975-79	1960-74	0	0*
B	1975-79	1975-79	7117	18
C	1960-64	1960-74	1179	1
C	1960-64	1975-79	552	1
C	1965-69	1960-74	781	0
C	1965-69	1975-79	676	1
C	1970-74	1960-74	783	6
C	1970-74	1975-79	1948	2
C	1975-79	1960-74	0	0*
C	1975-79	1975-79	274	1
D	1960-64	1960-74	251	0
D	1960-64	1975-79	105	0
D	1965-69	1960-74	288	0
D	1965-69	1975-79	192	0
D	1970-74	1960-74	349	2
D	1970-74	1975-79	1208	11
D	1975-79	1960-74	0	0*
D	1975-79	1975-79	2051	4
E	1960-64	1960-74	45	0
E	1960-64	1975-79	0	0*
E	1965-69	1960-74	789	7
E	1965-69	1975-79	437	7
E	1970-74	1960-74	1157	5
E	1970-74	1975-79	2161	12
E	1975-79	1960-74	0	0*
E	1975-79	1975-79	542	1

* Necessarily empty cells for ships not (yet) in action.

The company has to balance financial risk and is interested in a statistical analysis that allows a distinction between categories of safe ships and others.

Damage Rates in $^0/_{00}$				
	year of construction			
ship type	60-64	65-69	70-74	75-79
A	0.0	3.2	4.9	4.9
B	1.1	2.3	2.3	2.5
C	1.2	0.7	2.9	3.6
D	0.0	0.0	8.3	2.0
E	0.0	11.4	5.1	1.8

Lowest risk is observed for ship types B and C and highest for type E. Unexpectedly, the oldest ships appear to be the safest, whereas those built in 1965-1974 seem to yield highest risk. To validate this conjecture we define an appropriate generalized linear model (GLM, see section 1) satisfying all assumptions made in the previous section. The ML-estimation of the model parameters was done with the SQP-algorithm described in the previous section. The reported numerical results can be found in [Klinger (1992)].

$T = 40$ design vectors

$$\boldsymbol{z} = (1, z_{\cdot 1}, \ldots, z_{\cdot 8}, z_{\cdot 9})^T$$

are defined with effect coding for the factors. For instance, reference category $60 - 64$ yields

$(z_{\cdot 1}, z_{\cdot 2}, z_{\cdot 3})$	year of construction
$(1, 0, 0)$	$60 - 64$
$(0, 1, 0)$	$65 - 69$
$(0, 0, 1)$	$70 - 74$
$(-1, -1, -1)$	$75 - 79$

Reference category ship type E defines $z_{\cdot 4} \cdots z_{\cdot 7}$ for the remaining 4 categories A,B,C,D. $z_{\cdot 8} = 1$ is for period of operation in 75-79, whereas $z_{\cdot 8} = -1$ means 60-74. Finally, $z_{\cdot 9}$ describes the monthly service hours of the individual ship.

We follow [McCullagh & Nelder (1983)] and view the number y_t of damage incidents as independent Poisson variables. This gives a log-linear GLM with parameter vector $\boldsymbol{\beta} = (\beta_0, \beta_1, \ldots, \beta_9)^T$

$$\ln(E[y_{\cdot}]) = \beta_0 + \sum_{j=1}^{9} z_{\cdot j}\beta_j \, , \tag{46}$$

which yields the log-likelihood function

$$l(\beta) = \sum_{i=1}^{N} [\, y_i z_i^T \beta - \exp(z_i^T \beta)] \,. \tag{47}$$

The analysis of interactions is omitted here. Unrestricted ML-estimation gives

$$\hat{\beta} = (\underset{\beta_0}{-5.175}, \; \underbrace{\underset{60-64}{-0.448}, \; \underset{65-69}{0.214}, \; \underset{70-74}{0.312}}_{\text{year of construction (75-79 is reference category)}}, \; \underset{\dots}{\dots}, \; \underset{\text{service hours}}{0.905})$$

Due to effect coding the reference category parameter becomes

$$\hat{\beta}_{75-79} = -\sum_{j=1}^{3} \hat{\beta}_j = -0.078 \,. \tag{48}$$

Again, the conjecture that older ships are safer than new ones can be expressed in terms of the parameters of (46). Statistically spoken, we want to know whether the parameters corresponding to earlier construction periods are significantly smaller than the others. This can be formulated in the following hypotheses

$$H_0 : \beta_1 = \beta_2 = \beta_3$$

versus

$$H_1 : \beta_1 \leq \beta_2 \leq \beta_3 \,.$$

$\hat{\beta}$ satisfies restriction H_1. Under the equality constraints of H_0 we get

$$\tilde{\beta} = (\underset{\beta_0}{-3.940}, \; \underbrace{\underset{65-69}{0.084}, \; \underset{70-74}{0.084}, \; \underset{75-79}{0.084}}_{\text{year of construction}}, \; \underset{\dots}{\dots}, \; \underset{\text{service hours}}{0.759})$$

Following the theory of [Wollak (1987)], [McDonald & Diamond (1990)], and [Fahrmeir & Klinger (1994)] the corrected likelihood ratio statistic lq^* is distributed like a mixture of χ^2-variables with specific weights. [Klinger (1992)] reports $lq^* = 19.28$ with a "p-value" of 0.0001. Hence H_1 is accepted, and the year of construction has an highly significant influence on the number of damage incidents.

One possibility to check the significance of the negative influence of β_{75-79} is to recode the variables for the construction period with new reference category 60 – 64. We obtain

$$\hat{\beta} = (\underbrace{-5.175}_{\beta_0},\ \underbrace{\underbrace{0.214}_{65-69},\ \underbrace{0.312}_{70-74},\ \underbrace{-0.078}_{75-79}}_{\text{year of construction (60-64 is reference category)}},\ \ldots,\ \underbrace{0.905}_{\text{service hours}})$$

The statistical hypothesis to test is now

$$H_0 : \beta_1 \leq \beta_2 \leq \beta_3$$

versus

$$H_1 : \beta_1, \beta_2, \beta_3 \quad \text{arbitrary (not in } H_0\text{)}.$$

With

$$\boldsymbol{A} = \begin{pmatrix} 0 & -1 & 1 & 0 & 0 & 0 & 0 & 0 & 0 & 0 \\ 0 & 0 & -1 & 1 & 0 & 0 & 0 & 0 & 0 & 0 \end{pmatrix}, \boldsymbol{b} = \begin{pmatrix} 0 \\ 0 \end{pmatrix}$$

H_0 is equivalent with $\boldsymbol{A}\boldsymbol{\beta} \leq \boldsymbol{b}$. Under these restrictions we obtain with the algorithm of section 2 the ML-estimate

$$\tilde{\beta} = (\underbrace{-5.366}_{\beta_0},\ \underbrace{\underbrace{0.159}_{65-69},\ \underbrace{0.180}_{70-74},\ \underbrace{0.180}_{75-79}}_{\text{year of construction}},\ \ldots,\ \underbrace{0.935}_{\text{service hours}})$$

Again, according to [Klinger (1992)], the corrected likelihood ratio statistic is $lq^* = 3.04$ with a p-value of 0.0804. So hypothesis H_0 is accepted, which finally confirms the conjecture that in the considered data set older ships are more robust than newer ones.

Examples with linear parameter restrictions for binomial credit scoring and endodontric risk data are discussed in [Fahrmeir & Klinger (1994)].

4 Conclusion

SQP-methods are appropriate for a wider class of parameter restricted estimation problems than the here described generalized linear models with canonical parameters. If analytic first derivatives are available the FORTRAN-algorithm of [Schittkowski (1981)] will usually produce good results.

Sometimes the exact model is not known a priori and alternatives are to be checked before the best model is found. In this case the derivative coding for each new model is not only time consuming but also a source of many errors. Numerical difference quotients instead of analytic derivatives are an obvious but frequently not numerically stable alternative. In that situation automatic differentiation is a good way out. The PC-program PADMOS, cf. [Greiner, Kredler & Wagenpfeil (1992)], has been developed especially for such purposes. Text files contain the Pascal-like description of log-likelihood function and constraints. In the data file for each observation an extra line is reserved. A comfortable user interface with built-in editor enables a convenient input and modification of ML-problems. Models with some hundred observations and up to 15 parameters can be analyzed. Box, linear and even nonlinear type constraints are accepted. The execution time of PADMOS is about 10 times of that needed by analytic derivative procedures. But for any chosen model first **and** second derivatives are computed automatically, and hence the broad variety of powerful optimization algorithms, including methods with directions of negative curvature, is applicable, and ensures good convergence properties.

Acknowledgement

I wish to thank Joachim Klinger and Stefan Wagenpfeil for helpful comments and discussions.

References

[Albert & Anderson (1984)] Albert A. and Anderson J.A.: On the existence of maximum likelihood estimates in logistic regression models. Biometrica **71**, 1-10.

[Best & Ritter (1988)] Best M.J. and Ritter K.: A quadratic programming algorithm. ZOR **32**, 271-297.

[Fahrmeir & Klinger (1994)] Fahrmeir L. and Klinger J.: Estimating and testing generalized linear models under inequality restrictions. Statistical Papers **35**, 211-229.

[Fahrmeir & Kredler (1984)] Fahrmeir L. und Kredler Ch.: Verallgemeinerte lineare Modelle. In: Fahrmeir L. und Hamerle A. (Hrsg.): Multivariate statistische Verfahren. De Gruyter, Berlin.

[Gill et al. (1989)] Gill P.E., Murray W., Saunders M.A. and Wright M.H.: Constrained nonlinear programming. In: G.L. Nemhauser et al. (eds.): Handbooks in Operations Research and Management Science. North-Holland, Amsterdam.

[Greiner, Kredler & Wagenpfeil (1992)] Greiner M., Kredler Ch. and Wagenpfeil St.: Nonlinear optimization with PADMOS: User's guide and algorithms. Report **366**, DFG Schwerpunkt: Anwendungsbezogene Optimierung und Steuerung, TU München.

[Kaufmann (1988)] H.: On existence and uniqueness of maximum likelihood estimates in quantal and ordinal response models. Metrica **35**, 291-313.

[Klinger (1992)] J.: Tests für Ungleichungsrestriktionen in generalisierten linearen Modellen. Diplomarbeit (Referent. Prof. L. Fahrmeir), LMU München.

[Kredler (1986)] Ch.: Behaviour of third order terms in quadratic approximations of LR-statistics in multivariate generalized linear models. The Annals of Statistics **14**, 326-335.

[McCullagh & Nelder (1983)] McCullagh P. and Nelder J.A.: Generalized linear models. Chapman and Hall, London.

[McDonald & Diamond (1990)] McDonald P. and Diamond I.: On the fitting of generalized linear models with nonnegativity parameter constraints. Biometrics **46**, 201-206.

[Nelder & Wedderburn (1972)] Nelder J. and Wedderburn R.W.M.: Generalized linear models. J. Roy. Statist. Soc. **A 135**, 370-384.

[Nyquist (1991)] H.: Restricted estimation of generalized linear models. Applied Statistics, **40**, 133-141.

[Piegorsch (1990)] W.: One-side-significance tests for generalized linear models under dichotomous response. Biometrics **46**, 309-316.

[Ritter (1982)] K.: Numerical methods for nonlinear programming problems. In: Korte B. (ed.): MODERN APPLIED MATHEMATICS — Optimization and Operations Research. North-Holland Publ. Comp. Amsterdam.

[Robinson (1974)] S.M.: Perturbed Kuhn-Tucker points and rates of convergence for a class of nonlinear programming algorithms. Mathematical Programming **7**,1-16.

[Schittkowski (1981)] K.: The nonlinear programming method of Wilson, Han and Powell with an augmented Lagrangian type line search function. Part 1: Convergence analysis. Numer. Math. **38**, 83-114.

[Wedderburn (1976)] R.W.M.: On the existence and uniqueness of the maximum likelihood estimates for certain generalized linear models. Biometrika **63**, 27-32.

[Wilson (1963)] R.B.: A simplicial algorithm for concave programming. PhD thesis, Harvard University, Cambridge.

[Wollak (1987)] F.A.: An exakt test for multiple inequality and equality constraints in the linear model. JASA **82**, 782-793.

Machine Learning via Polyhedral Concave Minimization

O. L. Mangasarian*

Mathematical Programming Technical Report 95-20

Abstract. Two fundamental problems of machine learning, misclassification minimization [10, 24, 18] and feature selection, [25, 29, 14] are formulated as the minimization of a concave function on a polyhedral set. Other formulations of these problems utilize linear programs with equilibrium constraints [18, 1, 4, 3] which are generally intractable. In contrast, for the proposed concave minimization formulation, a successive linearization algorithm without stepsize terminates after a maximum average of 7 linear programs on problems with as many as 4192 points in 14-dimensional space. The algorithm terminates at a stationary point or a global solution to the problem. Preliminary numerical results indicate that the proposed approach is quite effective and more efficient than other approaches.

1 Introduction

We shall consider the following two fundamental problems of machine learning:

Problem 1.1 Misclassification Minimization *[24, 18] Given two finite point sets $\mathcal{A}$ and $\mathcal{B}$ in the n-dimensional real space R^n, construct a plane that minimizes the number of points of $\mathcal{A}$ falling in one of the closed halfspaces determined by the plane and the number of points of $\mathcal{B}$ falling in the other closed halfspace.*

Problem 1.2 Feature Selection *[4, 3] Given two finite point sets $\mathcal{A}$ and $\mathcal{B}$ in R^n, select a sufficiently small number of dimensions of R^n such that a*

*Computer Sciences Department, University of Wisconsin, 1210 West Dayton Street, Madison, WI 53706, email: *olvi@cs.wisc.edu*. This material is based on research supported by Air Force Office of Scientific Research Grant F49620-94-1-0036 and National Science Foundation Grants CCR-9322479.

plane, constructed in the smaller dimensional space, optimizes some separation criterion between the sets $\mathcal{A}$ and $\mathcal{B}$.

We immediately note that the misclassification minimization problem is NP-complete [6, Proposition 2]. But, effective methods for its solution have been proposed in [18] and implemented in [1]. An approximate technique [6] has also been implemented. The formulation that we propose in this work terminates in a finite number of linear programs (typically less than seven) at a vertex solution or stationary point of the problem.

We outline the contents of the paper now. In Section 2 we give a precise mathematical formulation of the misclassification minimization and feature selection problems and indicate how they can be set up as linear programs with equilibrium constraints and indicate some of the difficulties attendant this formulation. We then introduce in Section 3 a simple concave exponential approximation of the step function, similar to the classical sigmoid function of neural networks [28, 11, 17], but with the significant difference of concavity of the proposed approximation which is not shared by the sigmoid function. This concavity is possible, because the step function is applied here to nonnegative variables. This leads to a finite successive linearization algorithm (SLA) without a stepsize procedure that is described in Section 4 of the paper. Section 5 gives very encouraging results on numerical tests on the misclassification minimization and feature selection problems. Section 6 gives a concluding summary of the paper.

A word about our notation now. For a vector x in the n-dimensional real space R^n, x_+ will denote the vector in R^n with components $(x_+)_i := \max\{x_i, 0\}$, $i = 1, \ldots, n$. Similarly x_* will denote the vector in R^n with components $(x_*)_i := (x_i)_*$, $i = 1, \ldots, n$, where $(\cdot)_*$ is the step function defined as one for positive x_i and zero otherwise, while $|x|$ will denote a vector of absolute values of components of x. The base of the natural logarithm will be denoted by ε and for $y \in R^m$, ε^{-y} will denote a vector in R^m with component ε^{-y_i}, $i = 1, \ldots, m$. The norm $\|\cdot\|_p$ will denote the p norm, $1 \leqq p \leqq \infty$, while $A \in R^{m \times n}$ will signify a real $m \times n$ matrix. For such a matrix, A^T will denote the transpose, and A_i will denote row i. For two vectors x and y in R^n, $x \perp y$ will denote $x^T y = 0$. A vector of ones in a real space of arbitrary dimension will be denoted by e. The notation $\arg\min_{x \in S} f(x)$ will denote the set of minimizers of $f(x)$ on the set S. Similarly $\arg \operatorname{vertex} \min_{x \in S} f(x)$ will denote the set of vertex minimizers of $f(x)$ on the polyhedral set S. By a separating plane, with respect to two given point sets $\mathcal{A}$ and $\mathcal{B}$ in R^n, we shall mean a plane that attempts to separate R^n into two half spaces such that each open halfspace contains points mostly of $\mathcal{A}$ or $\mathcal{B}$. The symbol ":=" defines a quantity appearing on its left by a quantity appearing on its right. For $f : R^n \longrightarrow R$ which is differentiable at x, the notation $\nabla f(x)$ will represent the $1 \times n$ gradient vector. R^n_+ will denote the nonnegative orthant.

2 The Misclassification Minimization and Feature Selection Problems

We consider two nonempty finite point sets $\mathcal{A}$ and $\mathcal{B}$ in R^n consisting of m and k points respectively that are represented by the matrices $A \in R^{m\times n}$ and $B \in R^{k\times n}$. The objective of both problems here is to construct a separating plane:

$$P := \{x \mid x \in R^n,\ x^T w = \gamma\}, \tag{1}$$

where $w \in R^n$, $\gamma \in R$, such that some error criterion is minimized. Thus in the exceptional case when the convex hulls of $\mathcal{A}$ and $\mathcal{B}$ do not intersect, a single linear program [2] will generate a plane P that strictly separates the sets $\mathcal{A}$ and $\mathcal{B}$ as follows:

$$Aw \geqq e\gamma + e,\ Bw \leqq e\gamma - e \tag{2}$$

Our concern here is with the usually occurring case when *no* plane P exists satisfying (2). A desirable objective for such a case [24, 18, 1, 6] is to minimize the number of points of $\mathcal{A}$ lying in the complement of the closed halfspace reserved for it, that is, minimize the number of elements of $\mathcal{A}$ in:

$$\{x \mid x^T w < \gamma + 1\}, \tag{3}$$

as well as the number of points of $\mathcal{B}$ lying in the complement of the closed halfspace reserved for it, that is, minimize the number of elements of $\mathcal{B}$ in:

$$\{x \mid x^T w > \gamma - 1\} \tag{4}$$

Thus, if we introduce the nonnegative slack variables $y \in R^m$ and $z \in R^k$ and make use of the step function $(\cdot)_*$, the misclassification minimization problem can be stated as follows:

$$\min_{w,\gamma,y,z} \left\{ e^T y_* + e^T z_* \;\middle|\; \begin{array}{c} y \geqq -Aw + e\gamma + e,\ y \geqq 0, \\ z \geqq Bw - e\gamma + e,\ z \geqq 0 \end{array} \right\} \tag{5}$$

Note that without the step function $(\cdot)_*$ in (5), the problem becomes a linear program (essentially the robust linear program [2, Equation (2.11)], but without averaging over m and k), in which case y and z of (5) become:

$$y = (-Aw + e\gamma + e)_+,\ z = (Bw - e\gamma + e)_+ \tag{6}$$

Thus, problem (5) ***without*** the step function $(\cdot)_*$ is equivalent to:

$$\min_{w,\gamma} \left\| \begin{pmatrix} -Aw + e\gamma + e \\ Bw - e\gamma + e \end{pmatrix}_+ \right\|_1 \tag{7}$$

The objective of (7) measures sums of *distances* (assuming each row of A and B has unit 2-norm) of points of $\mathcal{A}$ in the open halfspace (3) from the plane

$x^T w = \gamma + 1$ as well as points of $\mathcal{B}$ in the open halfspace (4) from the plane $x^T w = \gamma - 1$. By contrast the objective of problem (5) is to *count* the points of $\mathcal{A}$ contained in the open halfspace (3) and the points of $\mathcal{B}$ contained in the open halfspace (4) and attempt to minimize the totality of such points. To show indeed that problem (5) minimizes the total number of misclassified points we state the following simple lemma.

Lemma 2.1 *Let $a \in R^m$. Then*

$$r \in \arg\min_r \{e^T r_* \mid r \geqq a,\ r \geqq 0\} \Rightarrow r_* = a_* \tag{8}$$

Proof If r is a solution of the indicated minimization problem then for $i = 1, \ldots, m$:

$$(r_i)_* = \left\langle \begin{array}{l} 0 \text{ if } a_i \leqq 0 \\ 1 \text{ if } a_i > 0 \end{array} \right\rangle = (a_i)_*$$

□

By using this lemma on problem (5) we obtain the following proposition, which shows that any solution of (5) (and we will show in Proposition 2.4 below that (5) is always solvable) generates a plane that minimizes the number of misclassified points, that is points of $\mathcal{A}$ in (3) and points of $\mathcal{B}$ in (4).

Proposition 2.2 *Let $(\bar{w}, \bar{\gamma}, \bar{y}, \bar{z})$ solve (5), then*

$$e^T \bar{y}_* + e^T \bar{z}_* = \min_{w,\gamma} e^T(-Aw + e\gamma + e)_* + e^T(Bw - e\gamma + e)_* \tag{9}$$

Proof For a fixed (w, γ), let

$$(y(w,\gamma), z(w,\gamma)) \in \arg\min_{y,z} \left\{ e^T y_* + e^T z_* \;\middle|\; \begin{array}{l} y \geqq -Aw + e\gamma + e, \\ y \geqq 0 \\ z \geqq Bw - e\gamma + e, \\ z \geqq 0 \end{array} \right\} \tag{10}$$

By Lemma 2.1 we have that

$$\begin{aligned} (y(w,\gamma))_* &= (-Aw + e\gamma + e)_* \\ (z(w,\gamma))_* &= (Bw - e\gamma + e)_* \end{aligned} \tag{11}$$

Since $(\bar{w}, \bar{\gamma}, \bar{y}, \bar{z})$ solves (5) we have by (10)-(11) that

$$e^T \bar{y}_* + e^T \bar{z}_* = \min_{w,\gamma} e^T(-Aw + e\gamma + e)_* + e^T(Bw - e\gamma + e)_* \tag{12}$$

□

To establish the existence of solution to problem (5) and to relate it to a linear program with equilibrium constraints (LPEC) [18, 19, 16, 15], we state the following lemma.

Lemma 2.3 *Let $a \in R^m$. Then*

$$r = a_*,\ u = a_+ \Leftrightarrow (r,u) = \arg\min_{r,u} \left\{ e^T r \,\middle|\, \begin{array}{l} 0 \leq r \perp u - a \geqq 0, \\ 0 \leqq u \perp -r + e \geqq 0 \end{array} \right\} \tag{13}$$

Proof The constraints of the minimization problem constitute the Karush-Kuhn-Tucker conditions for the dual linear programs:

$$\max_r \{a^T r \mid 0 \leq r \leqq e\},\ \min_u \{e^T u \mid u \geqq a,\ u \geqq 0\} \tag{14}$$

which are solved by:

$$r_i = \left\langle \begin{array}{l} 0 \text{ for } a_i < 0 \\ r_i \in [0,1] \text{ for } a_i = 0 \\ 1 \text{ for } a_i > 0 \end{array} \right. ,\quad u = a_+ \tag{15}$$

The objective function $e^T r$ minimized in (13) renders the solution r of (15) unique by making $r_i = 0$ for $a_i = 0$, thus giving $r = a_*$. □

By using Lemma 2.3, problem (5) can be written in the following equivalent form as an LPEC:

$$\min_{w,\gamma,y,z,r,u,s,v} \left\{ e^T r + e^T s \,\middle|\, \begin{array}{c} 0 \leqq r \perp u - y \geqq 0 \\ 0 \leqq s \perp v - z \geqq 0 \\ 0 \leqq u \perp -r + e \geqq 0 \\ 0 \leqq v \perp -s + e \geqq 0 \\ y \geqq -Aw + e\gamma + e; \\ z \geqq Bw - e\gamma + e \\ y \geqq 0, z \geqq 0 \end{array} \right\} \tag{16}$$

Since the nonempty (take $w = 0,\ \gamma = 0,\ y = e,\ z = e,\ r = e,\ s = e,\ u = e,\ v = e$) feasible region of (16) is the union of a finite number of polyhedral sets over which the linear objective function $e^T r + e^T s$ is bounded below by zero, it follows that $e^T r + e^T s$ attains a minimum on each of these polyhedral sets. The minimum of these minima is a solution of (16). Since (16) is equivalent to (5), we have the following.

Proposition 2.4 *The misclassification minimization problem (5) has a solution.*

We turn our attention to our second problem, the feature selection problem. The problem again is to separate the finite point sets $\mathcal{A}$ and $\mathcal{B}$ in R^n, *but* with the additional requirement of using as few of the dimensions of R^n as possible. If we take as our point of departure the robust linear program [2, Equation (2.11)], which is very effective in discriminating between sets arising from real world problems [21], and motivate our formulation by the perturbation results of linear programming [20] to suppress as many of the coefficients w of the separating plane $\{x \mid x^T w = \gamma\}$ as possible, we obtain the following problem for a suitably chosen $\lambda \in [0,1]$:

$$\min_{w,\gamma,y,z} \left\{ (1-\lambda)\left(\frac{e^T y}{m} + \frac{e^T z}{k}\right) + \lambda e^T v_* \;\middle|\; \begin{array}{c} Aw - e\gamma + y \geqq e, \\ -Bw + e\gamma + z \geqq e, \\ y \geqq 0,\ z \geqq 0 \\ -v \leqq w \leqq v \end{array} \right\} \tag{17}$$

For $\lambda = 0$, we obtain the robust linear program of [2]. For $\lambda = 1$, all components of w are suppressed yielding no useful result. For λ sufficiently small, the program (17) selects those solutions of the robust linear program, that is (17) with $\lambda = 0$, that minimize $e^T \mid w \mid_*$. This in effect suppresses as many components of w as possible. Computationally, one obviously varies λ until some "best" value of (w, γ) is obtained as evinced by a cross-validating procedure [30].

By using an identical technique to that used to establish the existence of a solution to problem (5), we can similarly replace problem (17) by an LPEC and establish existence of a solution to it. We thus can state the following result.

Proposition 2.5 *The feature selection problem (17) has a solution to each* $\lambda \in [0,1]$.

We turn our attention now to algorithmic considerations by first approximating the step function $(\cdot)_*$, which appears in both problems (5) and (17), by a smooth concave approximation.

3 Concave Approximation of the Step Function

One of the most common and useful approximations in neural networks [28, 11] is the *sigmoid* function approximation of the step function ζ_* defined as

$$s(\zeta, \alpha) := \frac{1}{1 + \varepsilon^{-\alpha\zeta}}\ , \quad \alpha > 0 \tag{18}$$

Here ε is the base of the natural logarithm. For moderate values of α, the sigmoid is a very adequate approximation of the step function ζ_*. A shortcoming of the sigmoid is that it is neither convex nor concave. This prevents us from invoking some of the fundamental properties of these functions. In the two applications of this paper, it turns out that the variables to which the step function is applied are nonnegative: y and z in problem (5) and

v in problem (17). Consequently, we propose the following simpler concave approximation of the step function for nonnegative variables

$$t(\zeta, \alpha) := 1 - \varepsilon^{-\alpha\zeta} \quad \alpha > 0,\ \zeta \geqq 0 \tag{19}$$

Two important consequences of this simpler concave approximation of the step function are: first, an existence proof to both the smooth concave approximation of the misclassification minimization problem (5) as well as to the smooth concave approximation of the feature selection problem (17) (Proposition 3.1 below), and second, a finite termination theorem (Theorem 4.2 below) for the successive linearization algorithm (SLA Algorithm 4.1 below). We now state the smooth approximations of the misclassification minimization and the feature selection problems.

3.1 Smooth Concave Misclassification Minimization Problem (5) Let $\alpha > 0$.

$$\min_{w,\gamma,y,z} \left\{ m + k - e^T\varepsilon^{-\alpha y} - e^T\varepsilon^{-\alpha z} \;\middle|\; \begin{array}{c} y \geqq -Aw + e\gamma + e, \\ y \geqq 0, \\ z \geqq Bw - e\gamma + e, \\ z \geqq 0 \end{array} \right\} \tag{20}$$

We note immediately that the concave objective function is bounded below by zero on the set $R^{n+1} \times R_+^{m+k}$ which contains the feasible region. Furthermore, this lower bound is attained by $y = 0$, $z = 0$, and some infeasible (w, γ) in general. The zero minimum is attained at a feasible point if and only if the convex hulls of $\mathcal{A}$ and $\mathcal{B}$ do not intersect. Otherwise the minimized objective of (20) approximates from below (for moderate values of α) the smallest number of misclassified points by any plane $x^T w = \gamma$.

3.2 Smooth Concave Feature Selection Problem (17) Let $\lambda \in [0, 1]$ and $\alpha > 0$.

$$\min_{w,\gamma,y,z} \left\{ \begin{array}{l} (1-\lambda)\left(\frac{e^T y}{m} + \frac{e^T z}{k}\right) + \\ + \lambda(n - e^T\varepsilon^{-\alpha v}) \end{array} \;\middle|\; \begin{array}{c} Aw - e\gamma + y \geqq e, \\ -Bw + e\gamma + z \geqq e, \\ y \geqq 0, z \geqq 0 \\ -v \leqq w \leqq v \end{array} \right\} \tag{21}$$

Again for this problem, the concave objective function is bounded below by zero on the feasible region. For various values of the parameter $\lambda \in [0, 1]$, emphasis of separation by the plane $x^T w = \gamma$ is balanced against suppression of as many coefficients of w as possible, with the term $(n - e^T\varepsilon^{-\alpha v})$ giving an approximation (from below) to the number of nonzero coefficients of w.

By making use of [27, Corollary 32.3.3] which implies that a concave function, bounded from below on a nonempty polyhedral set, attains its minimum on that set, we can state the following existence results for the two smooth problems above.

Proposition 3.1 *The smooth concave misclassification minimization problem (20) and the smooth concave feature selection problem (21) have solutions.*

We turn our attention to algorithmic considerations.

4 Successive Linearization of Polyhedral Concave Programs

By replacing the variables (w, γ) by the nonnegative variables (w^1, γ^1, ζ^1) using the standard transformation $w = w^1 - e\zeta^1$, $\gamma = \gamma^1 - \zeta^1$, the smooth problems (20) and (21) can be transformed to the following concave minimization problem:

$$\min_x \{f(x) \mid Ax \leqq b,\ x \geqq 0\}, \tag{22}$$

where $f: R^\ell \to R$, is a differentiable, concave function bounded below on the nonempty polyhedral feasible region of (22), $A \in R^{p\times\ell}$ and $b \in R^p$. By [27, Corollary 32.3.4] it follows that f attains its minimum at a vertex of the feasible region of (22). We now prescribe a simple finite successive linearization algorithm (essentially a Frank-Wolfe algorithm [9] without a stepsize) for solving (22) that appears to give good computational results. (See Section 5.) Other more complex computational schemes for this problem are given in [13, 12].

4.1 Successive Linearization Algorithm (SLA) Start with a random $x^0 \in R^n$. Having x^i determine x^{i+1} as follows:

$$\begin{array}{l} x^{i+1} \in \arg\text{vertex}\min_{x\in X}\ \nabla f(x^i)(x - x^i) \\ X = \{x \mid Ax \leqq b,\ x \geqq 0\} \end{array} \tag{23}$$

Stop if $x^i \in X$ and $\nabla f(x^i)(x^{i+1} - x^i) = 0$.
Comment: The condition $x^i \in X$ takes care of the possibility that x^0 may not be in X.

We show below that this is a finite algorithm which generates a strictly decreasing finite sequence $\{f(x^i)\}$, $i = 1, 2, \ldots, \bar{i}$, which terminates at an $x^{\bar{i}}$ that is a stationary point that may also be a global minimum solution.
Remark: SLA may be started from many different random starting points. This was not necessary in the present applications.

4.2 SLA Finite Termination Theorem Let f be a differentiable concave function on R^n that is bounded below on X. The SLA generates a finite sequence of iterates $\{x^1, x^2, \ldots, x^{\bar{i}}\}$ of strictly decreasing objective function values: $f(x^1) > f(x^2) > \ldots > f(x^{\bar{i}})$, such that $x^{\bar{i}}$ satisfies the minimum principle necessary optimality conditions

$$\nabla f(x^{\bar{i}})(x - x^{\bar{i}}) \geqq 0, \quad \forall x \in X. \tag{24}$$

Proof We first show that SLA is well defined. By the concavity of f and its boundedness from below on X, we have that

$$-\infty < \inf_{x \in X} f(x) - f(x^i) \leqq f(x) - f(x^i) \leqq \nabla f(x^i)(x - x^i), \quad \forall x \in X.$$

It follows for any $x^i \in R^n$, even for an infeasible x^i such as x^0, that $\nabla f(x^i)(x - x^i)$ is bounded below on X. Hence the linear program (23) is solvable and has a vertex solution x^{i+1}. It follows for $i = 1, 2, \ldots,$ that $\forall x \in X$:

$$\begin{array}{rcl} \nabla f(x^i)(x - x^i) & \geqq & \min_{x \in X} \nabla f(x^i)(x - x^i) \\ & = & \nabla f(x^i)(x^{i+1} - x^i) \left\langle \begin{array}{l} < 0 \ \text{(a)} \\ = 0 \ \text{(b)} \end{array} \right. \end{array} \tag{25}$$

We note immediately that because $x^i \in X$ for $i = 1, 2, \ldots,$ it follows that $\nabla f(x^i)(x^{i+1} - x^i) \leqq 0$. Hence only two cases, (a) or (b), can occur, as indicated above. When case (a) above occurs, the algorithm does not stop at iteration i, and we have from the concavity of f and the strict inequality of case (a) that:

$$f(x^{i+1}) \leqq f(x^i) + \nabla f(x^i)(x^{i+1} - x^i) < f(x^i)$$

Hence $f(x^{i+1}) < f(x^i)$, for $i = 1, 2, \ldots$. When case (b) occurs we then have that:

$$\forall x \in X: \ \nabla f(x^i)(x - x^i) \geqq 0, \tag{26}$$

and the algorithm terminates (provided $x^i \in X$, which may not be the case if $x^i = x^0 \notin X$), and set $\bar{i} = i$. The point $x^{\bar{i}}$ thus satisfies the minimum principle necessary optimality conditions (26) with $x^{\bar{i}} = x^i$, and $x^{\bar{i}}$ may be a global solution. Furthermore, since X has a finite number of vertices, $\{f(x^i)\}$ is strictly decreasing and $f(x)$ is bounded below on X, it follows that case (b) must occur after a finite number of steps. □

We turn our attention to some computational results.

5 Numerical Tests

The proposed approach was tested numerically on publicly available databases from the University of California Repository of Machine Learning Databases [22] as well as the Star/Galaxy database collected by Odewahn [26]. For all the numerical results reported, the value of α used in the concave

Data Set	m k n	Percent of Correctly Classified Points Time Seconds SPARCstation 20 Average No. of LPs over 10 Runs PMM	 SLA
WBC Prognosis	28 119 32	95.92 10.65	93.2 0.86 3.0
WBCD	239 443 9	98.57 24.65	97.6 9.47 5.7
Cleveland Heart	216 81 14	91.43 17.46	89.3 2.69 4.3
Ionosphere	225 126 34	98.42 27.26	97.0 10.30 4.0
Liver Disorders	145 200 6	74.85 18.51	71.4 1.09 5.5
Pima Diabetes	268 500 8	80.55 51.40	78.3 14.33 6.5
Star/Galaxy(Dim)	2082 2110 14	96.52 1122.70	96.1 779.89 5.9
Star/Galaxy(Bright)	1505 957 14	99.89 266.13	99.8 69.48 3.2
Tic Tac Toe	626 332 9	69.12 46.45	66.6 6.44 3.3
Votes	168 267 16	98.82 14.76	96.9 1.56 3.4
Total Times		1599.97	896.11

Table 1: Comparison of Successive Linearization Algorithm (SLA) Algorithm 4.1 for the Smooth Misclassification Minimization Problem (20) with the Parametric Minimization Method (PMM) [18, 1]. SLA was coded in GAMS [5] utilizing the CPLEX solver [7]. PMM was coded was coded in AMPL [8] utilizing the MINOS LP solver [23].

exponential approximation $t(\zeta, \alpha) = 1-\varepsilon^{-\alpha\zeta}$ to the step function ζ_*, was five. This value of α allows $t(\zeta, \alpha)$ to capture the essence of of the step function ζ_* with sufficient smoothness to make the proposed algorithm work effectively without overflow or underflow.

The first test consisted in applying the SLA 4.1 to the smooth misclassification minimization problem (20). For this problem ten databases were used from the Irvine repository and the Star/Galaxy database. Table 1 gives the percent of correctly separated points as well as CPU times using an average of ten SLA runs on the smooth misclassification minimization problem (20). These quantities are compared with those of a parametric minimization method (PMM) applied to an LPEC associated with the misclassification minimization [18, 1]. Table 1 shows that the much simpler SLA algorithm obtained a separation that was almost as good as the parametric method for solving the LPEC at considerably less computing cost. Each problem was solved using no more than a maximum average of 7 LPs over ten runs. Average of solution times of the SLA over all problems run was 56% of the average PMM solution times.

Our second test consisted of solving the smooth concave feature selection problem (21) by SLA 4.1. The test problem consisted of the Wisconsin Breast Cancer Database WBCD tested in the above set of tests, with one modification. Two new random features, uniformly distributed on the interval $[0, 10]$ were added to the problem, so that the problem space was R^{11} instead of the original R^9. With $\lambda = 0.05$ in problem (21), and by solving 6 successive linear programs, the SLA was able to suppress the effect of the random components x_{10} and x_{11} by setting w_{10} and w_{11} equal to zero, as well as some other components: w_3, w_4, w_5, w_7, and w_9. The resulting separation in R^4 correctly separated 97.1% of the points, which is almost as good as the 97.6% correctness obtained above without the feature selection option by solving the misclassification minimization problem (20) in R^9. This indicates that, for this problem, the stationary point obtained by the SLA algorithm in R^4 for the smooth feature selection problem (21) is almost as good as the stationary point obtained in R^9 for the smooth misclassification minimization problem (20). The key observation however, is that the feature selection approach proposed here, not only gets rid of extraneous random features, but also of unimportant features in the original problem.

6 Conclusion

We have formulated two important problems of machine learning: misclassification minimization and feature selection as the minimization of a simple concave function on a polyhedral set that is always solvable. A successive linearization algorithm that requires the solution of a few LPs in each instance appears to be a very effective method of solution.

Acknowledgement

I am indebted to my Ph.D. student Paul S. Bradley for the numerical testing of the proposed algorithm.

References

[1] K. P. Bennett and E. J. Bredensteiner. A parametric optimization method for machine learning. Department of Mathematical Sciences Math Report No. 217, Rensselaer Polytechnic Institute, Troy, NY 12180, 1994. ORSA Journal on Computing, submitted.

[2] K. P. Bennett and O. L. Mangasarian. Robust linear programming discrimination of two linearly inseparable sets. *Optimization Methods and Software*, 1:23–34, 1992.

[3] P. S. Bradley, O. L. Mangasarian, and W. N. Street. Feature selection via mathematical programming. Technical report, Computer Sciences Department, University of Wisconsin, Madison, Wisconsin, 1995. To appear.

[4] E. J. Bredensteiner and K. P. Bennett. Feature minimization within decision trees. Department of Mathematical Sciences Math Report No. 218, Rensselaer Polytechnic Institute, Troy, NY 12180, 1995.

[5] A. Brooke, D. Kendrick, and A. Meeraus. *GAMS: A User's Guide*. The Scientific Press, South San Francisco, CA, 1988.

[6] Chunhui Chen and O. L. Mangasarian. Hybrid misclassification minimization. Technical Report 95-05, Computer Sciences Department, University of Wisconsin, Madison, Wisconsin, February 1995. Advances in Computational Mathematics, to appear. Available from ftp://ftp.cs.wisc.edu/math-prog/tech-reports/95-05.ps.Z.

[7] CPLEX Optimization Inc., Incline Village, Nevada. *Using the CPLEX(TM) Linear Optimizer and CPLEX(TM) Mixed Integer Optimizer (Version 2.0)*, 1992.

[8] R. Fourer, D. Gay, and B. Kernighan. *AMPL*. The Scientific Press, South San Francisco, California, 1993.

[9] M. Frank and P. Wolfe. An algorithm for quadratic programming. *Naval Research Logistics Quarterly*, 3:95–110, 1956.

[10] David Heath. *A geometric Framework for Machine Learning*. PhD thesis, Department of Computer Science, Johns Hopkins University–Baltimore, Maryland, 1992.

[11] J. Hertz, A. Krogh, and R. G. Palmer. *Introduction to the Theory of Neural Computation*. Addison-Wesley, Redwood City, California, 1991.

[12] R. Horst, P. Pardalos, and N. V. Thoai. *Introduction to Global Optimization*. Kluwer Academic Publishers, Dodrecht, Netherlands, 1995.

[13] R. Horst and H. Tuy. *Global Optimization*. Springer–Verlag, Berlin, 1993. Second, Revised Edition.

[14] G. H. John, R. Kohavi, and K. Pfleger. Irrelevant features and the subset selection problem. In *Proceedings of the 11th International Conference on Machine Learning*, San Mateo, CA, 1994. Morgan Kaufmann.

[15] Z.-Q. Luo, J.-S. Pang, and D. Ralph. *Mathematical Programs with Equilibrium Constraints*. Cambridge University Press, Cambridge, England, 1996.

[16] Z.-Q. Luo, J.-S. Pang, D. Ralph, and S.-Q. Wu. Exact penalization and stationarity conditions of mathematical programs with equilibrium constraints. Technical Report 275, Communications Research Laboratory, McMaster University, Hamilton, Ontario, Hamilton, Ontario L8S 4K1, Canada, 1993. Mathematical Programming, to appear.

[17] O. L. Mangasarian. Mathematical programming in neural networks. *ORSA Journal on Computing*, 5(4):349–360, 1993.

[18] O. L. Mangasarian. Misclassification minimization. *Journal of Global Optimization*, 5:309–323, 1994.

[19] O. L. Mangasarian. The ill-posed linear complementarity problem. Technical Report 95-15, Computer Sciences Department, University of Wisconsin, Madison, Wisconsin, August 1995. Proceedings of the International Conference on Complementarity Problems, Johns Hopkins University, November 1-4, 1995, SIAM Publishers, Philadelphia, PA, submitted.

[20] O. L. Mangasarian and R. R. Meyer. Nonlinear perturbation of linear programs. *SIAM Journal on Control and Optimization*, 17(6):745–752, November 1979.

[21] O. L. Mangasarian, W. Nick Street, and W. H. Wolberg. Breast cancer diagnosis and prognosis via linear programming. *Operations Research*, 43(4):570–577, July-August 1995.

[22] P. M. Murphy and D. W. Aha. UCI repository of machine learning databases. Department of Information and Computer Science, University of California, Irvine, California, http://www.ics.uci.edu/AI/ML/MLDBRepository.html., 1992.

[23] B. A. Murtagh and M. A. Saunders. MINOS 5.0 user's guide. Technical Report SOL 83.20, Stanford University, December 1983. MINOS 5.4 Release Notes, December 1992.

[24] S. Murthy, S. Kasif, S. Salzberg, and R. Beigel. OC1: Randomized induction of oblique decision trees. In *Proceedings of the Eleventh National Conference on Artificial Intelligence*, pages 322–327, Cambridge, MA 02142, 1993. The AAAI Press/The MIT Press.

[25] P. M. Narendra and K. Fukunaga. A branch and bound algorithm for feature subset selection. *IEEE Transactions on Computers*, C-26(9):917–922, September 1977.

[26] S. Odewahn, E. Stockwell, R. Pennington, R. Hummphreys, and W. Zumach. Automated star/galaxy discrimination with neural networks. *Astronomical Journal*, 103(1):318–331, 1992.

[27] R. T. Rockafellar. *Convex Analysis*. Princeton University Press, Princeton, New Jersey, 1970.

[28] D. E. Rumelhart and J. L. McClelland. *Parallel Distributed Processing*. MIT Press, Cambridge, Massachusetts, 1986.

[29] W. Siedlecki and J. Sklansky. On automatic feature selection. *International Journal of Pattern Recognition and Artificial Intelligence*, 2(2):197–220, 1988.

[30] M. Stone. Cross-validatory choice and assessment of statistical predictions. *Journal of the Royal Statistical Society*, 36:111–147, 1974.

Optimization Concepts in Autonomous Mobile Platform Design

V.Mařík, L.Přeučil, P. Štěpán
AI Division, Department of Control Engineering
Czech Technical University in Prague*

Abstract. The contribution aims to introduce selected topics and possible approaches to application of optimization methods within the area of artificial intelligence (AI). There are mentioned some - from the mathematical point of view - atypical approaches to optimization together with examples of their use in the AI - namely in design of intelligent self-guided autonomous vehicles.

1 Introduction

Robotics is a research area which is aimed at design and development of robots, i.e. autonomous mobile/movable systems capable to operate independently and to solve complex tasks which require fast situation recognition, intelligent decision and subsequently an efficient physical action. The robotics research still follows the initial Čapek's vision [4] of robots which are considered as physical entities substituing human beings in carrying out of some specific tasks. Whereas the leading background vision remains the same, robotics explores and integrates up-to-date results of many scientific disciplines currently.

To analyze and categorize the desired functionality of a robot, let's consider a more precise definition of a robot [6]:

"A robot is a computer-controlled integrated system capable of autonomous and problem-oriented interactions with a real environment according to the instructions given by a human being. A robot should be able:

- *To percept and recognize the environment,*
- *To create and continuously modify the internal representation (model) of the environment,*
- *To deciside about its own activity: based on environment representation and in accordance with the required goals,*
- *To influence the environment: to manipulate objects and accomplish movements,*
- *To communicate with a human being in a natural or artificial language."*

This definition formulates quite explicitly three main subsystems of each robot, namely:

* The research has been supported by the EU grant PECO Copernicus'93, No. 5855 "Information Processing for Active CIM Subsystems" coordinated by Prof. Dr. K. Ritter, Institute for Applied Mathematics and Statistics, TU Munich.

- Perception subsystem responsible also for world model creation,
- Decision making subsystem, and
- Action subsystem.

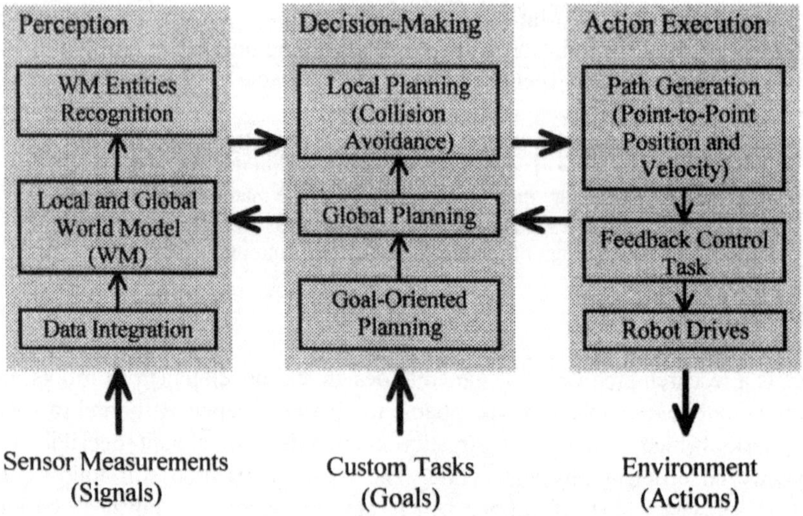

Fig. 1 Main functional components of an intelligent robot

The perception subsystem is responsible for preprocessing raw sensor data, in order to recognize basic elements of the higher level description of the environment and for integrating the partial descriptions into a (more or less complete and consistent) model of the environment (so called world model WM).

The decision making subsystem is responsible for formulating of global goals and for the global planning of activities to reach them. This subsystem widely explores the world model created by the sensor subsystem.

The action subsystem performs the detailed planning of actions, generates instructions for the motion system and directly controls the motion. It explores the world model and exceptionally some sensor data directly (e.g. tactile information for assembling can be used directly - without any complex processing in the decision making subsystem - to improve the finest motions and to make them very accurate).

Each function block mentioned in the Fig.1 is - more or less - an analogy to some biological subsystem which results form a long of evolution and which is naturally optimized to perform the required activity.

That is why the design or activity of each of the function blocks of each subsystem mentioned above is connected with specific optimization tasks.

An optimization task is - in principle - given by formulating:

- Goal(s)
- Optimization criterion/criteria
- Constraints (if any).

2 Types of Optimization Tasks

Two categories of optimization tasks should be considered:

- Tasks where an exact mathematical description of goals, criteria and constraints in the form of formulas is available. In such cases mathematical solutions aimed at computationally optimal approximation of the precisely expressed formulas are saught for.
- Tasks which are described by a performance-based specification of the desired system (either goals or criteria or constraints or of all of them are defined vaguely). We are facing to the problem of the best approximation of a loosely defined optimization criterion.

In the latter case, the classical methods of numeric optimization are not applicable, but the solutions can also be based on explicitly expressed pieces of problem-oriented knowledge. Rather than classical methods, the AI techniques can be explored with advantage.

The paradigm of "classical" versus AI-based optimization approaches is quite analogical to the "classical" software engineering versus knowledge engineering paradigm [11]. As a matter of fact, the roots of both these paradigms are the same: The methods for handling knowledge are quite different from those for exact handling of precisely described numerical objects. The knowledge-based methods can be conveniently used just in the tasks which cannot be specified quite exactly, i.e. in the form of precise analytical terms. The exploitation of knowledge (which is very often vague, uncertain, foggy, partially missing etc.) seems to be quite adequate and efficient (and natural) in the case of inexact, performance-oriented specification of optimization tasks.

There is one very good advantage of the performance-based specification: Such systems/methods based on this specification can never be verified from the point of view of their mathematical exactness/correctness, their performance may be only evaluated and compared to that which is (vaguely) intended. The process of designing/development of such a system may be recognized as everlasting incremental improvements of the system performance based on new knowledge which is added/modified either by a system designer or by own experience gathered by the system (learning processes). The system methodologies based on life-cycle models containing many various iterative feedbacks (e.g. POLITE-VRC or spiral prototyping methodologies [24], etc.) seem to represent the situation at best.

Two classes of - in principle - optimization tasks can be distinguished in the area of AI, namely:

Diagnostic tasks. The goal is to classify the entities represented by numerical or non-numerical descriptions into one of the pre-defined classes. The criterion is to achieve minimal average error when classifying a set of unknown patterns. Constraints usually don't occur or are expressed in the form of explicit knowledge (rules).

Planning tasks. The goal is to find an optimal sequence of operations to transform the system from an initial state to a final state in the state space. The optimization criterion is usually a function combining the costs of individual operations. Constraints play a very important role (even crucial for reduction of the state space search extent) and are usually explicitly expressed as a specific problem-domain knowledge.

A classical tool for robot planning task solving is the situation calculus [22]. Unfortunately, from the point of view of robustness, this technique is not directly applicable in practical navigation tasks.

The diagnostic tasks can be solved by the following AI methods:

Pattern Recognition - Feature Approach. The entities to be recognized are described by a vector of numeric features. There is a wide variety of machine learning techniques helping to setup statistically optimal probabilistic distributions representing individual classes (e.g. minimum error method, minimum distance method etc.).

Pattern Recognition - Structural Approach. The description of the entities has symbolic nature (symbols are used for both description primitives and relations among them). At the end of the inductive machine learning process, the classes are represented by an optimal finite set of structural rules (grammar rules) or structures (relational structures or automata): The optimization task is the task of the best approximation of the structural patterns in the training set by the induced finite structural descriptions.

Diagnostic Expert Systems. It is dealt with either rule-based systems (containing knowledge represented by production rules) or frame-based systems (in this case knowledge is stored in hierarchically organized chunks of knowledge - frames). The pieces of knowledge are either given explicitly by human experts (in this case, the optimization task is already solved by the expert who provides the knowledge) or learned inductively from examples (by inductive learning programs like ID3, AQ11 etc.). The ES shells are tools for applying the knowledge.

Neural Nets. Neural nets for solving the diagnostic tasks can be considered as specific pattern recognition systems based on feature approach: The only difference is that they use special architecture (network arrangement of uniform units) and approximate learning algorithms.

The goals of planning tasks are expressed in a much more vague manner. These tasks can be solved e.g. by the following AI methods:

State-space search reduced by heuristic knowledge. This group of methods includes e.g. A* based search, search by using meta-knowledge or task decomposition, search in a hierarchically organized state space, means-and-goals analysis approach etc. Also case-based reasoning can be included into this category. In all of these methods, the optimization task results as well as the optimality of the search process itself are much more dependent on quality of knowledge than on the search algorithms themselves.

Evolutionary computing. The genetic algorithms can be used for generation and selection of optimal plans. Whereas the optimization algorithms are not very complicated and well-known from the evolution in nature, the main problems are to find an appropriate coding of the problem into the genes' strings and definition of the corresponding optimization criteria. Both these problems to be solved require a very detailed analysis of the task in hand, as well as intensive experimental testing/verification.

Logic programming. Logic programming, and especially the up-to-date methods of the constrained logic programming (CLP) and inductive logic programming (ILP) represent very robust tools for automatic planning. The main problem of these methods is how to represent the real-life tasks of robotics in the formal language of predicate logic.

2.1 Typical Tasks to be Solved by Self-Guided Vehicles

To demonstrate the nature of the optimization techniques used in robotics, let's focus to a specific class of intelligent robots which are *self-guided autonomous vehicles* (SGV). Respecting the application field, the common task for this sort of robot is to conduct movements through 2D environments. The robot's goal is typically to *reach a destination* or to *cover a region*. All such activities are closely related to the following functions of the intelligent vehicle:

In the first place there is the *navigation task* which provides the vehicle's position and orientation at any world location. The self-navigation incorporates data acquisition by various sensor systems and their interpretation with respect to the level of a certain internal knowledge contained within an internal world model (WM). This results in two facts - the vehicle's position and orientation in the 2D world and into hypotheses of the world's properties expressed by the means of obstacle occurrence [10, 16,18, 25].

Knowing the actual position of the vehicle at any one time, and having previously mentioned knowledge about the environment, *the activity planning* and *plan execution* can be resolved [2, 7, 10, 20].

Generally, both categories of the processes - the world data acquisition and planning - are simultaneously operating functions, where each of them influences interactively the other one. This is a typical application of reactive planning [11].

In the case that the vehicle's environment is not completely known in advance before the execution of the planning task, the plan can only be estimated. Afterwards, during the plan execution and using interpreted measurements of the sensor system, the original plan estimate can be (and typically will have to be) modified.

As it can be clearly seen, nearly all the tasks connected with intelligent acting of a self-guided vehicle are described by performance-based specification. Therefore exploitation of the AI-based methods as well as methods of knowledge engineering seems to be quite natural there.

3 Main Functional Components of an Intelligent Mobile Robot

Succeeding to the basic concept introduced above a global functional block scheme can be defined. One of the possible architectures of an SGV which is fully compatible with basic concepts mentioned above, can consist of:

- Sensor preprocessing, fusion, and world model creation (perception subsystem)
- Activity planner (decision-making subsystem)
- Plan interpreter (action subsystem)

The *sensor processing and fusion* fulfills two basic tasks. It is responsible for low level preprocessing operations of the raw sensor data such as noise reduction, dropout detection, etc. More important is it's fusion function which deals with the task of the integration of measurements of position and orientation obtained via multiple measurements and/or sensors. The subsystem concentrates data having the highest reliability and passes it to a local world model. Then, the performance-based optimization task can be found - on the first place - in minimizing task-relevant information loss, and in the most cases computational costs as well. The process of sensor fusion uses - besides knowledge of sensor behavior - hypotheses on data validity, provided by understanding the content of the local map.

The *world model* as a whole consists of two data-storing parts, the already mentioned *local world map* and the *global world map*. The related *position location* provides information on the coordinates and orientations of the vehicle via local to global map data matching. As the local map contains temporary data about the vehicle's neighborhood, the global map contains long-time valid data of the whole work area.

The *scene feature finder* serves the local map data understanding by the means of search for basic elements - features which set up the scene description. This function controls updating of the global and local maps with already extracted scene features and provides adaptability to changes in the environment.

The main subject to optimization for the above mentioned is an optimal splitting of both knowledge representations: the local and global world maps. Typical objective criterion for doing this should respect e.g. minimizing of used storage or maximizing data compression, and maximizing data access speed, etc.

The *path planner* subsystem explores knowledge on the work area structure and topology to solve the task(s) given to the vehicle. The completed path plan is converted by the *plan interpreter* subsystem into a sequence of commands - driving parameters, that feed directly the vehicle drives. The plan interpreter also tightly cooperates with the *collision avoidance* system, so that all the commands during the plan execution can be fulfilled. If not, the planner is reactivated and local replanning is invoked.

The planning part of the systems requires - from the optimization point of view - minimization of path length, energy consumption, or time, elapsed to goal fulfilling. A typical feature of this part is it's performance over - more or less - uncertain data which leads mostly to dynamic (on-line) methods.

3.1 World Data Acquisition and the Sensor Subsystem

With respect to nature and performance of used sensors, all the data on the environment can be gathered within certain limits on their quantity and/or rate. The gathering process has to provide sufficient information but must not overload succeeding processing.

The behavior of sensors under various conditions brings another problem. With respect to their physical principles, more or less uncertainty in the measurements can be expected.

These two facts form the main task of the sensor subsystem - *data volume reduction* and *reliability improvement* - by preprocessing and fusion. The incoming information, provided by the sensors, is in a pure signal form, and can be either directly fused, or passed to further processing. This processing stage assigns a semantic context to raw signal data and converts it into higher level representations.

3.3 Data Fusion Methods

The methods used for information fusion range from simple statistical approaches to knowledge-based decision making techniques. The choice of a suitable approach depends in the first place on the level of information representation, introduced above, and on a prior knowledge of the sensor response.

Knowing the sensor principle more or less exactly the description of the sensor behavior - the sensor model - can be built up. Basically, the less accurate model is available, the more heuristic method has to be used. Unfortunately, in many cases no accurate model is known and therefore sensor behavior can be described only by typical failure situations. This can be described by heuristic rules, which drive a decision if the measurement has to be refused or accepted.

Signal-Level Fusion. The method is usually related to processing coming from multiple sensors that are sensing the same entity from the same aspect. The fusion should result in the same form of signals as the original ones but with improved quality. Basically, the signals can be seen as random variables corrupted by uncorrelated noise. The fusion can then be considered as an estimation process of the original variable [16]. The most common measure of the signal quality in this case is its variance.

One method of implementing the signal-level fusion is by calculating a *weighted average* of redundant information coming from the sensors. The weights are set up with respect to estimated variances of the signal following the rule: the greater the variance of a certain signal, the lower its weight. This technique can be also combined with outlying value selection by omitting measurements which are "very far" from the mean value.

If an estimate of signal statistics is known, the *Kalman filter*, instead of the previous method, can be used. Having a linear system with Gaussian error distribution the Kalman filter will provide statistically optimal estimates for the fused data. Besides current estimates of the current values, the Kalman filter also provides a prediction of the future values. Beside the above mentioned main approaches a variety of modifications can be used [1].

Pixel-Level Fusion. This is used to improve the information content of 2D data - pixels of an image obtained by a combination of multiple images. A typical example of that is the assignment of depth information to the intensity of images, with respect to their pixels. This can be done, for example, by scene depth reconstruction from two different views (stereo vision) or by utilizing direct range images from a range-finding device. The improvement in data quality via pixel-level fusion can be typically noted through the improvements in performance of the succeeding image processing tasks as segmentation, feature extraction, etc.

The fused image can be created either through pixel-by-pixel fusion or through the fusion of associated local neighborhoods of corresponding pixels. The former case is mostly used when fusing raw 2D signals. The latter one can already be used for higher-level information representation for fusion of pixel-level feature extraction (e.g. an edge image). It is possible to use many kinds of pixel-level fusion methods, which generally belong to the area of computer vision [1, 12]. Beside these approaches, general methods for data fusion such as the Bayesian approach, Kalman filtering, etc. can be extended to 2D signals as well.

Feature-Level Methods. This group of methods is dedicated to two basic purposes - to increase accuracy and reliability of a feature, and to create additional composite features. A primary feature is created by the assignment of some semantic meaning to the result of processing [6] of a certain portion of sensory data. As an example, this might again be the case of edge pixel extraction, where each data is assigned its measure of edginess.

Composite features are then created making-use of the previously extracted ones, but bringing a new quality. In our example, as a composite feature can be considered complete edge detection, e.g. every complete edge consists of a set of edge pixels which are connected to neighbors with a similar direction.

Symbol-Level Methods. The supreme level methods enable us to put together multiple sensor information on the top level of data abstraction. Basically, the symbol level fusion allows to join information from extremely different sensors, being of very different nature, and the outputs of which can not be easily unified. A model of the world serves for coupling element. The represented data - symbols - originate from two sources:

- From processing of sensory data
- As prior information from a world model

A symbol calculated from the sensory data is matched to the world model content. This can be done by matching features derived from the sensory data to features derived from the world map. The obtained degree of similarity expresses whether the hypothesis on the world state (features derived from sensory data) was a mismatch and/or uncertainty infiltrated from the sensor itself. Frequent successful matches can be used for world model updating.

The other fundamental approaches to data fusion incorporate on this level: Bayesian estimation, Dempster-Shafer evidential reasoning, and production rule-based systems with confidence factors [1].

3.4 Sonar Data Fusion - a Case Study

To give a good idea, let's illustrate the above approach by an example of data (signal) fusion obtained by a sonar range-finder, as a very frequently used technique in the area of mobile robotics.

The basic method for recovery of spatial information from range data is based on extraction of a geometric model directly from sensor readings. Another alternate approach uses a grid map as a low level representation, avoiding early commitment to the geometric description. This can be expressed by identifying the following components of the data interpretation and the integration processes:

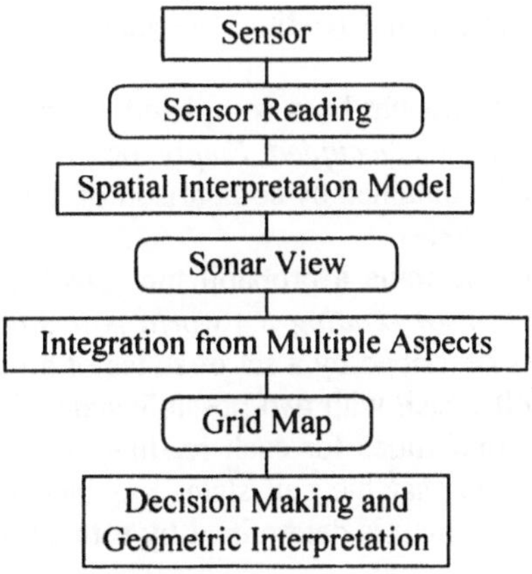

Fig. 2 Range data interpretation process

- *Spatial Interpretation Model* This model converts the measured distance in a given azimuth into probability values of occupancy and emptiness for each spatial cell. This model is dependent on sensor hardware parameters.
- *Integration* of measurement from different positions. The probability values from Spatial Interpretation Model are integrated into a common grid using fusion methods. The most common approaches are: *probability addition* and *Bayesian fusion.*
- *Decision Making* is used at the stage of final labeling of cells as *Occupied, Empty* or *Unknown.*

Integration Method. A sonar map can be built by evaluation of the emptiness and occupancy probability distribution for each sonar range reading. The probabilities are projected onto a discrete grid with succeeding integration of different views within the sonar map.

Whereas the map contains information derived from the previous measurements it should allow an integration with other readings. In a general case, the internal map structure depends on the used integration method.

For probability addition case the sonar map consists of 2D array of cells corresponding to a horizontal grid imposed on the area to be mapped. The grid has $M \times N$ cells, each $\Delta \times \Delta$ sized. During the integration process each cell contains two values: the emptiness probability p_E and the occupancy probability p_O, both initialized to zeros, and ranging within an interval of *(0,1)*.

The fusion process itself is done in two steps by evaluating the emptiness and occupancy probability profiles of each reading (what is actually the sensor probability model). The next step projects probability profiles onto the grid map and updates the emptiness and occupancy values contained by the map. The map probabilities can be integrated/updated using probability addition formulas:

$$p_E = p_E + r_E - p_E \cdot r_E \qquad\qquad p_O = p_O + r_O - p_O \cdot r_O$$

Where r_O (probability that cell is occupied during actual reading) and r_E (probability that the cell is empty during actual reading) are taken from the sonar probability model.

Once the integration process has been completed, the content of each map cell is converted into the state attributes *Occupied, Empty*, or *Unknown.* The decision about the final state of a cell can be obtained by comparison of the relative strengths of the emptiness to the occupancy values.

To apply the Bayesian formula a probabilistic sensor model in the form of a conditional *distribution P(sensor reading R | world is in state S)* has to be defined. The state S of the world is described by a set of states of all cells in the map. In our case, for a map with n^2 cells, each with two possible states, it would be necessary to specify 2^{n*n} conditional probabilities for each reading. To avoid this combinatorial explosion, it can be assumed that the cell states are independent discrete random variables, and the state of the map is determined by estimating the state of each cell individually.

In our case, the evidence is given by a sonar range reading R and determination of the probabilities $P(O_c|R)$ and $P(E_m|R)$ is desired After application of an additional simplifying assumption that $P(R|O_c) = 1-P(R|E_m)$ and $P(O_c) = 1-P(E_m)$. The Bayesian theorem [3] changes into:

$$P(Oc|R) = \frac{P(R|Oc)P(Oc)}{P(R|Oc)P(Oc) + (1 - P(R|Oc))(1 - P(Oc))}$$

Under assumption that $P(O_c) = 1-P(E_m)$, each cell contains only one probability value $P(O_c)$. At the beginning the value is set to *0.5* (represent unknown state). The value below 0.5 means the cell is empty, the value over *0.5* means the cell is occupied. The value $P(R|O_c)$ is obtained from the sonar model mentioned below.

Sonar Model. The sonar model is essential for transformation of sensor readings into probabilistic values. A single range reading is interpreted as making an assertion about two volumes: one is probably empty and the other one is probably occupied. For this purpose, the sonar beam can be simplified and modeled by the two probability profiles f_E, f_O, defining the mentioned volumes.

The Fig. 3 shows a possible emptiness and occupancy probability distributions of the sonar sensor using exponential model [5, 12, 21]. The plane defines the *unknown* level, the values above the plane represent the *occupied* probabilities, and the values below represent the *empty* probabilities.

In other words: featuring of the exponential model ensures that the amount of reasonable information provided by the sensor is relatively constant for every measured distance. That means that numerous wrong measurements in greater distances are spread over many cells but are "removable" by a small number of correct measurement.

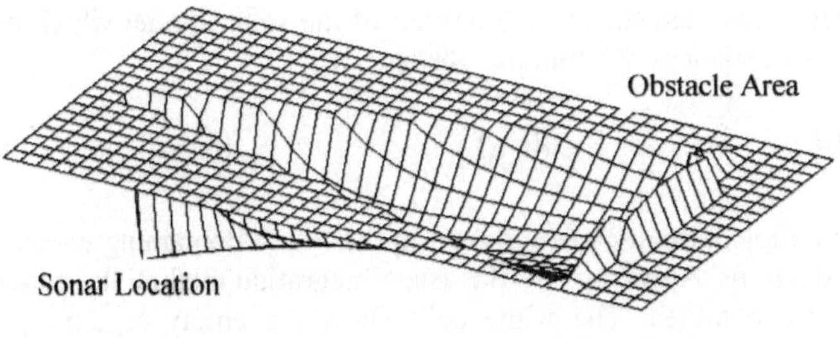

Fig. 3 Sonar model with exponential probability distribution

Multiple Reflections Removal. Existence of multiple reflections belongs to the main problems of sonar data processing. If the sensor provides measurements corrupted by multiple reflections, such readings have to be filtered out. Having only a single scan (a set of range measurements over the whole horizon) it is generally not possible to recognize presence of the mentioned phenomenon.

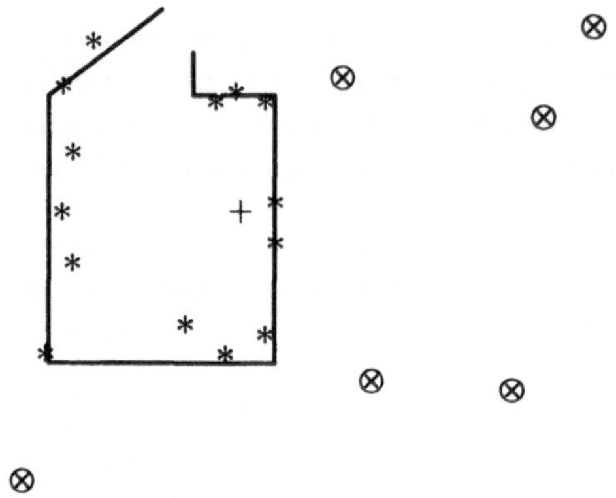

Fig. 4 Multiple reflections. Symbols * and ⊗ denote correct measurement, respectively measurement spoiled by multiple reflections (gathered at the Prof. Ritter's meeting room, Institute for Applied Mathematics and Statistics, TU Munich)

This problem can be approached by making-use of a presumption on the minimum of the measured distances from one certain position. This minimum distance defines a safe area where an obstacle cannot exist. The mentioned heuristic can be applied to suppress bad measurements by redefinition of the sonar model which has emptiness probability dependent on the minimal distance M:

$$E_r(\delta) = e^{-\frac{\delta}{M}} \quad \text{for } \delta \in (0, R) \qquad\qquad E_r(\delta) = 0 \quad \text{otherwise}$$

Let's assume typical input data as shown in the Fig. 4 containing measurements with multiple reflections. Applying the Bayesian integration method the patterns shown in the Fig. 5 are obtained. The white cells show the empty regions, gray cells are unknown areas, and the black ones stand for occupied regions. All examples use the Bayesian integration method.

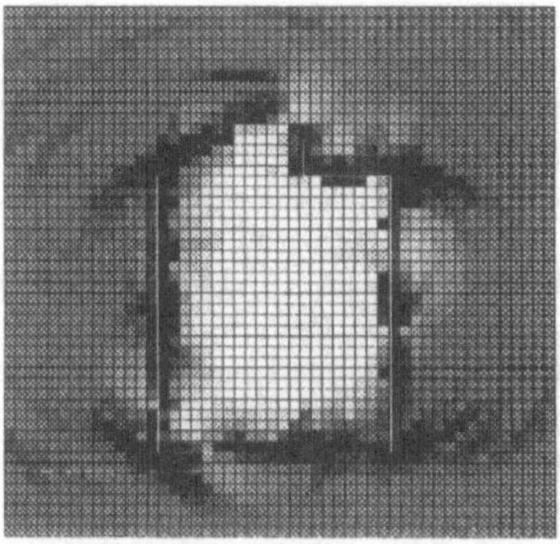

Fig. 5 Bayesian fusion with exponential model and multiple reflection removal. Dark areas denote regions with higher measure of occupancy, the light regions stand for higher value of emptiness. (data source as by Fig. 4)

5 Activity Planning

Two classes of tasks are to be solved: the *service trip* and *region filling*. The former task which solves the motion of the robot from the current to a goal position, is considered for the generic case. A slightly more complicated task is the latter problem of region filling. For this specification is not as unambiguous as for the former. The task might be given simply by list of regions to be filled (irrespective of order!), or by the environment area, which is to be filled. The system then has to split the filling area into regions - relatively compact subareas - which will be ordered in a sequence and filled by the vehicle's path, both following some certain optimal criterion.

The area for autonomous activity of the robot can be seen in *how* the given goal can be reached making-use of the given resources [6, 8, 9]. The resources are, in this example, considered to be admissible movements of the vehicle through environment under constrains being set up by:

- Limits on the radii of turns, and vehicle speed (maneuvering capabilities).
- The planned trajectory should safely avoid all obstacles (obstacle avoidance).

This process is also called *activity (path) planning* .

Moreover for our case, under a certain simplification of the output of the planning process - a path - could be defined as a sequence of places going from the start point to the goal point along an arc. The start position is given, the sequence of arcs is to be planned.

5.1 Path Planning in Graph-Based Models

Assuming that the end-points of arcs are vehicle states, and the admissible arc of the same length between two states defines a transition between states (an edge), we obtain a *state-space graph.* A planning algorithm provides the sequence of system states which represents an optimal path with minimum cost. This can be based on state-space graph search techniques [2, 10, 18, 20]. Each transition between two states is accompanied by quantitative features, for example: distance of the states, energy consumption, minimal collision risk, or the time elapsed by performing the transition, etc. These serve to evaluate the price of transitions between states, via a chosen *cost function* which enables the employment of *heuristic search* techniques.

Although considerable progress has been made using heuristic approaches, these alternative methods do not fully achieve satisfactory performance, and have limitations of their applicability caused by the build-up of heuristic rules. If we omit blind-search methods, let us mention the purely formal approach of *dynamic programming* [15, 20], which can be used for optimal path search. The dynamic programming (backward algorithm) consists of two stages:

The first one generates a tree subgraph of the full state graph which contains all the possible ways from the initial point to the goal.

The second stage then evaluates the cost function step by step from the goal backwards to the initial point following the iterative formula:

$$C(0)=0, \quad C(i)=min_{i-1,i}(C(i-1)+c(i-1,i))$$

Where $C(i)$ denotes the total cost of the path in distance of i steps from the goal. The value $c(i-1,i)$ stand for the price of the transition from $i-1^{th}$ to i^{th} state. At this point it is necessary to show that this sort of optimization method explores in single steps only the information contained in the costs of transitions.

Another classic technique used for finding an optimal path in a graph is the A^* algorithm [13, 17, 18]. In addition to processing the information in the graph itself, as in the previous approach, A^* provides a prescription of how to use additional information on the situation from which the graph was derived. As a result, the algorithm uses far less computational effort than standard algorithms which achieve the same results.

Besides graph nodes, edges and their costs that create a standard graph, A^* uses one more sort of data - a number $h(i)$ denoting an estimate of a lower limit on the cost of moving from the ith node to a goal node. For the most common case of shortest path search the $h(i)$ might be shortest (straight line) distance to the goal. This estimate is a way of incorporating additional information which is usually based on knowledge which is not directly represented within the graph.

The A^* performs search forwards from the start to the goal. The initial path cost is equal to the estimate $h(i)$ between the start and the goal nodes. Let $g(n)$ represent the cost of the selected path from start to successor node n, where $g(start)=0$. Let the current node be the first. Then the algorithm can be sketched:

Choose node N from the set of successors of current node. If the term $g(N)+h(N)$ is not minimal over the whole set of successors, repeat the term evaluation for other successors until the smallest value is found. Consider the node having a minimal term value for a new current node and repeat them all until the goal node is achieved. Total (minimal) path cost is then equal to *g(goal)*.

Less computational effort than A^* can be required by the IDA^* algorithm [13] which performs depth-first iterative deepening with heuristic cutoffs.

Search Speedup. There is a couple of ways of how to improve the efficiency of search algorithms. The main idea behind these is that the search can be performed in two directions simultaneously. In the simplest case a search can be performed from the goal (backward chaining) and from the start (forward chaining). Completion of the search is reached, whenever both processes meet each other.

These *bi-directional* algorithms for a single processor [13] are of two kinds:

The *non-wave-shaping* methods aim to find the solution as fast as possible by eliminating the nodes which probably do not lead to a solution. Both searches intersect at a place between the start and the goal.

The *wave-shaping algorithms* always expand the nodes which are found to be the closest ones to the nodes of the opposite search. This ensures that the opposing searches meet in the middle of the path between the start and the goal, this is paid for by the higher computational effort when nodes for expansion are being selected.

As for another way the *island search* can be considered the in which the state space is broken into many smaller (and spatially consistent) subspaces along the estimate of the optimal path from the start to the goal. Each of these subgraph contains a source node (the island) which is assumed to be on the optimal path. The neighboring source nodes therefore simultaneously serve as a start and goal states for the succeeding search processes.

To illustrate the profit of the previous approach, let us assume having N islands placed at the same distance on the path, the average depth for the search from any island is n/N, where n stands for the length of the solution (number of states which have to be passed from the start to the goal). If the original complexity of the used algorithm without island is, say for A^* case $O(b^n)$, where b denotes the branching factor, the use of N islands requires only $N.O(b^{n/N})$.

The great advantage of the island-approach is that it is suited for implementation on a multiprocessor machine, what can bring an exponential speedup ratio with the number of used processors [13], although uniprocessor implementation is also profitable.

The only limiting factor is whether desired island can be determined in advance or not. Their definition can be done either by making use of some prior knowledge of

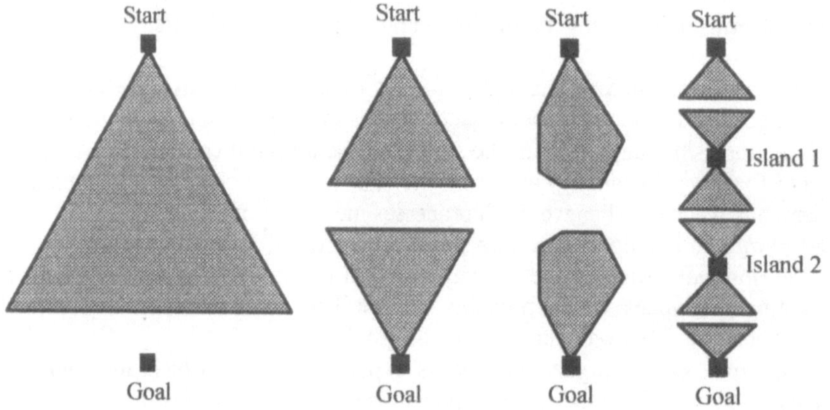

Fig. 6. Various search techniques. From the left: Unidirectional search, bi-directional search, bi-directional search with wave-shaping, and bi-directional search with islands.

the environment (certain locations which the robot must pass through) or via hierarchical planning , where the islands states are outputs of the prior upper-level planning stage.

Planning over Visibility Graphs. Another wide-spread technique uses modified representation of the WM. The path planning, respectively the minimum-path search which is conducted not over a standard fixed state-space graph, but a dynamically generated *visibility graph* [14]. The method of creating the graph explores the further knowledge of the physical properties of the world and the vehicle, in order to restrict its complexity. This fact could be expressed in rules guiding the graph build-up:

- Create a node near the object corner only, and only if, the node is visible from the current node, and no real edge exists between them.
- Do not expand the node to a new position, which is directly unreachable because of the kinematics of the vehicle.

The obtained graph is substantially less complex than the appropriate state-graph, containing all transitions among possible neighboring positions. The important property of the visibility graphs stands in the possibility of dynamic (on-line) generation of the graph during a planning process.

But savings in computational effort when building WM on-line also bring a problem. Having an incomplete WM, there exists no planning algorithm which would guarantee a global optimum of the found solution. A modified A* algorithm performing planning over visibility graphs has been introduced in [14].

5.2 Path Planning in Grid-Based Models

Another method for path planning in grid-based models of the environment is commonly called *distance transform*. This method is basically suitable for global path planning under the assumption of complete information of the presence of obstacles in the environment. The approach of distance transform [14, 25] considers the planning task as backward path finding from the goal to the initial point.

This method covers the environment by a uniform grid (see the certainty grid world model) and then propagates distances from the through free grid elements from the goal to the initial point. The same distance value is assigned to all pixels, which are found in the same distance from the goal. In this manner the distance marking step by step is spread all over the free grid elements unless one of the markings reaches the initial point of the robot.

The shortest path from the initial point to the goal can be then found by tracing downhill the steepest descent path.

Fig. 7. Distance marking in the grid-based WM and distance-transform path planning. Minimal paths from the Start to the Goal are: DDDRRRRDRD and DDDRRRRRDD, where D, and R denote the transitions to "down" and "right" pixels.

If there is no downhill path, e.g. the current pixel is on a "distance plateau" (all of the grid cells, including obstacles are initialized to high value), it can be concluded that no path from the initial point to the goal exists.

The optimum criterion as minimization of the path length was intuitively taken in this case. If additional minimization of some other features (e.g. consumed energy, number of turns, maximum velocity, etc.) are required, their influence has to be incorporated into the evaluation of the distance between cells.

6 Plan Execution Task

Although the path execution task is not a purely performance oriented optimization but more a numerical one, a the used method of plan execution and the vehicle control is dependent on the kinematics model of the used vehicle (tricycle, car-like). A path plan found by a planner is mostly represented by either *symbolic command language* or *sequence of succeeding positions.*

The symbolic language representation describes the plan as a sequence of driving commands. Such macro-commands can define vehicle's actions as in the following example:

(1) DEFINE start position, speed, and heading
(2) MOVE straightforward for 2 meters, accelerate to speed 0.5 ms^{-1}
(3) MOVE 60 degrees turn having diameter of 3 meters, no acceleration.
(4) MOVE straightforward for 5 meters, decelerate to zero speed

As the shown commands can not directly control drives of the vehicle, they have to be interpreted first. The interpretation process, done by the driving subsystem, translates the macro-commands into a smooth trajectory. The converted trajectory, typically represented by a dense sequence of desired positions and headings of the vehicle, can then be directly fed to vehicle drive controllers (point-to-point control).

The second approach uses sequences of positions (and headings) which the robot has to attain. The robot path between two succeeding points is not defined and is subject to interpolation, which also provides smooth trajectories.

The only difference between the both approaches lies in the level on which the plan has been created. The former one is better suited for symbolic planners or planners working over network-based WMs. The latter one seems to be more practical for use with grid-based WMs.

6.1 Control Problem Solution

From the theoretical point of view, the control problem between two points can be classified as multidimensional multicriterial nonlinear one. Such task can't be solved by a simple feedback controller efficiently - the system has too complicated and nonlinear structure. Therefore a hierarchical approach has to be used:

- The top level control system sets trajectory parameters for the last point of the control horizon which is assumed to be fixed.
- The medium level control system parametrizes local segments of the given trajectory.
- At the bottom level, a simple local PID controller compares the desired and actual local trajectory parameters and provides a control action.

The central problem to be solved is caused by different number of freedom degrees of the physical system and of the approximation curve. The trajectory planning problem has degree of freedom equal to 9, while the most 2D curve parametrizations consider even number of parameters. There are two basic ways how to approach the problem:

- Application of a curve with higher number of parameters with fixation of free parameters. The vehicle then follows such trajectory exactly.
- Application of a curve described by smaller number of parameters fitting exactly the final state and approximately the initial state. The vehicle uses a controller to minimize the difference between actual trajectory and trajectory estimate (defined by the plan).

None of these methods is superior to the other and both can be implemented in practice.

Kinematics and Dynamics Separation. For every type of approximation curve holds, that the control problem can be split into two subproblems which can be solved separately: trajectory planning (kinematics problem) and planning of the speed of the vehicle along the trajectory (dynamic problem).

Typical solutions to the former problem are path approximations by various types of curves like polynomial, circular, and Bezier curve fittings [19]. The latter can be neglected with respect to low speed of vehicle motions, having sufficiently fast dynamics of the used vehicle drives.

As it can be seen, the control problem is the only one connected with the self-guided vehicle control which can be solved quite exactly by classical numerical optimization methods. The main reason for it is the fact that the problem is well and completely defined from the viewpoint of numerical mathematics

7 Conclusion

From the point of view of the system theory, the human beings are systems which have passed through a long way of evolution. This evolution can be considered as a many-fold optimization procedure. On the other hand, persons as individuals try to permanently optimize their behavior and the flexibility of doing this plays the role of a measure of their intelligence.

It was shown in this paper, that robots aiming to replicate some aspects of human behavior (incl. decision-making functionality and performing of intelligent actions in physical environment), have to be equipped by many different optimization techniques.

Under the term *optimization*, only numeric optimization techniques are usually considered. These are based on *precise mathematical descriptions* of goals, criteria and constraints. But, the optimization tasks in robotics are usually expressed by means of performance-based specifications of goals and/or vaguely defined heuristic criteria and/or not very precise constraints. Such problems require application of heuristic methods in iterative optimization cycles. At the first glance, it looks like the AI methods are also nothing else but numerical iterative optimization processes. But the precise mathematical models are precise models for handling vaguely defined data of vaguely specified optimization problems. This is, e.g. the problem of sensor data fusion where precise formulae are used to manipulate uncertain data; this is the case of heuristic functions which precisely manipulate roughly, heuristically estimated costs etc.

The main goal of the paper was just to stress the fact that the up-to-date optimization problems require methods which are very far from the purely numerical optimization techniques. The paper should help on one hand to understand the optimization problems from new, wider perspectives, on the other hand to demonstrate a strong optimization nature of AI-based techniques.

References

1. Abidi M.A., Gondzales R.C. (Eds.): Data Fusion in Robotics and Machine Intelligence, Academic Press, 1992

2. Allen J., Hendler J., Tate A. (Eds.): Readings in Planning, Morgan Kaufmann Publishing Inc., Palo Alto, 1990

3. Berger J.O.: Statistical Decision Theory and Bayesian Analysis, Springer-Verlag, Berlin, 1985, Second Edition

4. Čapek K.: R.U.R., Prague, 1929

5. Elfes A.: Sonar-Based Real World Mapping and Navigation, IEEE Journal on Robotics and Automation, Vol.3, No.3, June 1987

6. Havel M.: Robotics - an Introduction to Theory of Cognitive Robots, SNTL Publ. Prague, 1980 (in Czech).

7. Kanayama Y., Yuta S.: Vehicle Path Specification by a Sequence of Straight Lines, IEEE Journal of Robotics and Automation, Vol.4, No.3, June 1988, pp. 256-276

8. Kohout J.: A Perspective on Intelligent Systems, Chappman and Hall Publ., U.K., 1990

9. Lažanský J.: Practical Application of Planning Tasks, Advanced Topics in Artificial Intelligence (V. Mařík, O. Štěpánková, R. Trappl, Eds.), LNAI No.617, Springer Verlag, Prague, 1992, pp. 238-244

10. Lozano-Perez T., Wesley M.A.: An Algorithm for Planning Collision Free Paths Among Polyhedral Obstacles, Communication of the ACM, Vol.22, No.10, 1979, pp. 560-570

11. Mařík V., Vlček T.: Some Aspects of Knowledge Engineering, in: Advanced Topics in Artificial Intelligence (V. Mařík, O. Štěpánková, R. Trappl eds.), LNAI No.617, Springer Verlag, 1992, pp. 316-337

12. Matthies L., Elfes A.: Integration of Sonar and Stereo Range Data Using a Grid-Based Representation, Proc. of the IEEE International Conference on Robotics and Automation , 1988, pp. 727-733

13. Nelson P.C., Toptsis A.A.: Unidirectional and Bidirectional Search Algorithms, IEEE Software, March 1992, pp. 77-83

14. Oubrecht P.: Planning Algorithms for Robotics, Diploma Thesis, Dept. of Control Eng., Czech Technical University, Prague, 1994

15. Pierre A.D.: Optimization Theory with Applications, John Willey and Sons Publ., 1969

16. Přeučil L., Mařík R., Šára R., et.al: Vision-Based Robot Navigation in the 3D World, in Proc. of IEEE Conference on Artificial Intelligence, Simulation, and Planning in High Autonomy Systems, Tucson, Arizona, September 1993

17. Rich E., Knight K.: Artificial Intelligence, Mc Graw-Hill Publ., 1991.

18. Shapiro S.C. (Ed.): Encyclopedia of AI, John Wiley & Sons Publ., New York, 1990

19. Sládek B.: 2D Vehicle Control, Reseach Report 9502, Institute for Applied Mathematics and Statistics, TU Munich and Department of Control Eng., CTU Prague, 1995, pp. 1-16

20. Suh S., Shin K.: A Variational Dynamic Programming Approach to Robot-Path planning With a Distance-Safety Criterion, IEEE Journal of Robotics and Automation, Vol.4, No.3, June 1988, pp. 334-349

21. Štěpán P., Přeučil L.: Statistical Approach to Range-Data Interpretation, in Proceedings of Artificial Intelligence Techniques AIT'95, Brno, Sept. 18-20, 1995, pp. 315-323

22. Štěpánková O., Havel I.M.: A Logical Theory of Robot Problem Solving, Artificial Intelligence 7, 1976, pp.129-161

23. Takahashi M., Oono J., Saitoh K.: Manufacturing Process Design by CBR with Knowledge Ware, IEEE Expert, Vol. 10, No. 6, Dec. 1995, pp. 74-80

24. Vasseur H.A., Pin F.G., Taylor J.R.: Navigation of a Car-Like Mobile Robots in Obstructed Environments Using Convex Polygonal Callus, journal on Robotics and Autonomous Systems, Elsevier Science Publ., Vol. 10, 1992, pp. 123-146

25. Zelinski A., Yuta S.: Reactive Planning for Mobile Robots Using Numeric Potential Fields, Intelligent Autonomous Systems IAS-3, 1993, pp. 84-93

A Fuzzy Set Approach for Optimal Positioning of a Mobile Robot Using Sonar Data

Martin Pfingstl
Department of Mathematics,
Technical University of Munich, Germany

and

Stefan Schäffler
Department of Mathematics
Technical University of Munich, Germany

Abstract: A procedure for optimal positioning of a mobile robot based on sonar data is presented. This procedure uses simulated sonar scan data in order to gain step by step information about the unknwon room, which is modelled as a fuzzy set. The definition and application of a special measure allows the calculation of a new position, which promises the maximal amount of additional information about the unknown room, if the new position is interpreted as the next scan position. The efficiency of the algorithm is shown by examples.

Key Words: fuzzy sets, mobile robot, sonar data, simulation, world mapping

1 Introduction

One of the most interesting areas in robotics is the optimal positioning of a mobile robot within an unknown work space ([LDW92]). We assume that at each position of the mobile robot additional information about the work space can be obtained by sonar measurements, which are often used in robotics. In this situation the following questions arise:

1. How should the measured data be interpreted?
2. Which new position, where new measurements will be done, provides the maximal additional information about the work space?
3. How should the different measurements be combined?

4. How can real measurements be replaced by simulated data?
In this paper we present a fuzzy set approach for the optimal positioning of a mobile robot using simulated sonar data. Details about fuzzy sets are given in ([KGK95]). The optimal position is computed as a point in the work space, where the expected information represented by new measurements at this point is maximal. The computation of this new point is based on a special information measure, which is available by a fuzzy set approach. The efficiency of the algorithm is shown by examples.
Chapter 2 deals with the simulation of sonar data. In chapter 3, the necessary background of fuzzy sets are given. An interpretation of sonar measurements in order to find a mapping of the unknown work space is described in chapter 4. Using this interpretation the next position of the robot is computed based on a special information measure (chapter 5). Finally, in chapter 6 examples are documented.

2 Simulation of Sonar Data

This chapter deals with the simulation of sonar devices. The practical evaluation of sonar data firstly requires several sonar senders and receivers. But these of course have to be installed and to be connected with a computer in order to work with them. Additionally, one has to solve other problems like the positioning of the mobile robot. To avoid this, it is required to have a simulation program for the evaluation of sonar data (see [BMW91], [RT95] and [Rei95]).
The input of the program should consist of the description of a room in a specific manner and of a position of the robot equipped with sonar devices and then should just simulate a complete sonar scan at the given location.
The program simulates a single sonar sensor, which at the beginning acts as a transmitter and then, after a short relay time, as a receiver. The room is assumed to be a 2 dimensional object and is supposed to be known to the simulation program. The sensor device is rotated around its own axis in N steps. At each step a sonar scan is carried out.
This simulation is based on a ray-tracing algorithm. A fine bunch of rays is created at the front of the sonar sensor for each single measurement. Each such ray is examined, until it - after one or more reflections - again hits the front of the sonar device or until it is so weak that the sonar sensor can not recognize it any more. The runtime t between transmitting of the rays and the return of the first ray, which is strong enough to be recognized by the sensor, is measured. With the speed of sound c in the air a reflecting object with the distance

$$z = \frac{c \times t}{2}$$

is assumed.

As a matter of course, one of the most important things in the simulation is this bunch of rays, which is created at the beginning at the front of the sensor. In reality, infinite many rays are created, but in a simulation only a finite number of rays can be created and examined. Another difficulty of the simulation are the walls. In reality there do not exist perfect walls, i.e. walls, which are totally plain and straight. Every wall has a certain level of roughness, which can not be modelled exactly. Leaving this problem aside it could happen, that the simulated sonar data contain more precise information than the scans obtained in reality. Due to this it was useful not only to reflect rays at the walls, but also to transmit a diffuse part of the strenght of the ray in all directions in order to recognize walls, which behave like real walls. This method has lead to great success.

3 Fuzzy Sets

In this chapter we give a short introduction into the world of fuzzy sets. Our goal is to make the reader familiar with the notions and definitions from the area of fuzzy sets, so one can follow our descriptions without using further knowledge and literature.

Classical sets are defined by their characteristic function. Given a universe U, which contains all elements, a set M is defined by a function f:U $\longrightarrow$ {0;1}.

f(x) is equal to 1 for all elements x, which are contained in M, and f(x) is equal to 0 for all elements x, which are not contained in M. One can also think of the values 0 and 1 as grades of membership: 0 means, the element is surely not contained in M, 1 means, the element surely belongs to M.

Now let us consider the set M of all young people for instance. It is obvious that somebody, who is 2 years old, belongs to M. On the other side, somebody who is 75 years old does not belong to M. But it does not make much sense to say that somebody who is 35 years old belongs to M or does not belong to M. It makes much more sense to assign a value of perhaps 0.7 between 0 and 1 in order to express, that one is unsure, whether this person belongs to M or not. The value 0.7 could mean, that this person should be considered as a young person.

A function $\mu : U \longrightarrow [0, 1]$ with a reference set U is called a fuzzy set. In order to work with fuzzy sets, operations like complement, conjunction and disjunction have to be generalized to fuzzy sets. There are several ways of defining these operations. Each kind of definition has its advantages and disadvantages. The chosen definition mainly depends on the application, in which context the fuzzy sets are used. The most common type of definition, which we will use here, is as follows: The complement of a fuzzy set $\mu : U \longrightarrow [0, 1]$ is $\mu^c : U \longrightarrow [0, 1]$ with $\mu^c(x) = 1 - \mu(x) \quad \forall x \in U$ (see Fig. 1). This definition corresponds with the well known complement of classical sets.

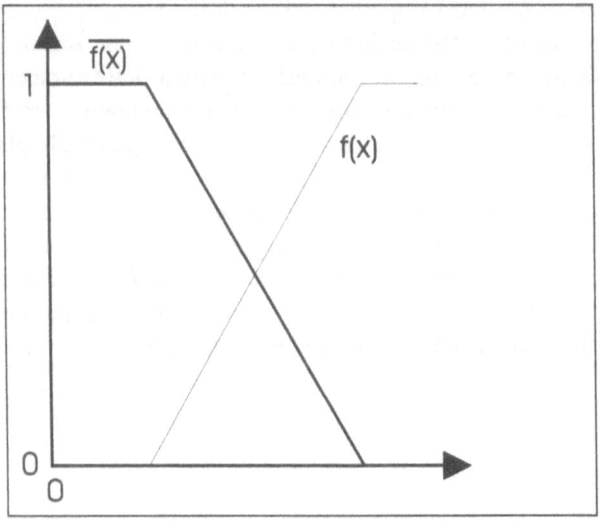

Fig. 1

The disjunction of two sets $f: U \longrightarrow [0,1]$ and $g: U \longrightarrow [0,1]$ is defined as

$$h: U \longrightarrow [0,1], \text{ h(x)=min\{ f(x) , g(x) \} } \forall x \in U$$

(see Fig. 2)

If applied to classical sets, this again is the already known disjunction.

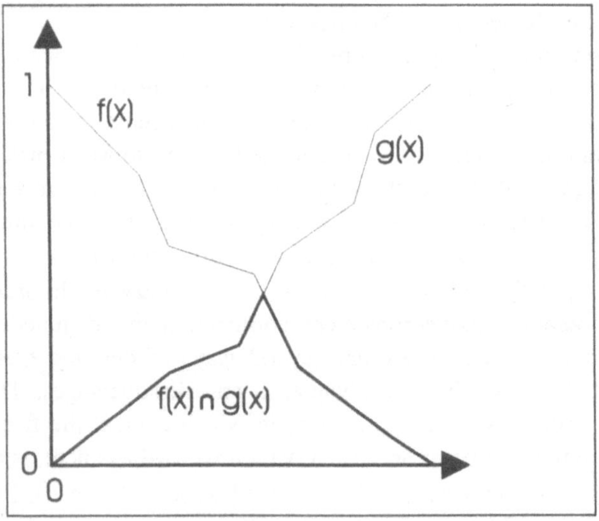

Fig.2

The conjunction of two sets $f: U \rightarrow [0,1]$ and $g: U \rightarrow [0,1]$ is defined as

$$h: U \rightarrow [0,1]\ ,\ \mathrm{h(x)=max}\{\ \mathrm{f(x)}\ ,\ \mathrm{g(x)}\ \}\ \ \forall x \in U$$

(see Fig. 3)

Yet again, if applied to classical sets, this is the already known conjunction.

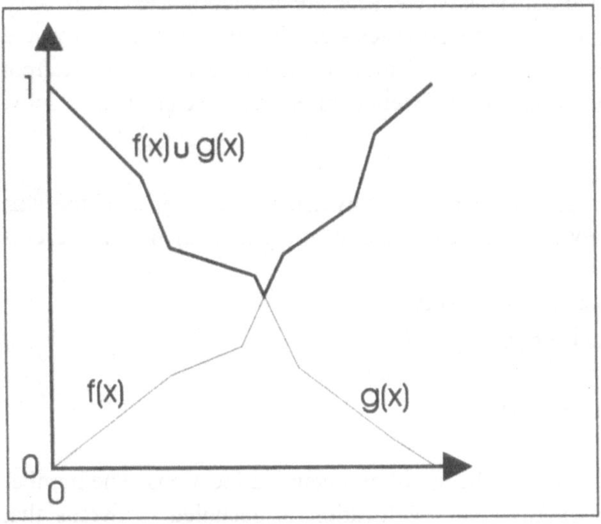

Fig. 3

Now we need another operation, which is specific for fuzzy sets. In almost all applications one has to decide, whether an element belongs to a set or not. For example, in the world mapping problem there is to decide, whether a wall is at a specific location or not. The name for this kind of operation is the a-cut or a-niveau-set. Let $f: U \rightarrow [0,1]$ be a fuzzy set, then the a-cut of f is given by the set $g: U \rightarrow [0,1]$ with

$$\begin{array}{ll} \mathrm{g(x)=1} & \text{if } f(x) \geq a \\ \mathrm{g(x)=0} & \text{if } f(x) < a \end{array}$$

Hence, the a-cut of f is a classical set.

4 The World Mapping Problem

An important problem using sonar data is the fact, that this data contain correct information only partially. Therefore, the most important task is to recognize, which part of information is correct and which part of information is incorrect and hence has to be eliminated. For example, due to multiple reflections of sonar rays on walls one eventually gets wrong information about the distance of a wall. In connection with sonar data it is only possible to locate an object with a special uncertainty. Hence, a modelling with the use of fuzzy sets is appropriate. One simply tries to find out from the sonar data the set of all walls and the set of all positions, which can be entered. These two sets are modelled as fuzzy sets, because one never - except in some special cases - can be sure, that the expected situation - free or not free - is correct. Therefore, sonar scans from different locations are made and the room is remodelled after every scan. This is now explained in more detail.

The discretisized room is assumed to be the universe U, i.e. a grid of the size *SizexSize* is laid over the room. A single element of the grid array then represents a small part of the room.
Now we are trying to find out two fuzzy sets

$$f_o : [0...Size]x[0...Size] \rightarrow [0,1]$$

and

$$f_f : [0...Size]x[0...Size] \rightarrow [0,1]$$

f_o is the set, which tries to represent the set of all occupied locations. The number between [0;1] shows the certainty that the location is occupied. 0 means that nothing is known,1 means that this location is apparently occupied.
f_f is the set, which indicates all the locations, which are free and can be entered. The number between [0;1] indicates, how sure the program is, that the location is free in fact.

At the beginning the robot is assumed to be at a certain location of the grid, perhaps at position (Size/2,Size/2). Then a scan is made from this position. Now two new fuzzy sets g_o and g_f are built, which describe the room using the information from the previous scan and which use informations from f_o and f_f. After that, g_o with f_o and g_f with f_f are combined to get a better model of the room.
At first, all the information from f_o and f_f, which seems to be correct, has to be copied to g_o and g_f, i.e. all elements, which have a component greater than the global program constant *GoodInfo*. This leads to the following piece of pseudo code:

```
for x=0 to Size
  for y=0 to Size
    g_o[x][y]=g_f[x][y]=0
    if f_o[x][y]>GoodInfo then g_o[x][y]=f_o[x][y]
```

if f_f[x][y]>*GoodInfo* then g_f[x][y]=f_f[x][y]
next y
next x

Now the previous scan is integrated into g_o. Such a scan contains the following kind of information:

Object at location (x_1, y_1)
Object at location (x_2, y_2)
...
Object at location (x_{n-1}, y_{n-1})
Object at location (x_n, y_n)

Using this information it is possible to construct a new fuzzy set, which contains more precise information. $g_o[x_i][y_i]$ is a number, which represents the certainty of a wall at this location and which depends on the distance of a singular scan point from the current scan position. It is 1 at the current position of the robot and then goes down to 0 at a distance of the global program constant $MaxDist_o$. The function used is:

$$certainty = \sqrt{1 - \frac{Dist}{MaxDist_o}} \qquad \text{if } Dist < MaxDist_o$$

$$certainty = 0 \qquad \text{otherwise}$$

This of course means that all gathered data from the last scan, which represent objects in a distance of more than $MaxDist_o$ are disregarded.

Now, the information about the free room has to be updated. To do this, all positions (x_i/y_i) are examined, whose distance from the current robot position is less than $MaxDist_f$.
After that, it has to be examined, whether it is legal to mark this position in the grid as free and whether this position is in the shadowed region of a wall. In this case it does not make any sense to mark this position as free, whereas it is probable, that positions before a recognized wall are free - otherwise there would have been another recognized obstacle and therefore the position would again lie in the shadowed region of a wall.
Such positions, which do not lie in shadowed regions of any object get a number, which is 1 at the position of the robot and 0 in a distance of $MaxDist_f$ using the same function as before, but with another global constant $MaxDist_f$:

$$certainty = \sqrt{1 - \frac{Dist}{MaxDist_f}} \qquad \text{if } Dist < MaxDist_f$$

$$certainty = 0 \qquad \text{otherwise}$$

Unfortunately, another problem arises from the fact, that by the use of sonar measurements only point like data are gathered, i.e. walls are not recognized as walls but are recognized as many points, which perhaps form a shape similar to a wall.
So the problem occurs very often, that a wall is stored with many gaps. If one then calculates the shadowed regions of this wall, many incorrect information is built into g_f, because free positions are marked and they do not lie in the shadowed regions of the few points, which form the wall.
Due to this it is necessary to enlarge the walls artificially in order to reduce this nasty effect and to reduce the amount of incorrect information in g_f.

After this step one gets a model of the walls and the free positions in g_o and g_f using the last scan and the (hopefully) right information from previous scans.
f_o and f_f contain the last modelling of the room. Now these fuzzy sets have to be combined, f_o with g_o and f_f with g_f. This is done through the conjunction of f_o with g_o and of f_f with g_f, i.e. for all components the maximum of the two sets is taken and stored in f_o or f_f:

```
for x=0 to Size
  for y=0 to Size
    fo[x][y]=max(go[x][y], fo[x][y])
    ff[x][y]=max(gf[x][y], ff[x][y])
  next y
next x
```

The last step is the restauration of the data consistancy. It is very probable that contradictory information is available in f_o and f_f.
What should it mean that f_o[x][y] and f_f[x][y] are both different from 0? It would mean that this position is occupied by a wall with some certainty and at the same time this position is free with another certainty. This is somehow contradictory. Either there is a wall or there is a free position with any certainty.
In all such cases simply the smaller component is subtracted from the bigger one and the smaller one is set to 0. This now enables recognition of false data. One scan could show, that there is an object, but there is none. The next scan now shows, that there is no object and hence, there are contradictory data in f_o and f_f.

The two data are compared and the certainty of the wall is reduced or set to 0 depending on the given numbers.

5 Optimal Positioning of a Mobile Robot

The previous chapter described how to recognize the unknown room from different scans. Now one problem is left: What position should be the next scan position? What position promises the largest amount of additional information? The purpose is to recognize the room with as few scans as possible. So, a measure for the expected amount of additional information has to be calculated for every position (x/y), which is a candidate for the position of the robot. This problem is solved in the following way:
The program uses a simulation method. The robot is positioned at the location (x/y) and it is assumed that a scan made at this position lead to exactly the same values as already available in f_o. The room is modelled as described in chapter 4. Then, g_o with f_o and g_f with f_f are combined, the result is stored in h_o and h_f. These two fuzzy sets contain more information than f_o and f_f: Additional information about free locations is available in h_f and additional information about the walls in h_o, because the certainty of the walls may have increased.
Of course, all of this is done under the assumption, that a scan at that location leads to data already given. Now, the sets h_o and h_f have to be compared with f_o and f_f. The differences in all components then have to be sumed up. This sum can be used as an indicator for the amount of additional information one gets from a scan at the location (x/y):

```
Infoamount=0
for x=0 to Size
  for y=0 to Size
    Infoamount=h_o[x][y]-f_o[x][y]+h_f[x][y]-f_f[x][y]
  next y
next x
```

If one applies this method to all locations, which are a candidate for the next scan position i.e., for all locations (x/y), which are free, then it is quite easy to find out the location, which promises the most amount of additional information. This position is then suggested as the next scan position.

6 Examples

This chapter shows the effectivity of the algorithm. We have chosen two rooms, the first being not too complicated, the other being quite complicated.
There are two kinds of pictures: One picture is the picture of the room recognized by the program. The black pixels are the walls, which have been recognized. Actually, there should be shades of black color according to the certainty of a calculated wall. But we colored it totally black, because otherwise there could be some confusion about the walls and the interior region of the room. The recognized interior region of the room is colored in shades of grey colors. Light grey means, the program is not very sure about this localtion to be free, dark grey means, the program is quite sure, that this location is free.
Two positions are marked in this picture, namely the current position of the robot, i.e. the last scan position. It is the bigger mark always in the middle of a dark circle. Furthermore, the next scan position calculated by the program is marked. This mark is obviously in a region, which is light grey.
The other kind of picture should make clear, why the program chose this next scan position. This picture shows the information measure at all possible next scan positions. The darker the location is, the more additional information the program expects from a scan at this position. Hence, the next scan position will always be in one of the darkest regions in this picture. Again, two positions are marked:
- The current position of the robot, now at a location, which is light grey. This is clear, because from a scan at the current position very few additional information can be obtained.
- The suggested next scan position, which of course lies in a dark region.

Now lets start with the examples. The room to be recognized looks as follows:

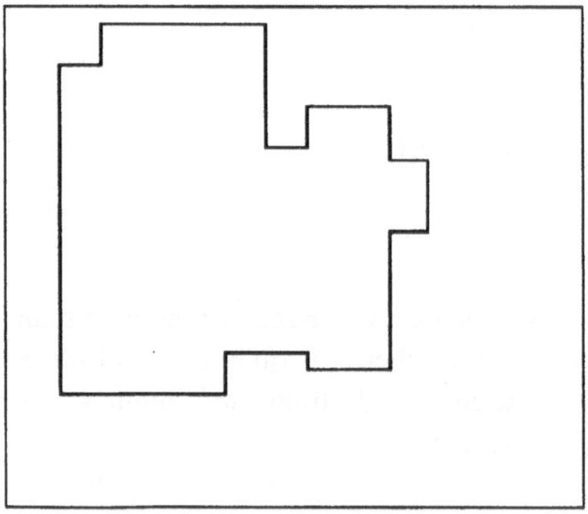

Room to be recognized

The size of the room is about 10m x 10m. The gridsize was 5cm x 5cm. $MaxDist_o$ was 250 cm, $MaxDist_f$ was 170 cm. The first scan produced the following result. The left picture shows the room known to the program, the right picture shows the amount of expected information:

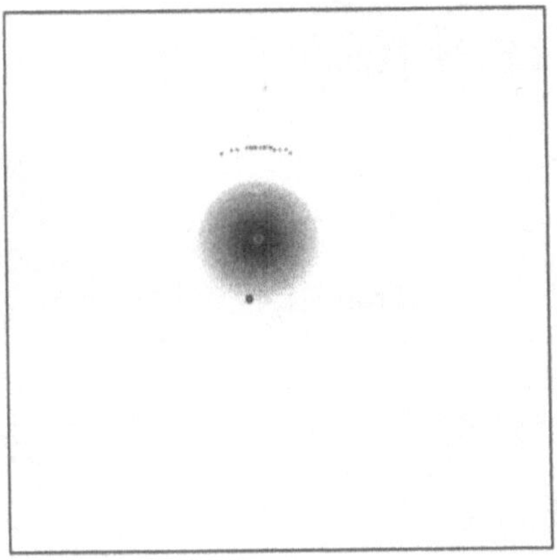

room 1.1

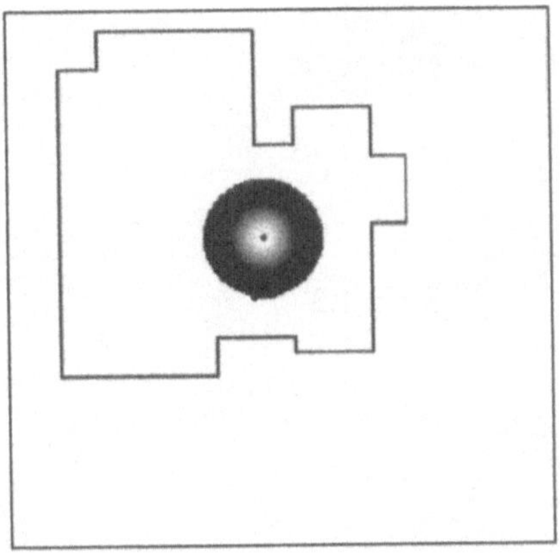

info 1.1

Almost no walls have been recognized. This is due to $MaxDist_o$, which sets the certainty of recognized walls in a distance of more than 250 cm to 0. The robot moves to the new position and performs a scan with the following result:

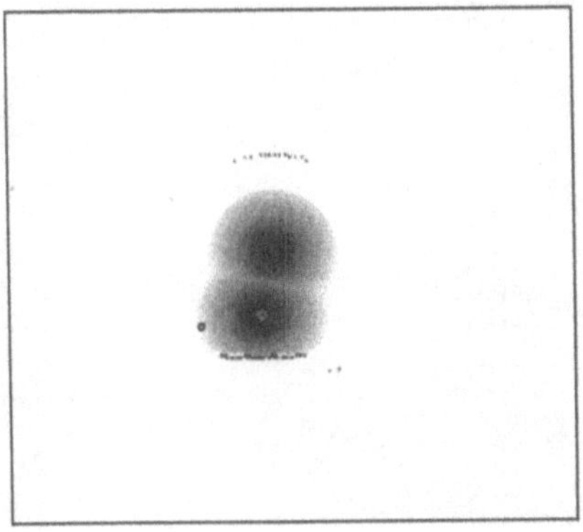

room 1.2

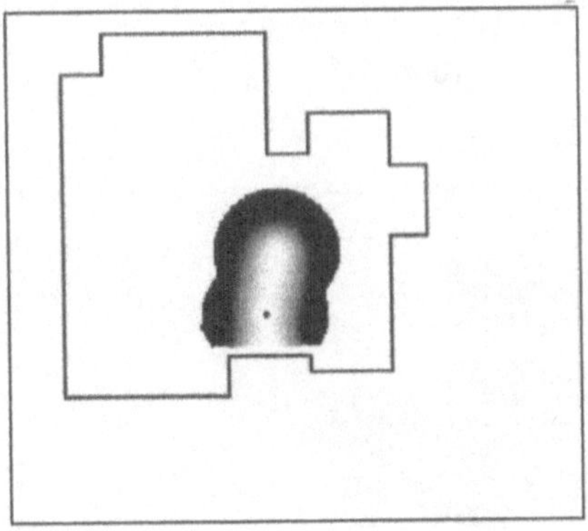

info 1.2

This approach leads to the following pictures:

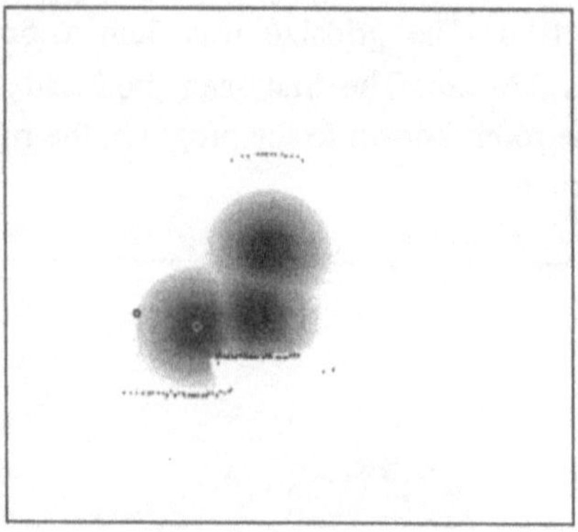

room 1.3

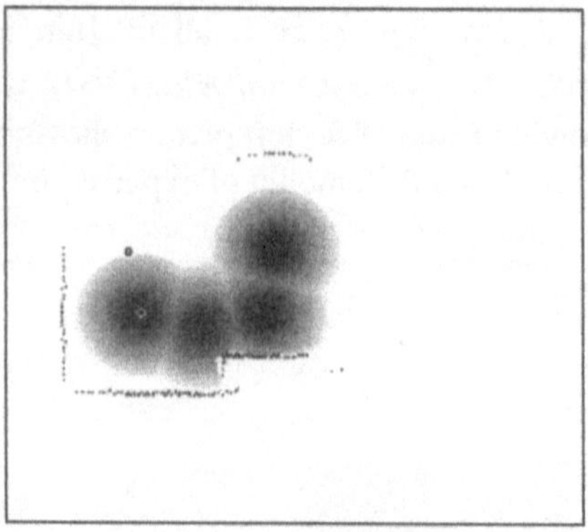

room 1.4

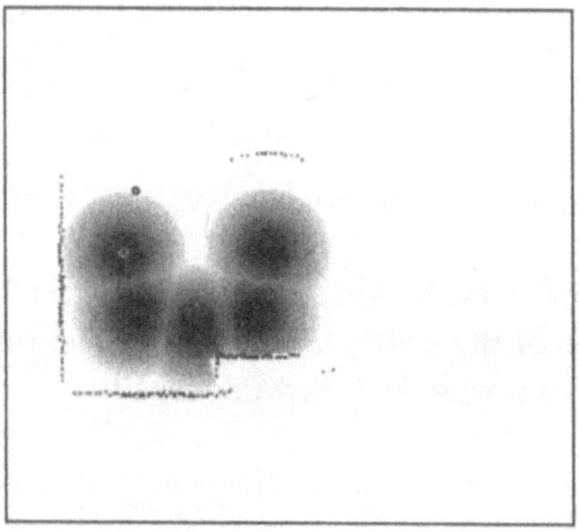

room 1.5

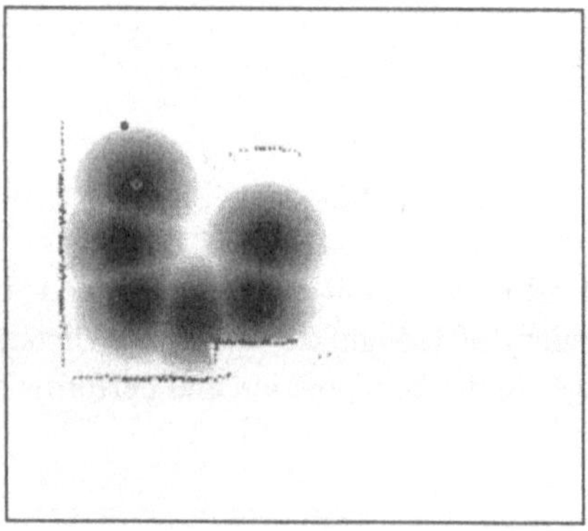

room 1.6

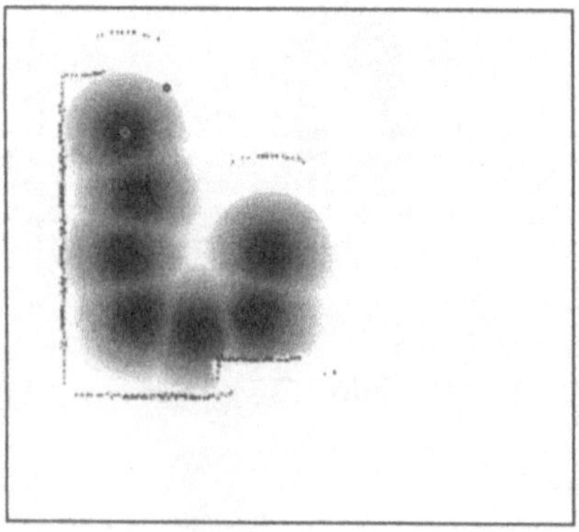

room 1.7

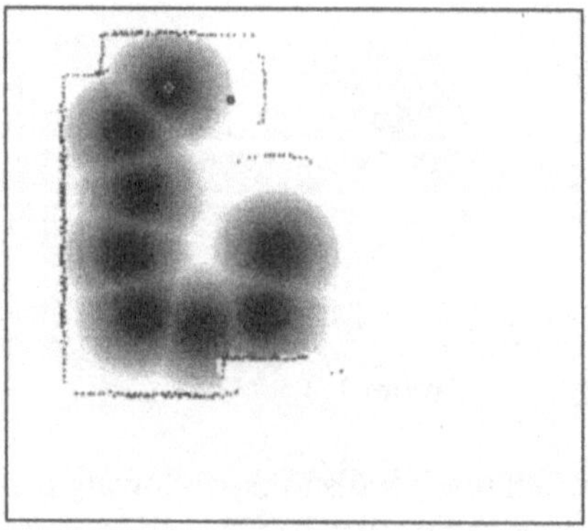

room 1.8

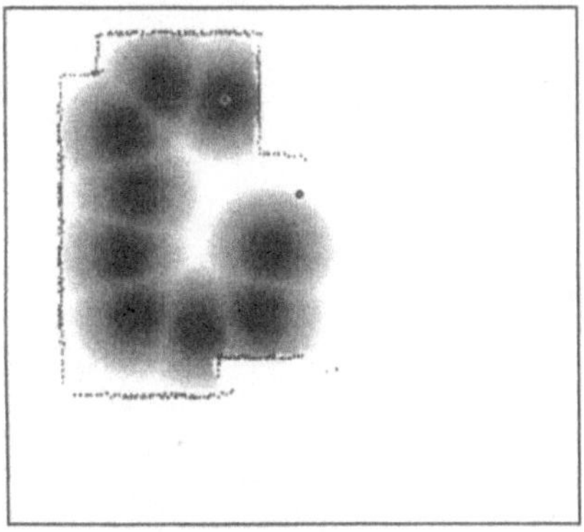

room 1.9

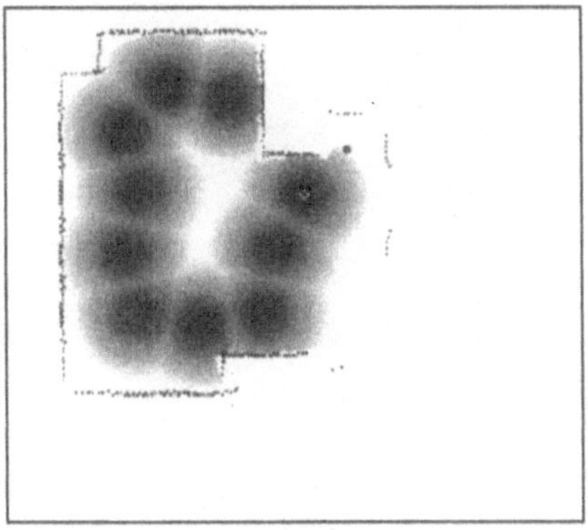

room 1.10

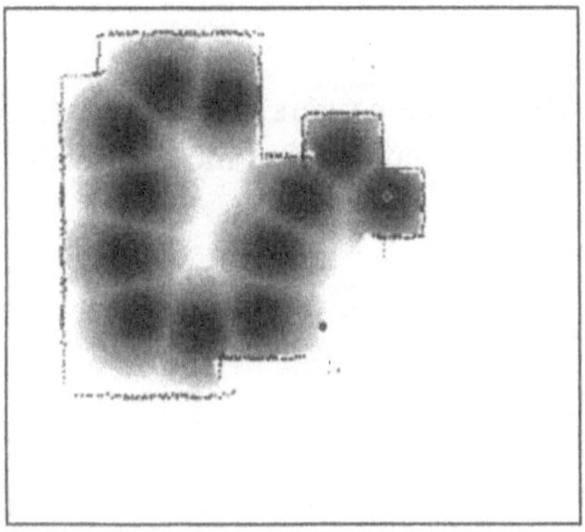

room 1.12

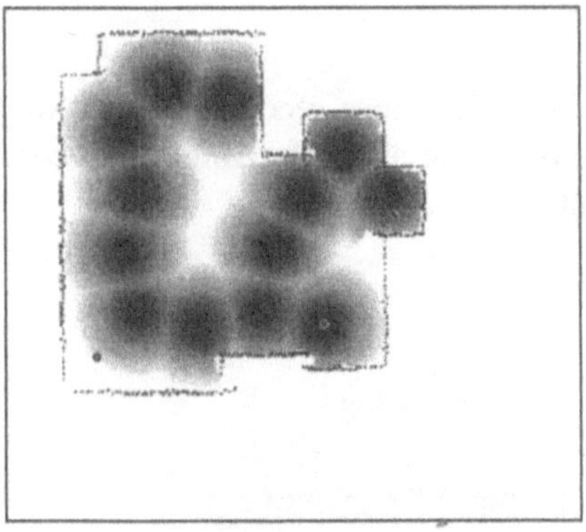

room 1.13

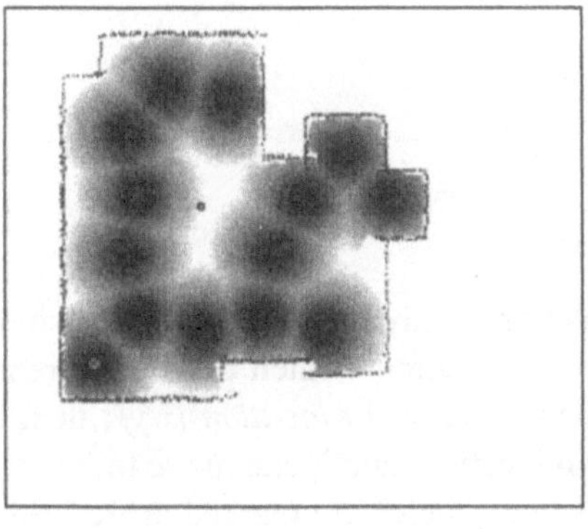

room 1.14

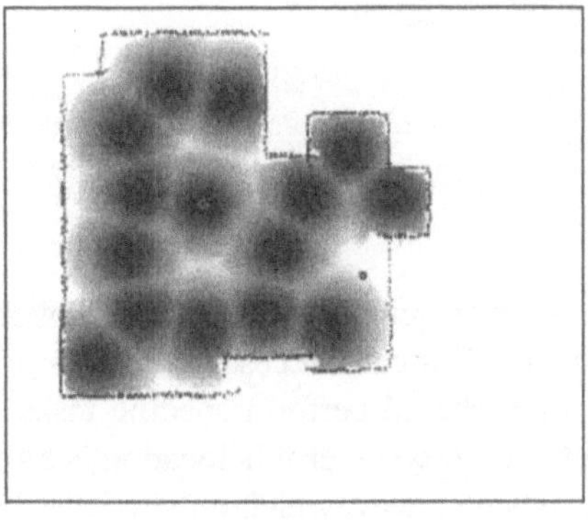

room 1.15

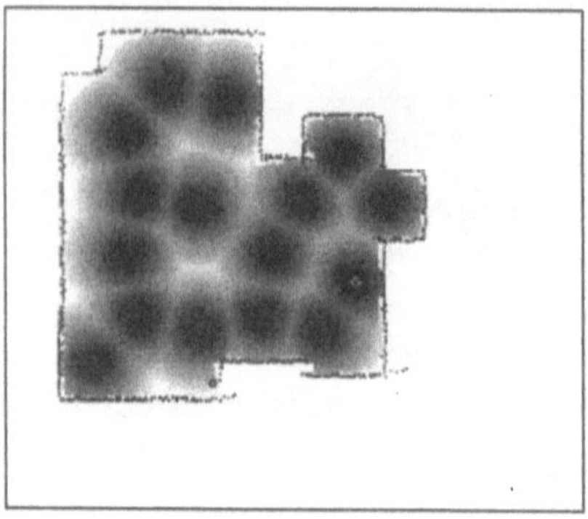

room 1.16

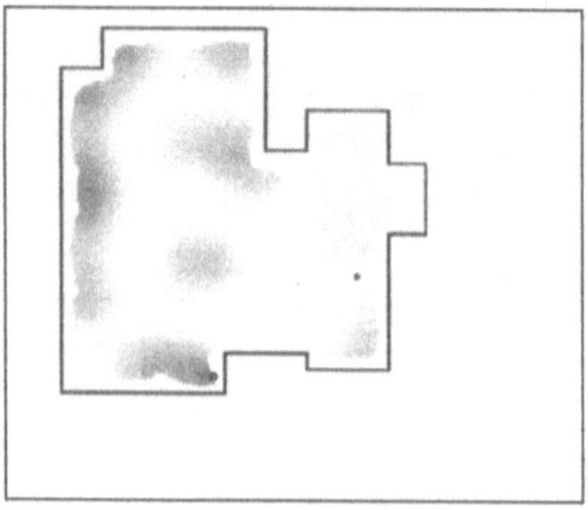

info 1.16

The room has been recognized very well. Only some corners are not recognized exactly. If we allow the program to continue, then - after some more scans - the result looks like this:

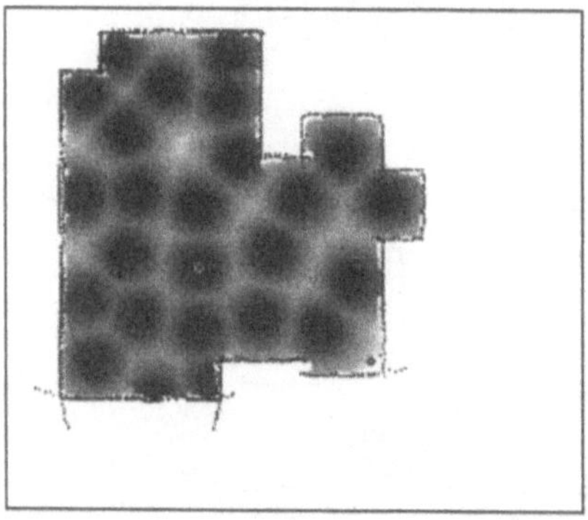

room 1.25

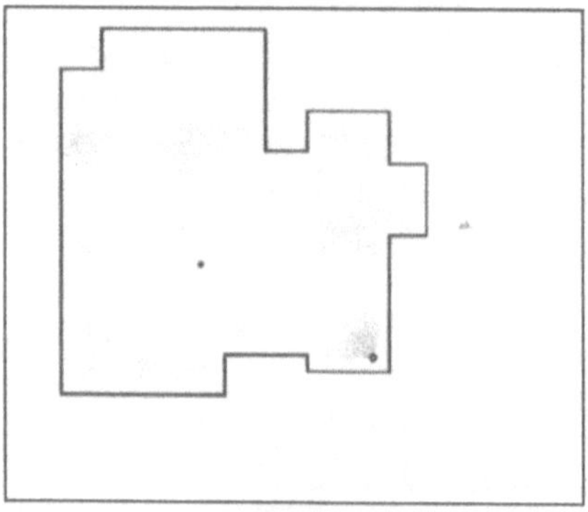

info 1.25

The result is quite satisfactory. The robot now not only knows the locations in the room which it can't access, but also knows all locations which can be entered. If the robot should perform specific tasks such as '*move to location (x/y)*', then the robot knows, whether this location is a valid location, and it can move to it simply by choosing a way, which only touches positions, which are marked as free. So no further room modelling is needed. The result can be used directly.

The program produced the following sequence of positions:

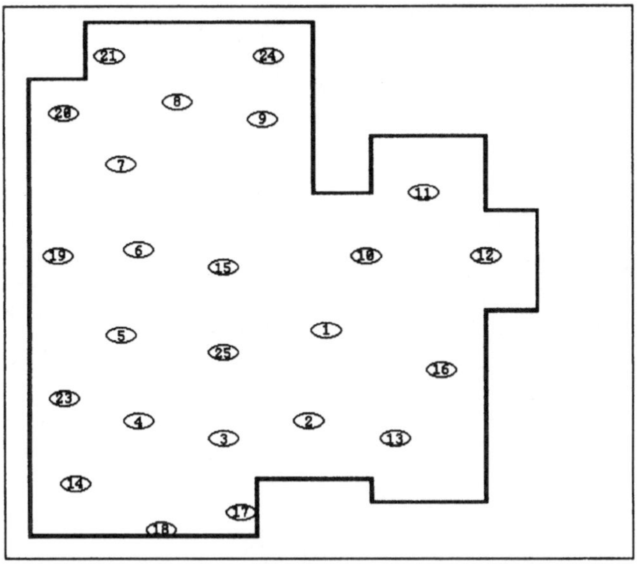

We now want to check, how the program works, if the room is more complicated. The following room is now considered:

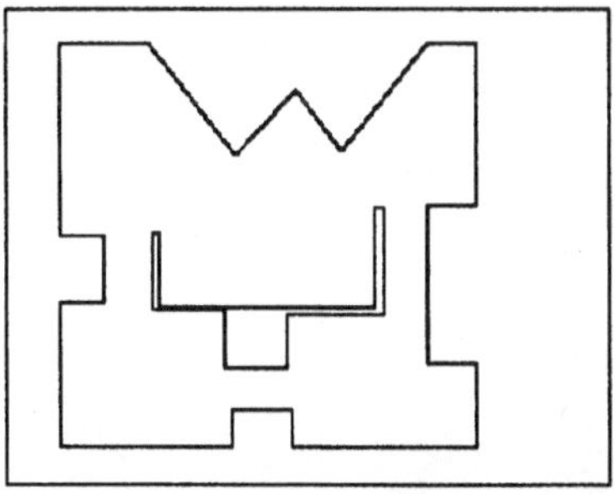

Room to be recognized

As one can see, the robot now has to 'see' some quite small floors in order to reach another part of the room. When we first started the program, we observed that the robot behaved in another way as a human being would do. Imagine two possible next scan positions, which lead to the same amount of additional information. A human being then would take the position, which is close to the current position. The robot did not. Hence, we implemented a routine for checking the path lenght to these positions and subtracted this lenght from the information measure. This lead to the following results, where the robot prefers near next scan positions.

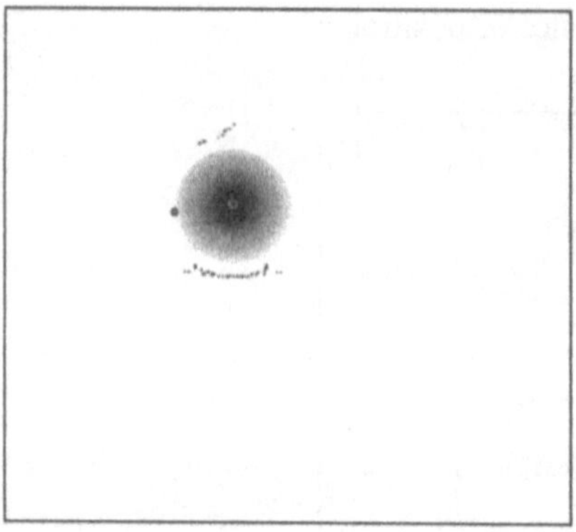

room 2.1

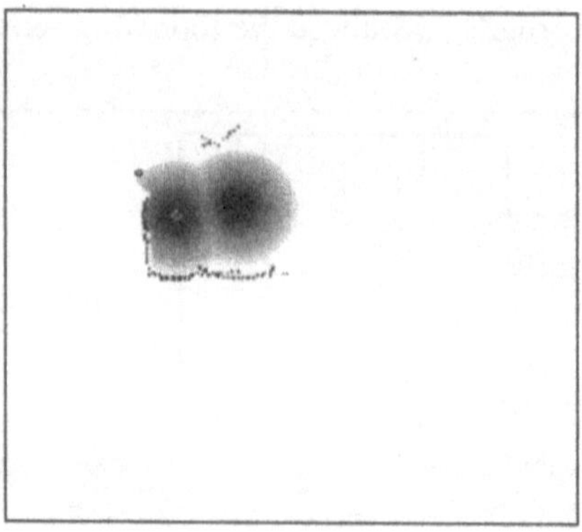

room 2.2

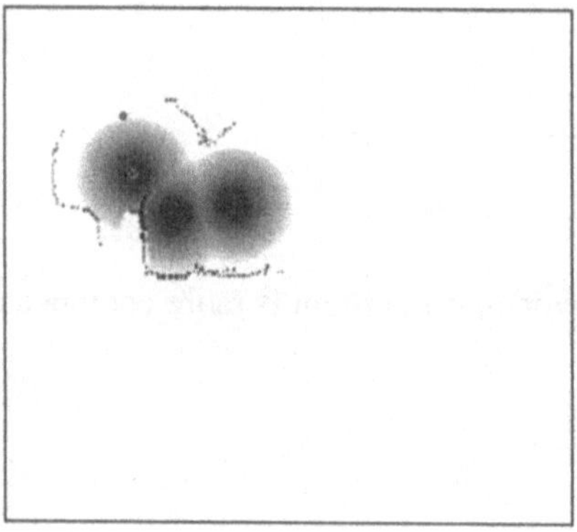

room 2.3

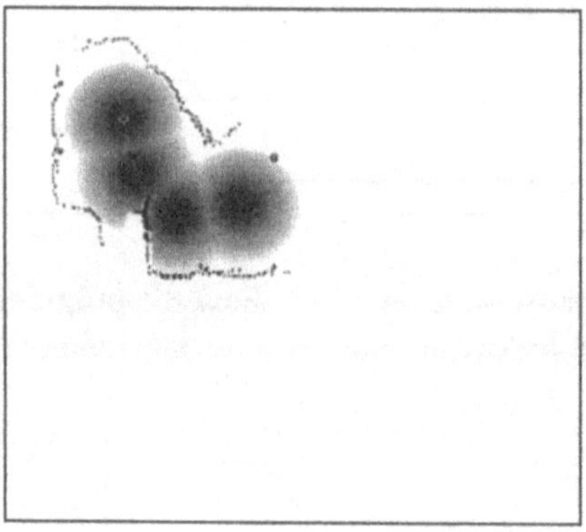

room 2.4

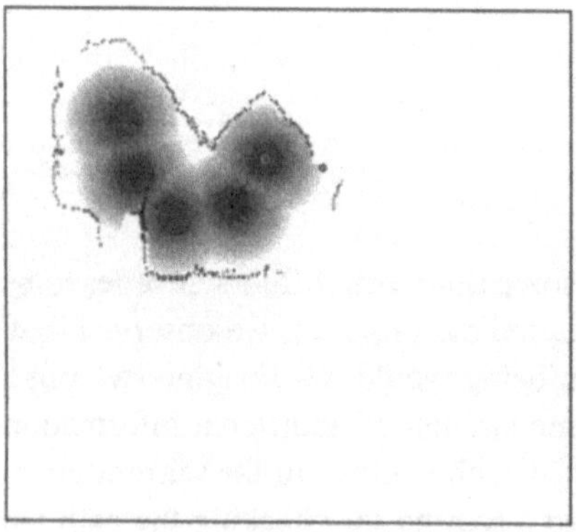

room 2.5

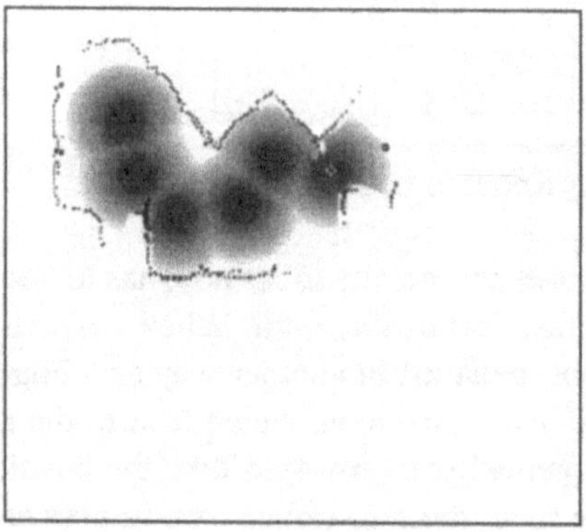

room 2.6

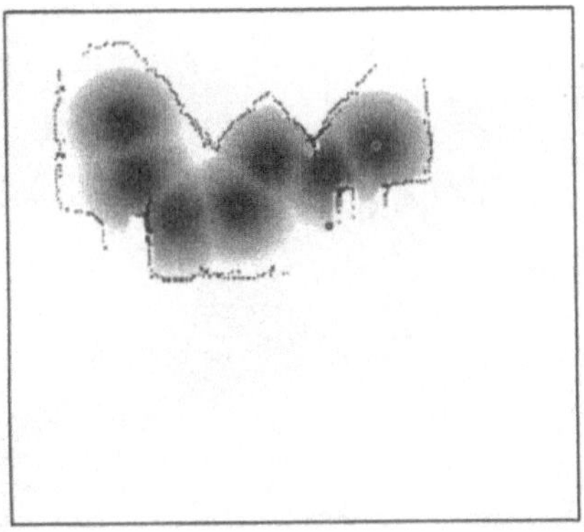

room 2.7

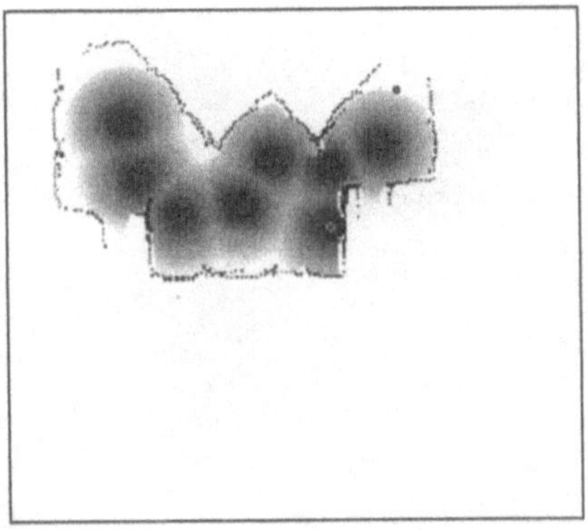

room 2.8

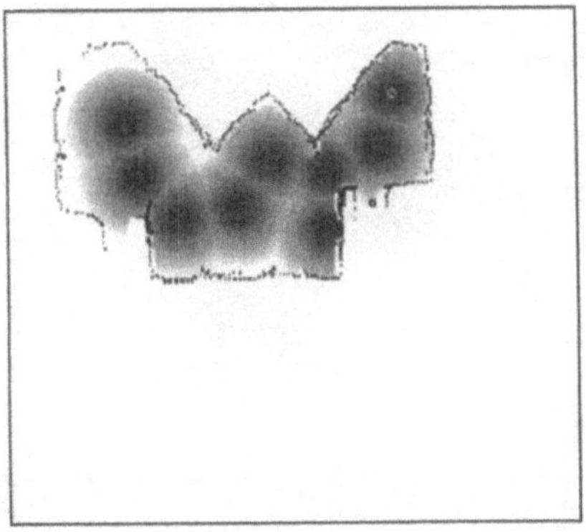

room 2.9

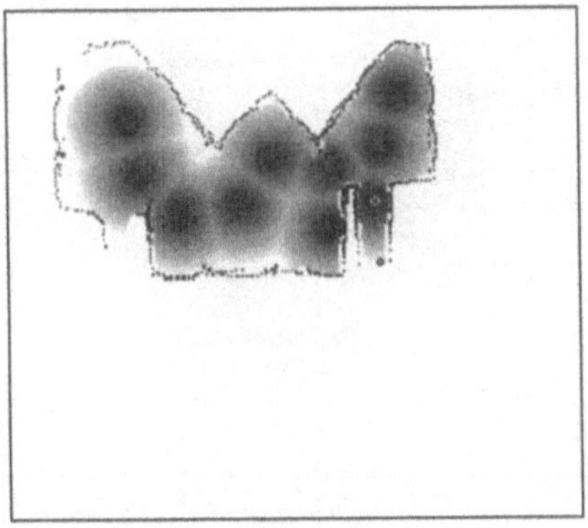

room 2.10

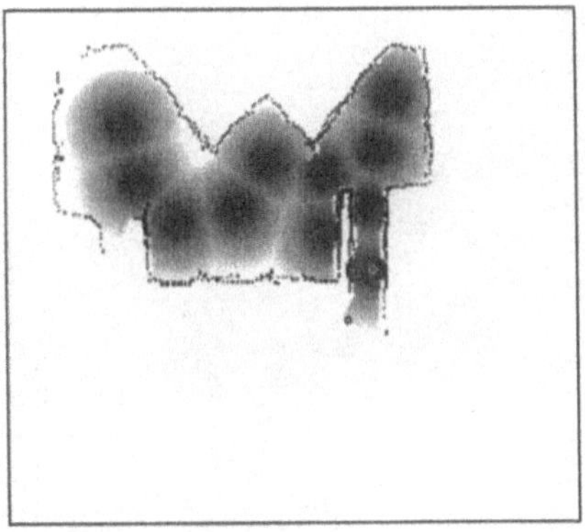

room 2.11

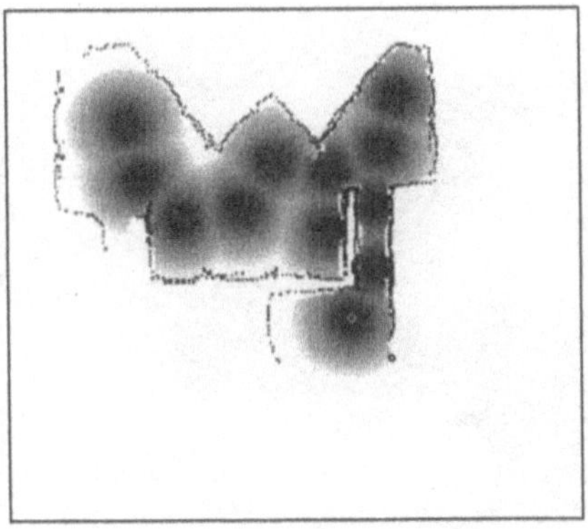

room 2.12

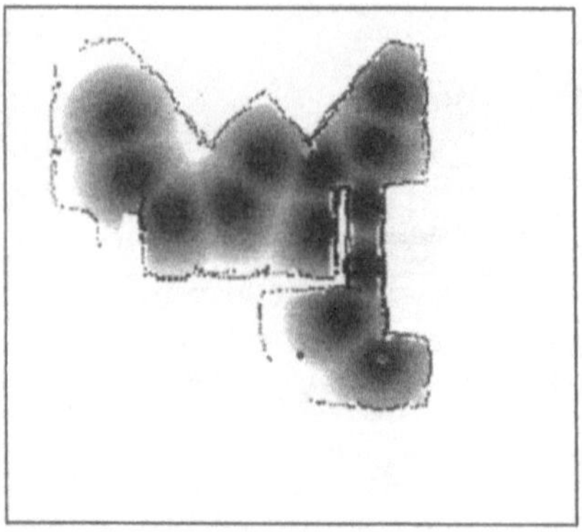

room 2.13

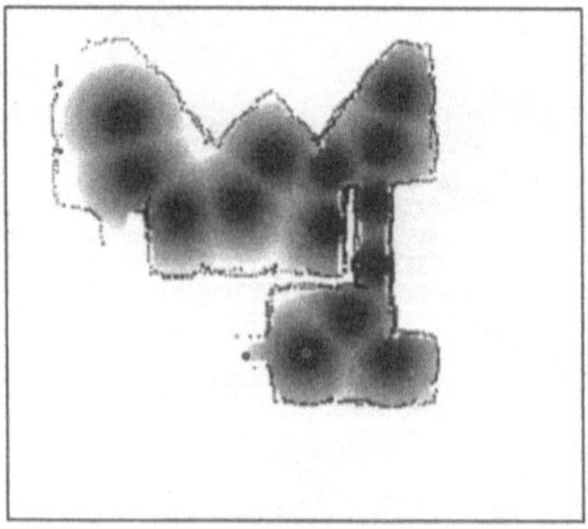

room 2.14

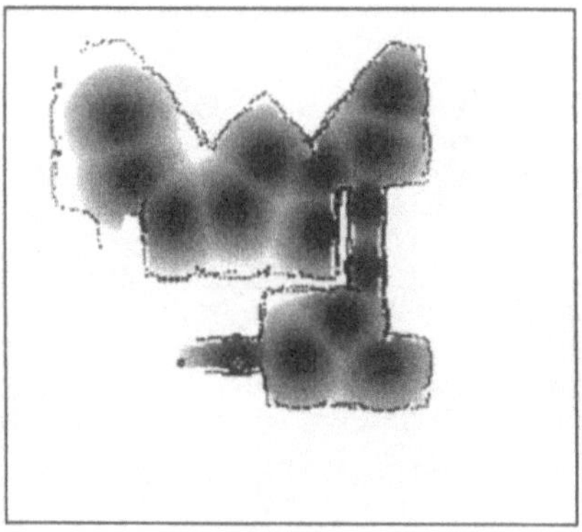

room 2.15

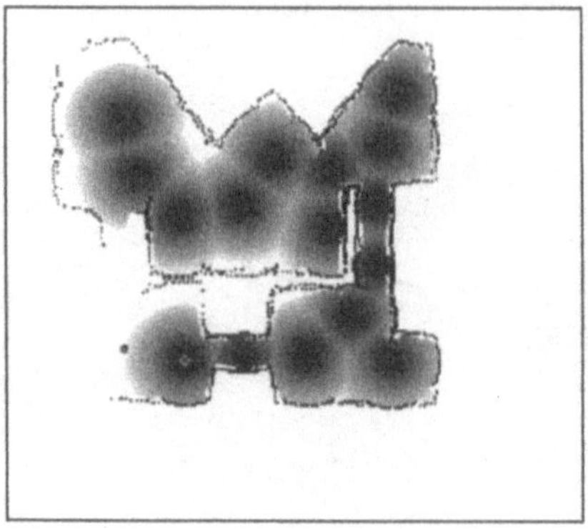

room 2.16

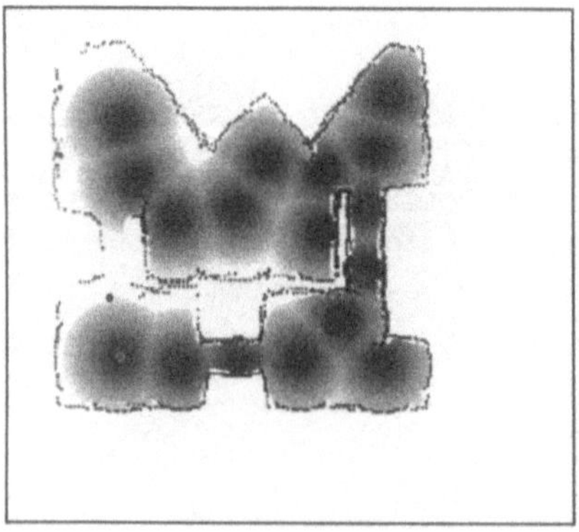

room 2.17

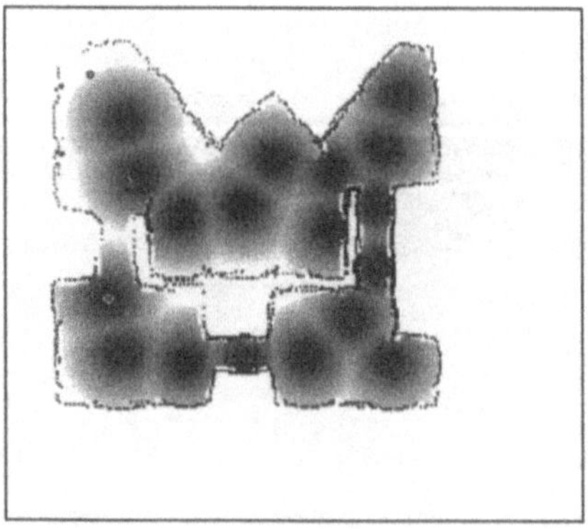

room 2.18

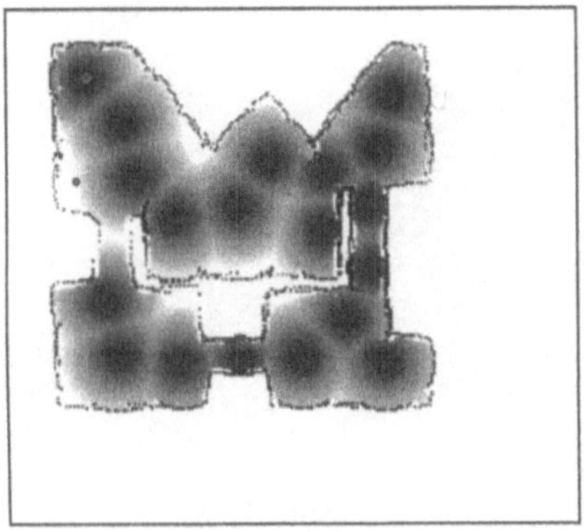

room 2.19

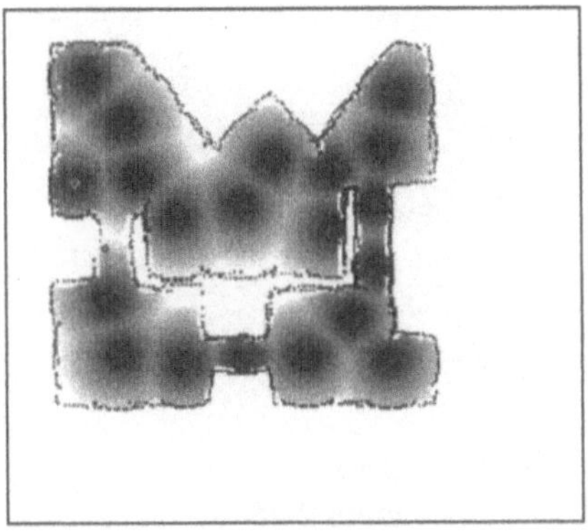

room 2.20

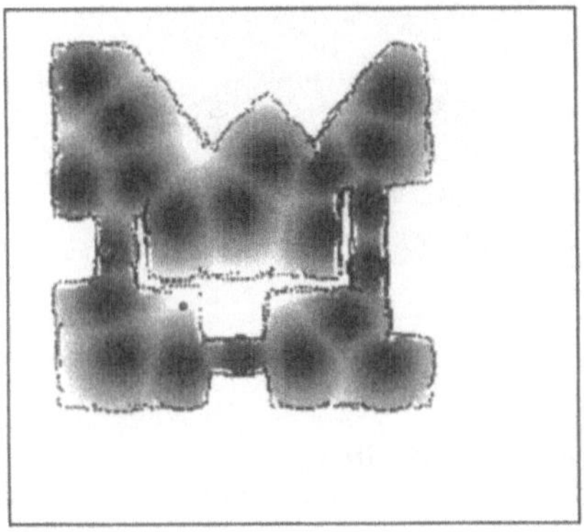

room 2.21

room 2.22

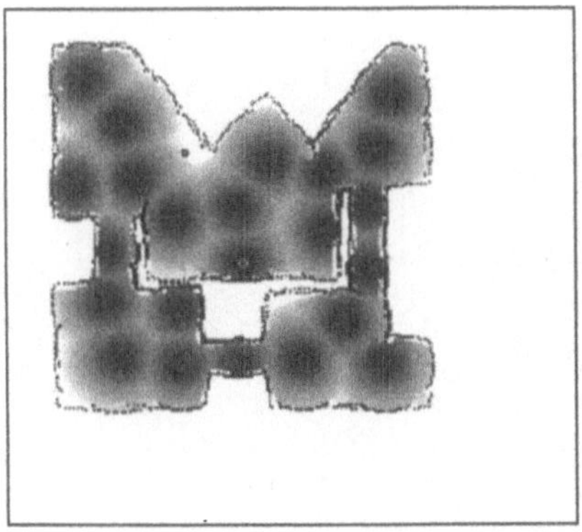

room 2.23

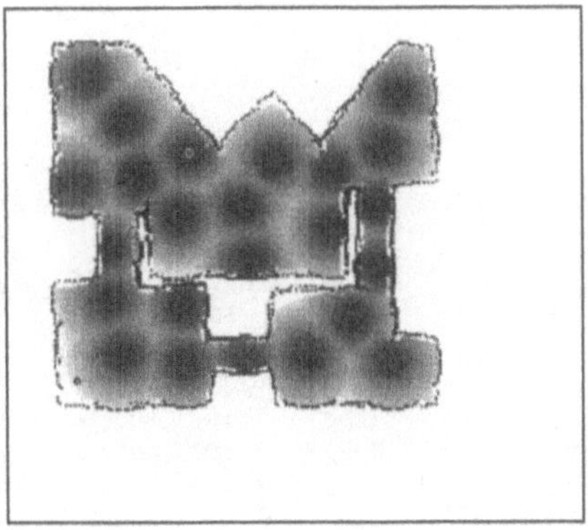

room 2.24

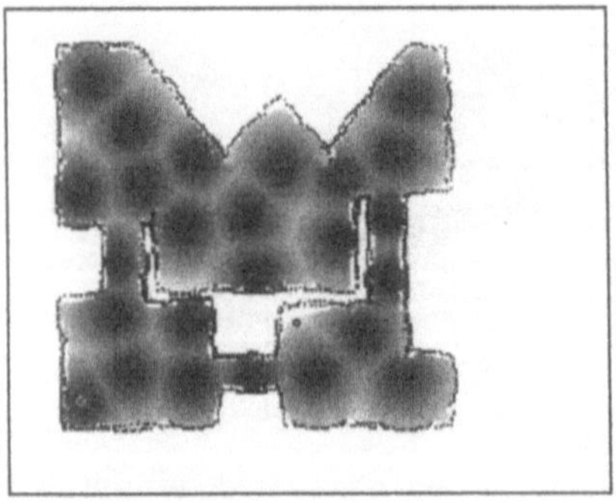

room 2.25

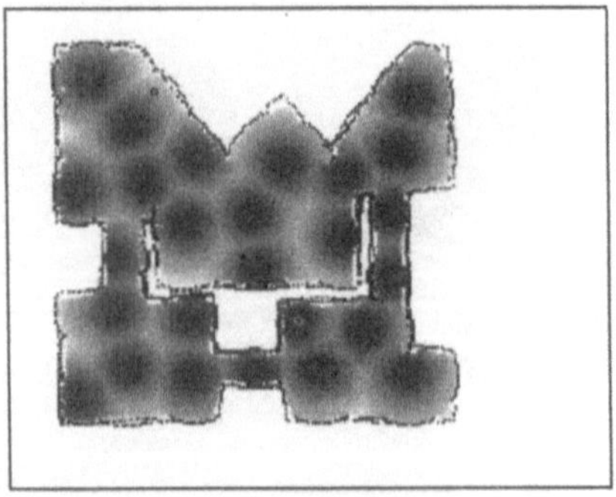

room 2.26

room 2.27

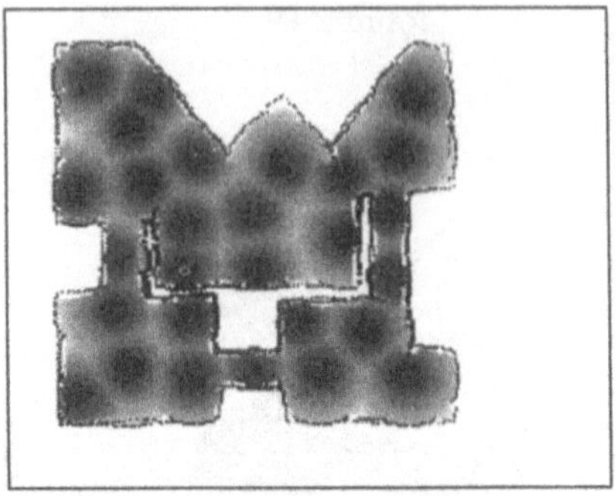

room 2.28

Now the program produced this sequence of scan positions:

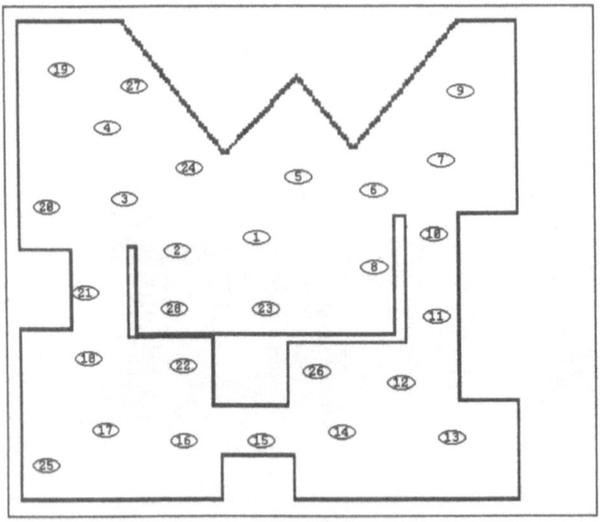

Again, the room has been recognized very well.

References

[BMW91] R. E. Blahut, W. Miller, Jr. and C. H. Wilcox. *Radar and Sonar Part I* Springer Verlag, 1991

[KGK95] R. Kruse, J. Gebhardt and F. Klawonn. *Fuzzy Systeme*. B.G. Teubner, Stuttgart, 1995

[LDW92] J. J. Leonard and H. F. Durrant-Whyte. *Directed Sonar Sensing for Mobile Robot Navigation*. Kluwer Academic Publishers, Boston, 1992

[Rei95] A. Reich. *Simulation der Entfernungsmessung anhand des Laufzeitverfahrens mit einem Ultraschallsensor*. Fortgeschrittenenpraktikum, Institut für Angewandte Mathematik und Statistik, TU München, 1995

[RT95] M. Rempter and M. Troll. *Interactive Simulation and Visualization of Sonar Data*. Information Processing for Active CIM Subsystems 8, TU München, 1995

Gradient Computation by Matrix Multiplication

Louis B. Rall
5101 Odana Road
Madison, Wisconsin 53711
U.S.A.

Key Words: Automatic Differentiation, Gradients, Sensitivities
MOS Classifications: 65H99, 65K99

Abstract. The components of the gradient of a function defined by a code list are components of the eigenvectors of a matrix which is the Jacobian of the code list. These eigenvectors can be computed by the power method, yielding algorithms equivalent to the forward and reverse modes of automatic differentiation for computation of gradients.

1. Code Lists. An important class of functions $t : D \subset \mathbf{R}^m \to \mathbf{R}$ (or $\mathbf{C}^m \to \mathbf{C}$) are those which can be represented by *code lists*

$$t = (t_1, \ldots, t_m, t_{m+1}, \ldots, t_n) \in \mathbf{R}^n, \tag{1.1}$$

where $\mathbf{t}_0 = (t_1, \ldots, t_m) \in \mathbf{R}^m$ is the *input* to the code list and $t_n \in \mathbf{R}$ is its *output*. The integer n is called the *length* of the code list. Once values are assigned to the input entries $t_1, \ldots, t_m$, the subsequent entries $t_{m+1}, \ldots, t_n$ of the code list are computed from previous entries by arithmetic operations

$$t_i = g_i(t_1, \ldots, t_{i-1}) = t_j \circ t_k, \quad j, k < i, \quad \circ \in \{+, -, \cdot, /\}, \tag{1.2}$$

or by

$$t_i = \phi(t_j), \quad j < i, \tag{1.3}$$

where $\phi : D_\phi \subset \mathbf{R} \to \mathbf{R}$ belongs to some set Φ of *library functions* and $t_j \in D_\phi$. In the case of (1.2), we say that t_i *depends* on t_j and t_k, and in (1.3) t_i depends on t_j. This process gives the value of the output

$$t_n = t(\mathbf{t}_0) = t(t_1, \ldots, t_m), \tag{1.4}$$

corresponding to the given input $\mathbf{t}_0 = (t_1, \ldots, t_m) \in D$, where dependence on the intermediate entries $t_{m+1}, \ldots, t_{n-1}$ has been suppressed. For example, the code list

$$(1.5)\qquad \begin{aligned} t_1 &= t_1, \\ t_2 &= t_2, \\ t_3 &= t_1 \cdot t_2, \\ t_4 &= \sin(t_1), \\ t_5 &= t_3 + t_4, \\ t_6 &= t_5 + 4, \\ t_7 &= \mathrm{sqr}(t_2), \\ t_8 &= 3 \cdot t_7, \\ t_9 &= t_8 + 6, \\ t_{10} &= t_6 \cdot t_9. \end{aligned}$$

represents the function $t : \mathbf{R}^2 \to \mathbf{R}$ which can also be expressed by the equivalent formula

$$(1.6)\qquad t_{10} = t(t_1, t_2) = (t_1 t_2 + \sin t_1 + 4)(3t_2^2 + 6).$$

The following sets of indices will be useful for analysis of the code list. First, $U = \{1, 2, \ldots, n\}$ denotes the "universe," and for $j > m$, $K_j \subset \{1, \ldots, j-1\}$ the set of indices i for which the entry t_j depends explicitly on t_i. It follows from the definition of code lists that K_j contains only one or two indices. For $j < n$, $I_j \subset \{j+1, \ldots, n\}$ denotes the set of indices i such that t_i depends explicitly on t_j, that is,

$$(1.7)\qquad I_j = \{i \mid j \in K_i\}.$$

Furthermore, we define $S_0 = \{1, 2, \ldots, m\}$, the set of indices of the inputs, and $D_0 \subset U \backslash S_0$ the set of indices j such that t_j depends *only* on t_i for $i \in S_0$, i.e., only on the inputs. To continue, define $S_{r+1} = S_r \bigcup D_r$, $D_{r+1} \subset U \backslash S_r$ to be the set of indices j such that t_j depends only on t_i for $i \in S_r$. It follows from the condition of nonemptiness of I_j that S_r is a proper subset of S_{r+1}, so $S_k = U$ for some integer k, which will be called the *effective length* of the code list. The effective length is equal to the number of steps required to evaluate the code list in parallel if sufficiently many processors are available.

For the example code list (1.5), one has

$$(1.8)\qquad \begin{aligned} &S_0 = \{1, 2\},\ D_0 = \{3, 4, 7\}, \\ &S_1 = \{1, 2, 3, 4, 7\},\ D_1 = \{5, 8\}, \\ &S_2 = \{1, 2, 3, 4, 5, 7, 8\},\ D_2 = \{9\}, \\ &S_3 = \{1, 2, 3, 4, 5, 6, 7, 8, 9\},\ D_3 = \{10\}, \end{aligned}$$

so $S_4 = U$ and $k = 4$ in this case.

2. Automatic Differentiation. Before compilers with formula translation were available, computer routines to evaluate functions were written in form similar to the code list (1.5). It was noted by Moore [5] and later by

Wengert [7], among others, that the rules of differentiation are simple to apply to the entries of a code list, provided the library functions involved have known derivatives, and consequently the derivatives of each entry in the code list can be evaluated using quantities already computed. It then follows from the chain rule that derivatives of the function defined by the code list can be evaluated along with the function. This process is often called *automatic differentiation* in order to distinguish it from *symbolic* differentiation, which has as its goal the production of formulas for the derivatives of the function, which in turn can then be evaluated.

Since the early days of the subject, striking advances have taken place in both computer programming and automatic differentiation. In order to program evaluation of a function such as (1.6), it is no longer necessary to form a code list, all one has to do is enter an expression which is well-formed according to the syntax rules of some high-level programming language being used, for example,

$$\texttt{f} \ := \ (\mathbf{x} * \mathbf{y} + \sin(\mathbf{x}) + \mathbf{4}) * (\mathbf{3} * \mathtt{sqr}(\mathbf{y}) + \mathbf{6}); \tag{2.1}$$

in Pascal [4]. When the program containing the expression (2.1) is compiled, a simple left-to-right parser can then produce the code list (1.5) which in turn can be used to generated actual computer code to perform the indicated operations [1]. The result of the assignments of the values $t_1 := x$, $t_2 := y$, will be $f := t_{10}$, subject of course to the unavoidable roundoff error inherent in computer arithmetic.

Automatic differentiation has also evolved considerably since its beginnings with the entry-by-entry computation of derivatives of functions defined by code lists to ones given by expressions and to the differentiation of entire computer programs [2]. In this paper, attention will be confined to evaluation of the gradient of the output of functions defined by code lists,

$$\nabla t_n = \left(\frac{\partial t_n}{\partial t_1}, \frac{\partial t_n}{\partial t_2}, \ldots, \frac{\partial t_n}{\partial t_m}\right). \tag{2.2}$$

The individual components

$$s_i(t_n) = \frac{\partial t_n}{\partial t_i}, \tag{2.3}$$

of ∇t_n, are often called the *sensitivities* of the output t_n to the inputs t_i, $i = 1, 2, \ldots, m$.

3. Forward and Reverse Modes. The forward mode of automatic differentiation, as the name implies, consists of application of the rules of differentiation to the entries of the code list in sequence. The gradients of the inputs are the m unit coordinate vectors $\mathbf{e}_1, \mathbf{e}_2, \ldots, \mathbf{e}_m$ of $\mathbf{R}^m$ or $\mathbf{C}^m$. The chain rule can then be applied to each subsequent entry in the form

$$\frac{\partial t_j}{\partial x_i} = \sum_{k \in K_j} \frac{\partial t_j}{\partial t_k}\frac{\partial t_k}{\partial x_i}, \quad i = 1, 2, \ldots, m, \tag{3.1}$$

for $j = m+1, \ldots, n$, to obtain the gradient of t_n. A compact way to formulate this process is in terms of *gradient arithmetic*,

$$
\begin{aligned}
\nabla(t_j \pm t_k) &= \nabla t_j \pm \nabla t_k, \\
\nabla(t_j t_k) &= t_j \nabla t_k + t_k \nabla t_j, \\
\nabla(t_j / t_k) &= (t_k \nabla t_j - t_j \nabla t_k)/t_k^2, \quad t_k \neq 0, \\
\nabla(\phi(t_j)) &= \phi'(t_j) \nabla t_j.
\end{aligned} \tag{3.2}
$$

The important thing to note is that $\nabla t_1, \ldots, \nabla t_{j-1}$ (or the corresponding partial derivatives) have been computed before they are needed for the evaluation of ∇t_j. The forward mode of automatic differentiation is thus easy to understand and implement, but ordinarily requires computational effort proportional to mn. As applied to the example code list, the forward mode gives

$$
\begin{aligned}
\nabla t_1 &= (1, 0), \\
\nabla t_2 &= (0, 1), \\
\nabla t_3 &= (t_2, t_1), \\
\nabla t_4 &= (\cos t_1, 0), \\
\nabla t_5 &= (t_2 + \cos t_1, t_1), \\
\nabla t_6 &= (t_2 + \cos t_1, t_1), \\
\nabla t_7 &= (0, 2t_2), \\
\nabla t_8 &= (0, 6t_2), \\
\nabla t_9 &= (0, 6t_2), \\
\nabla t_{10} &= \big(t_9(t_2 + \cos t_1), 6t_2 t_6 + t_1 t_9\big).
\end{aligned} \tag{3.3}
$$

An advance in the serial evaluation of gradients came about through the later introduction of the reverse mode [3]. This is another way in which to apply the chain rule, in which the output t_n is differentiated with respect to $t_n, t_{n-1}, \ldots, t_2, t_1$ after evaluation of all of the entries in the code list. Formally,

$$
\frac{\partial t_n}{\partial t_n} = 1, \quad \frac{\partial t_n}{\partial t_j} = \sum_{i \in I_j} \frac{\partial t_n}{\partial t_i} \frac{\partial t_i}{\partial t_j}, \quad j = n-1, n-2, \ldots, 1. \tag{3.4}
$$

Thus, the components of the gradient ∇t_n

$$
\frac{\partial t_n}{\partial t_m}, \frac{\partial t_n}{\partial t_{m-1}}, \ldots, \frac{\partial t_n}{\partial t_1},
$$

are obtained in reverse order. The computational effort for the reverse mode is proportional to n, the length of the code list. On the other hand, all entries in the code list have to be computed and stored before starting the computation of derivatives. The reverse mode applied to the example function gives

(3.5)
$$
\begin{aligned}
&\frac{\partial t_{10}}{\partial t_{10}} = 1, \quad \frac{\partial t_{10}}{\partial t_9} = t_6, \quad \frac{\partial t_{10}}{\partial t_8} = \frac{\partial t_{10}}{\partial t_9}\frac{\partial t_9}{\partial t_8} = t_6,\\
&\frac{\partial t_{10}}{\partial t_7} = \frac{\partial t_{10}}{\partial t_8}\frac{\partial t_8}{\partial t_7} = 3t_6, \quad \frac{\partial t_{10}}{\partial t_6} = t_9,\\
&\frac{\partial t_{10}}{\partial t_5} = \frac{\partial t_{10}}{\partial t_6}\frac{\partial t_6}{\partial t_5} = t_9, \quad \frac{\partial t_{10}}{\partial t_4} = \frac{\partial t_{10}}{\partial t_5}\frac{\partial t_5}{\partial t_4} = t_9,\\
&\frac{\partial t_{10}}{\partial t_3} = \frac{\partial t_{10}}{\partial t_5}\frac{\partial t_5}{\partial t_3} = t_9,\\
&\frac{\partial t_{10}}{\partial t_2} = \frac{\partial t_{10}}{\partial t_7}\frac{\partial t_7}{\partial t_2} + \frac{\partial t_{10}}{\partial t_3}\frac{\partial t_3}{\partial t_2} = 6t_2t_6 + t_1t_9,\\
&\frac{\partial t_{10}}{\partial t_1} = \frac{\partial t_{10}}{\partial t_4}\frac{\partial t_4}{\partial t_1} + \frac{\partial t_{10}}{\partial t_3}\frac{\partial t_3}{\partial t_1} = t_9(\cos t_1 + t_2).
\end{aligned}
$$

Because of the low dimensionality of the input space, the advantage of the reverse mode is not apparent in this example.

4. The Jacobian of the Code List and its Eigenvectors. A unified formulation of the forward and reverse mode algorithms can be made by considering the code list t to represent a mapping from $\mathbf{R}^n$ to $\mathbf{R}^n$ (or $\mathbf{C}^n$ to $\mathbf{C}^n$) instead of from $\mathbf{R}^m$ to $\mathbf{R}$. The Jacobian matrix

$$J = J(t) = \left[\frac{\partial t_i}{\partial t_j}\right]_{i,j=1,\ldots,n}$$

of this mapping has the structure $J = E + L$, with

(4.1)
$$E = \begin{bmatrix} I_m & 0 \\ 0 & 0 \end{bmatrix},$$

where I_m denotes the $m \times m$ identity matrix, and L is strictly lower triangular. Furthermore, the first m rows of L are zero, and L is sparse, each row contains at most two nonzero entries. For the example code list,

(4.2)
$$J = \begin{bmatrix}
1 & 0 & . & . & . & . & . & . & . & 0\\
0 & 1 & 0 & . & . & . & . & . & . & 0\\
t_2 & t_1 & 0 & . & . & . & . & . & . & 0\\
\cos t_1 & 0 & 0 & 0 & . & . & . & . & . & 0\\
0 & 0 & 1 & 1 & 0 & . & . & . & . & 0\\
0 & 0 & 0 & 0 & 1 & 0 & . & . & . & 0\\
0 & 2t_2 & 0 & 0 & 0 & 0 & 0 & . & . & 0\\
0 & 0 & 0 & 0 & 0 & 0 & 3 & 0 & . & 0\\
0 & 0 & 0 & 0 & 0 & 0 & 0 & 1 & 0 & 0\\
0 & 0 & 0 & 0 & 0 & t_9 & 0 & 0 & t_6 & 0
\end{bmatrix}.$$

It follows that $L^{k+1} = 0$ for some integer $k \leq n - m$. Furthermore, the eigenvalues of J are 1 with multiplicity m, and its remaining eigenvalues are 0. Since $E^2 = E$ and $EL = 0$,

$$J^k = (E + L)^k = E + LE + \cdots + L^{k-1}E + L^k, \tag{4.3}$$

and thus $JJ^k = (E + L)J^k = J^k E$, from which

$$JJ^k E = J^k E. \tag{4.4}$$

The first m columns of the matrix $J^k E$ are the same as the first m columns of J^k, and form an $n \times m$ matrix M of the form

$$M = \begin{bmatrix} I_m \\ G \end{bmatrix}.$$

Consequently, the columns of M are linearly independent and thus a complete set of eigenvectors of J corresponding to the eigenvalue 1. These eigenvectors can be computed by the algorithm

$$\begin{aligned} M_0 &= \begin{bmatrix} I_m \\ 0 \end{bmatrix}, \\ M_j &= JM_{j-1}, \quad j = 1, 2, \ldots, k-1, \\ M &= M_k = JM_{k-1}, \end{aligned} \tag{4.5}$$

which is simply the power method for computing eigenvectors of J starting with the first m coordinate vectors $\mathbf{e}_1, \ldots, \mathbf{e}_m$ of $\mathbf{R}^n$, respectively. Note that it is not necessary to compute M_1, which is simply the first m columns of J.

For the example code list, $k = 4$ and

$$M = \begin{bmatrix} 1 & 0 \\ 0 & 1 \\ t_2 & t_1 \\ \cos t_1 & 0 \\ t_2 + \cos t_1 & t_1 \\ t_2 + \cos t_1 & t_1 \\ 0 & 2t_2 \\ 0 & 6t_2 \\ 0 & 6t_2 \\ t_9(t_2 + \cos t_1) & t_1 t_9 + 6t_2 t_6 \end{bmatrix}. \tag{4.6}$$

Comparison of (4.6) with (3.3) reveals that the ith row of M is the gradient ∇t_i of the corresponding entry in the code list (1.5). This is a consequence of the forward chain rule (3.1) interpreted as matrix multiplication.

Theorem 4.1. Algorithm (4.5) computes the matrix

$$M = M_k = \begin{bmatrix} \nabla t_1 \\ \nabla t_2 \\ \vdots \\ \nabla t_n \end{bmatrix}, \tag{4.7}$$

where k is such that $S_k = U$.

Proof. A finite induction will be used. Note that the first m rows of M_0 are ∇t_i, $i = 1, 2, \ldots, m$, also that the columns of M_0 are of the form $[\nabla t_j \ 0]^T$. Consequently, by (3.1), the ith row of $M_1 = JM_0$ is ∇t_i for $i \in S_1$. Now suppose for $j < k$ that the ith row of M_j is ∇t_i for $i \in S_j$. It follows again from (3.1) that for $i \in D_j$, the ith row of M_{j+1} will be ∇t_i, and thus also for $i \in S_{j+1}$. Hence, (4.7) will hold for $j + 1 = k$, at which point the algorithm will terminate since $JM_k = JM_k$, so $M_k = M$. QED

The forward mode as implemented in Algorithm (4.5) provides a lot of information, namely the gradient of each entry in the code list as well as the gradient of the output. If only ∇t_n is required, then it can be obtained by computing only the last row of J^k by the following algorithm, noting that

$$[\nabla t_n \quad \cdots \quad] = [0 \quad \cdots \quad 0 \quad 1]\, J^k. \tag{4.8}$$

Thus, the computation

$$R_0 = [0 \quad \cdots \quad 0 \quad 1], \quad R_i = R_{i-1}J, \quad i = 1, 2, \ldots, k, \tag{4.9}$$

gives

$$R_k = [\nabla t_n \quad \cdots \quad]. \tag{4.10}$$

Again, R_1 is simply the last row of J, and thus does not need to be computed. Algorithm (4.9) is equivalent to the reverse mode of automatic differentiation. For the example code list, one has

$$\begin{aligned}
R_0 &= [0 \quad 0 \quad 0 \quad 0 \quad 0 \quad 0 \quad 0 \quad 0 \quad 0 \quad 1], \\
R_1 &= [0 \quad 0 \quad 0 \quad 0 \quad 0 \quad t_9 \quad 0 \quad 0 \quad t_6 \quad 0], \\
R_2 &= [0 \quad 0 \quad 0 \quad 0 \quad t_9 \quad 0 \quad 0 \quad t_6 \quad 0 \quad 0], \\
R_3 &= [0 \quad 0 \quad t_9 \quad t_9 \quad 0 \quad 0 \quad 3t_6 \quad 0 \quad 0 \quad 0], \\
R_4 &= [t_9(t_2 + \cos t_1) \quad t_1 t_9 + 6 t_2 t_6 \quad 0 \quad \cdots \quad 0].
\end{aligned} \tag{4.11}$$

The analysis of the forward and reverse modes can also be carried out on the basis of graph theory [3], the above provides an equivalent formulation in terms of matrix theory.

5. Implications for Parallel Computation. Although the reverse mode of automatic differentiation appears advantageous for serial computation of gradients, forward computation can be carried out effectively in parallel. For example, if enough processors are available, the computation of t_i and ∇t_i can be carried out simultaneously for all $i \in D_j$. The computation time is thus proportional to k, the effective length of the code list. If storage is a potential problem, the values of t_j and ∇t_j can be discarded as soon as I_j contains no indices greater than the current index i [6].

The Author is grateful to Prof. George F. Corliss for a number of helpful comments.

References

1. C. N. Fischer and R. J. LeBlanc, Jr., Crafting a Compiler, Benjamin/Cummings, 1988.

2. A. Griewank and G. F. Corliss (Eds.), Automatic Differentiation of Algorithms: Theory, Implementation, and Application, SIAM, 1991.

3. M. Iri, Simultaneous computations of functions, partial derivatives, and estimates of rounding errors—complexity and practicality, *Japan Journal of Applied Mathematics* **1**, pp. 223–252, 1984.

4. K. Jensen and N. Wirth, Pascal User Manual and Report, 3rd Ed., Springer, 1985.

5. R. E. Moore, Interval Arithmetic and Automatic Error Analysis in Digital Computing, Ph.D. Dissertation, Stanford University, 1962.

6. L. B. Rall, Automatic Differentiation, Techniques and Applications, Lecture Notes in Computer Science No. 120, Springer, 1981.

7. R. E. Wengert, A simple automatic derivative evaluation program, *Communications ACM* **7**, pp. 463–464, 1965.

Simulating Ultrasonic Range Sensors on a Transputer Workstation

MICHAEL REMPTER
Institut für Angewandte Mathematik und Statistik
Technische Universität München

Abstract. A simulation based model of sonar range sensing for robot navigation is presented that accounts for multiple reflections of the sonar signal. Incorporating methods of linear system theory and acoustics an approach somewhat similar to 'ray tracing' in computer graphics is used for the model. The high parallelism of the ray tracing approach is exploited to implement the algorithm on a transputer network. Finally, the encouraging results of our tests are presented. It is shown, that most of the typical sonar ranging errors are predictable features of the sensor.

Key Words: time of flight ranging, sensor simulation, linear system theory, transputer, parallel computing.

1 Introduction

Ultrasonic sensors provide an inexpensive means for determining the surroundings of an object, for example of a robot. So acoustic time of flight (TOF) systems are often used as range sensors in robot navigation. Unfortunately, range measurement with a sonar device is not as straightforward as one might expect. The main problem is that sonar ranging systems produce complex signals that require some interpretation to be used successfully. Our hope is that the simulation gives us a better understanding of the sonar range measurements. Besides, a good simulation is an important tool for developing computer algorithms in robotics for problems like world mapping, obstacle avoidance, and motion planning.

The time of flight ranging system is presented in chapter 2. In 1987 KUC and SIEGEL developed a physically based computer model to simulate acoustic sensors in a two-dimensional (2D) environment ([KS87]). Their model is presented in chapter 3. But this approach assumes right angled corners and does not handle multiple reflections. In 1992 WILKES ET. AL. combined the ideas of KUC and SIEGEL with an 'ray tracing'-approach ([W+92]). Our simulation algorithm is based on their ideas. It allows arbitrary room shapes, surfaces with various sonar reflection characteristics, and multiple reflection readings. The algorithm is presented in chapter 4. The computational effort of the ray tracing approach is enormous but the approach is also massively parallel. Hence we have implemented the algorithm on a workstation with up to 131 INMOS transputers. The parallelization is shortly described in chapter 5. The results of our simulation are presented in chapter 6.

2 A Time of Flight Ranging System

In the most common form of sonar range sensing the sonar range sensor (*transducer*) is both transmitter and receiver. The transducer emits a signal (*pulse*) and measures the time delay – the *time of flight* (TOF) – until it receives an echo. An example of an echo is shown in figure 1.

The TOF ranging system produces a reading whenever the echo amplitude

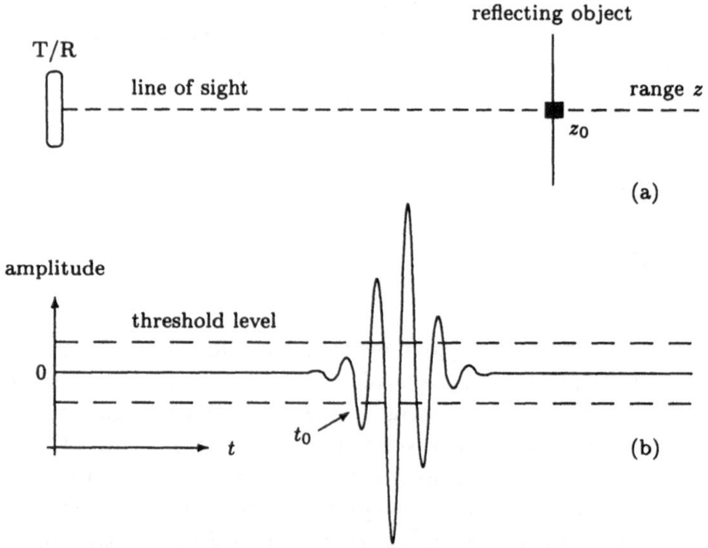

Figure 1: Components of a TOF ranging system: (a) physical configuration, (b) observed time waveform.

exceeds the present threshold level for the first time (fig. 1). If this occurs at time t_0 the detected object has the distance

$$z_0 = \frac{c\,t_o}{2}$$

where $c = c(T) = 331.4 \cdot \sqrt{\frac{T}{273\,\mathrm{K}}}\ [\frac{\mathrm{m}}{\mathrm{s}}]$ denotes the speed of sound in air at temperature T in Kelvin (K).

The shape of the radiated pulse is well known, a typical waveform is approximately given by

$$p(t) = \exp\left(\frac{-t^2}{2\,\sigma^2}\right)\sin(2\pi f_r t)\,\chi_{[-3\sigma,3\sigma]}(t)$$

where $\chi_A(.)$ is the characteristic function of the set A, f_r is the resonant frequency and σ is a measure of the pulse duration.
The emitted sound intensity is maximal in the direction which is normally inclined to the transducer and depends on the angle ϑ to that direction. It is given by the *directivity term* ([KF62])

$$P(\vartheta) = \frac{2\,J_1\,(k\,a\,\sin\vartheta)}{k\,a\,\sin\vartheta}\;\chi_{]-\frac{\pi}{2},\frac{\pi}{2}[}(\vartheta)$$

Here is $k = \frac{2\pi}{\lambda}$, λ the wavelength of the resonant frequency, a the radius of the receiver aperture and $J_1(.)$ the Bessel function of order 1. Figure 2 shows a typical sound intensity pattern given in decibel.

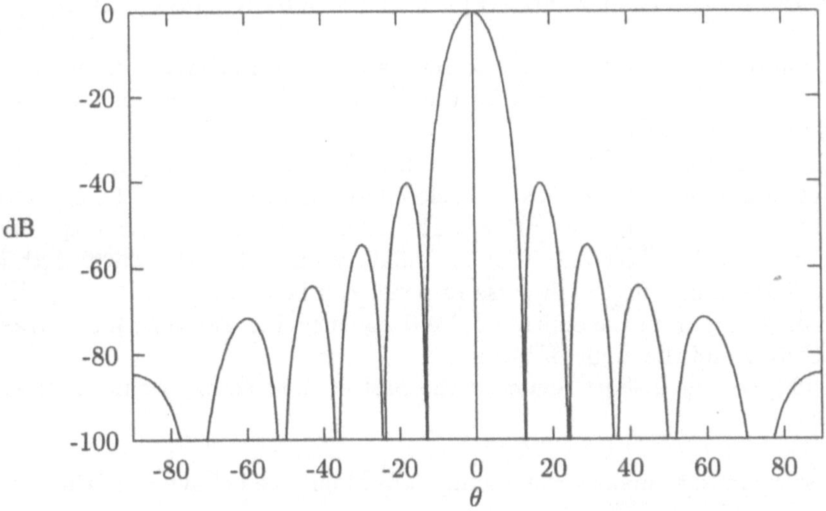

Figure 2: Radiated sound intensity pattern for a transducer with $a = 19\,\text{mm}$ and $\lambda = 6.95\,\text{mm}$.

The straightforward interpretation of a time of flight ultrasonic range sensor, i. e. the assumption that if a reading is produced then an obstacle must lie along the transducer line of sight at that range, often produces false range data. So for example the observed signal can be the result of multiple reflections. Another problem are so called *weak returns* that are reflection aside the aim direction of the transducer evoked by a side lobe of the signal (fig. 2) which occur at low threshold levels. Obstacles and walls which are very close to the transducer may not be detected because of the minimal range of the sensor. Of course, objects which are beyond maximal range can not be detected. A surface whose normal has a large slope to the signal propagation direction may not produce an echo, even if it is close to the sensor. On the other hand this last problem is a good motivation for ignoring floor and ceiling in the simulation.

3 Impulse Response Approach

For the simulation we need to know the time waveform $r(.)$ detected at the receiver output. Using a linear system model with the radiated pulse time waveform $p(.)$ as input and $r(.)$ as output, we can express it as

$$r(t) = \int_{-\infty}^{\infty} \hat{h}(t,\tau)\,p(\tau)\,d\tau = \int_{-\infty}^{\infty} h(t-\tau)\,p(\tau)\,d\tau \tag{1}$$

where the kernel $\hat{h}(t,\tau) = h(t-\tau)$ is called *impulse response* ([ZD63]).

This approach makes it possible to separate between effects of transmitter, reflector and receiver, and makes the ray tracing approach for the sonar simulation convenient.
The impulse response is not as easy to handle as input or output. KUC and SIEGEL published in 1987 ([KS87]) a closed form representation. Their idea is based on Huygens principle ([MI68]). The transmitting and receiving aperture are broken up into elements that are small compared to the wavelength of the radiated pulse. So the impulse response can be expressed in closed form depending only from the emitted pulse (frequency and waveform), the size of the receiver and the angle of incidence.
We will give now a short review of the part of their theory we need for our simulation.

At first we have to make some assumptions about the behaviour of the sonar sensor and of sonar signals:

- The reflection of the sonar signal from surfaces is predominantly specular.
- Floors and ceilings can safely be ignored. That is necessary, because the simulation only uses a two-dimensional map of the environment.
- The transducer is modeled as a plane circular piston in an infinite baffle ([KF62]). Some further simplification lead to the well known radiation sound intensity pattern of sonar transducers shown in figure 2.
- *Far-field* approximation: By the time the wavefront reaches the receiving element, the spherical wavefront can be approximated by a planar

wavefront.
This approximation simplifies the computation significantly and produces reasonable results at range $z > \frac{a^2}{\lambda}$ from the transmitter.

- The sonar system responds only to the first occurrence of a signal exceeding the threshold level.
- The range finder incorporates an amplifier whose gain increases linearly with time.
- *Narrow band* assumption: The gain of the amplifier is constant over the bandwidth of the transducer.

The standard ultrasonic range-finders fulfill the last three assumptions. Also for the typical frequencies reflection from the most surfaces is specular. An amplifier is necessary to reach a convenient range, because of the high attenuation of sound in air. Indeed, after a certain time the amplifier would gain the normal background roaring, so the sonar has a maximal range. There is also a positive minimal range, because of the far-field approximation and the transmitter-receiver characteristic of the transducer.

In our simulation a pulse is modeled by a fan of rays, which are assumed to be transmitted by an almost punctiform source. The receiver is a circular aperture. For a better understanding we prepare this case by studying the impulse response of two other systems.

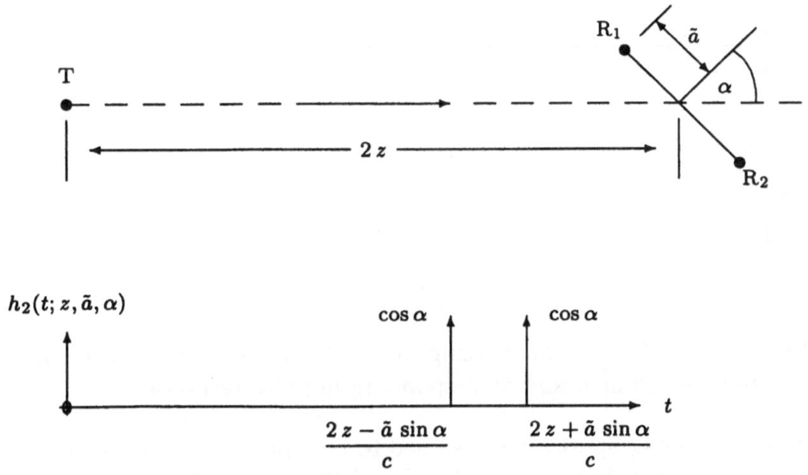

Figure 3: Single transmitting and two receiving elements and corresponding impulse response.

Huygens principle allows to handle arbitrary transducer shapes by breaking them up into elements, which are small compared to the wavelength. So we start with the case of a single transmitting and a parallel single receiving element. The output $r(.)$ is the input $p(.)$ only shifted by certain time. Hence the impulse response becomes

$$h_1(t; z) = \delta\left(t - \frac{2z}{c}\right),$$

where $\delta(.)$ is the Dirac delta function ([OS75]).
Now we assume that there is more than one punctiform receiver and that the wave propagation direction is no longer normally inclined with respect to the receivers. Without loss of generality we use two receivers (fig. 3). We have the sum of two linear systems, so the impulse response of the system is the sum of the impulse responses ([ZD63]). Typical sonar sensors are sensitive to normally incident pressures, so the impulse response becomes scaled with an obliquity factor $\cos\alpha$:

$$h_2(t; z, \tilde{a}, \alpha) = \cos\alpha \cdot \left(\delta\left(t - \frac{2z - \tilde{a}\sin\alpha}{c}\right) + \delta\left(t - \frac{2z + \tilde{a}\sin\alpha}{c}\right)\right)$$

where α is the angle of incidence and $\tilde{a}$ is half the separation of the elements.

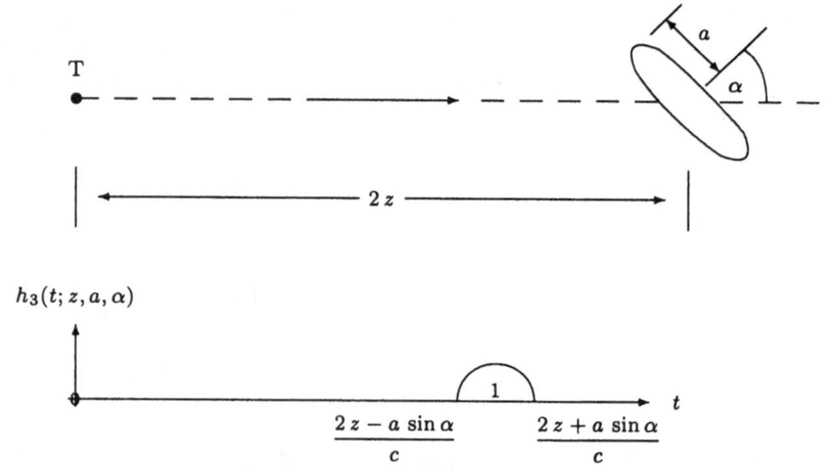

Figure 4: Single transmitting element and a circular receiving aperture of radius a and corresponding impulse response.

The circular receiver aperture is assumed to be a plane and approximated by arranging N elements in a rectangular grid. With the far-field approximation, all the elements detect the signal simultaneously when the aperture is normally inclined to the propagation direction of the pulse. Otherwise the signal needs some time to pass all the receiving elements, the impulse response is spread out in time. For a circular aperture the limiting process $N \to \infty$ delivers the following continuous impulse response (after scaling h_3 appropriately for $\int_{-\infty}^{\infty} h_3(t)\,dt = \cos\alpha$):

$$h_3(t;z,a,\alpha) = \begin{cases} \dfrac{c}{\pi a^2 \sin\alpha} \cdot \cos\alpha \cdot 2\sqrt{a^2 - \dfrac{(ct-2z)^2}{\sin^2\alpha}} & , \quad 0 < |\alpha| < \dfrac{\pi}{2} \\ \qquad \text{and} \quad \dfrac{2z - a\sin\alpha}{c} \le t \le \dfrac{2z + a\sin\alpha}{c} & \\ \delta\left(t - \dfrac{2z}{c}\right) & , \quad \alpha = 0 \\ 0 & , \quad \text{otherwise} \end{cases}$$

In the first case the first term is a scaling constant, the second term is the obliquity factor and the third one is the length of the cross-section of the sensor aperture the signal reaches at time t.

4 The Simulation Algorithm

Given a known two-dimensional environment, the task of the simulation algorithm is to produce realistic sonar scans. Following an idea from WILKES ET. AL. [W+92] our algorithm joins the model presented in the last chapter with an approach somewhat similar to ray tracing. The simulation generates a finely spaced fan of rays and computes the reflections and diffractions of these rays from surfaces until they are attenuated beyond detectability or return to the receiver.

The behaviour of a ray hitting an obstacle is very much depending on the kind of obstacle. The reflection at plane surfaces is mainly specular, the ray is reflected according to the well known formula $\alpha_i = \alpha_o$, where α_i is the angle of incidence. The amplitude of the ray is attenuated by a relative reflectance coefficient. But there is also a weak diffuse reflection at plane surfaces, some parts of the rays are reflected in every direction – especially back to the transducer.

Objects like corners with an extension shorter as the wavelength no longer

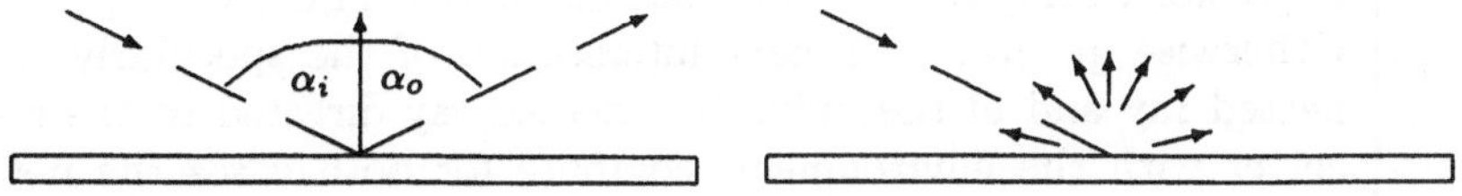

Figure 5: A ray specularly and diffusely reflected by a plane surface

reflect but diffract the sonar signal, it becomes a cylindrical wave ([KF62]).

Modelling those objects as line sources of echoes, the strength of the signal is additionally attenuated by $2\pi\sqrt{\frac{d}{\lambda}}$, where d is the distance from the corner.

So needed facts about the environment are a map of the room with the relative reflectance coefficients of each modelled surface, or even with a set of such coefficients per surface, the room temperature and the size of all the object in the environment.
We will now give a high level statement of the simulation algorithm. A more detailed description of it and of the simulation program is given by REICH in [Rei95] and by REMPTER and TROLL in [RT95].

Step 0: (initialization)

Transducer-parameters are the resonant frequency and the corresponding wavelength, the transducer-radius, minimal and maximal time of flight, the pulse waveform and the threshold level. Further parameters are the line-segment environment, the position and the facing direction of the transducer. The range is set to infinity.

Step 1: (heap initialization)

Generate a finely spaced fan of rays leaving the front of the transducer and compute the first intersection with the environment. Push the travelled distance, the new position, the angle of incidence and the attenuated signal strength on the heap, which is ordered by the distance travelled from the transducer.

Step 2: (range finding)

while the heap is not empty do

> Pop the intersection at the top of the heap from the heap.
> If the travelled distance is larger than the maximal range, then set the range to infinity and goto Step 3.
> If the intersection is at the front of the transducer, compute the output $r(.)$ with formula (1) and the impulse response $h(.) = h_3(.)$. If $|r(t)|$ exceeds the threshold level at any time t, set the range to half of the travelled distance and goto Step 3.
> If the intersection is next to an diffracting obstacle, try to generate an intersection of the diffracted ray with the receiver. If that is possible, compute the new data and sort it into the heap.
> Otherwise generate the next intersection of the specularly reflected ray and of the diffusely reflected ray directed to the receiver with the environment, compute for both cases the new data and sort it into the heap.

Step 3: (end)

Clear the heap and return the range.

The algorithm can simulate arbitrary ultrasonic time of flight sensors, because its free parameters are the radiated pulse time waveform, the sonar resonant frequency, the size of the transducer, the signal detection threshold and minimum and maximum range. The simulation also allows arbitrary room shapes, surfaces with various sonar reflection characteristics and multiple-reflection readings.
But there are also some limitations. So the simulation uses only a planar model of the world and signals arriving at the receiver from different directions at the same time are not added.

5 Parallelization

The Transputer T805-25 from INMOS is a 32 bit microprocessor with a 64 bit floating point unit and four bidirectional communication links, which allow networks of transputers constructed by direct processor to processor connections. A detailed description of our transputer workstation is given by SCHÖNE and SCHULTZ in this book ([SS96]).

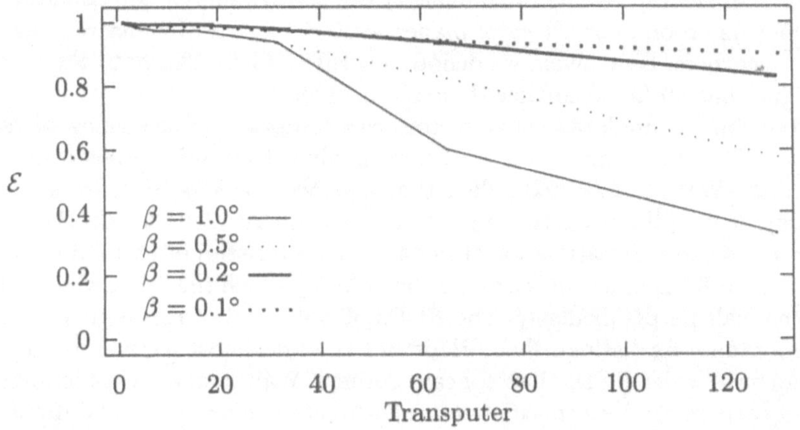

Figure 6: Efficiency of the simulation depending on the number of worker-transputers used and the angle β between adjacent rays

The ray tracing approach is massively parallel. The progress of each of the rays, which is generated in step 1 of the algorithm can be followed independently. Hence step 1 and step 2 become distributed about a larger number of so called *worker processes*. Each of these worker processes is running on

its own transputer. The *master process* who is also running on its own transputer only has to distribute the data in the beginning, to collect the ranges computed by the different workers and to determine the minimal range.
Let T_n be the execution time of the parallel program with n workers. We define the *efficiency* as the ratio

$$\mathcal{E}(n) = \frac{T_1}{n \cdot T_n} \, .$$

where only the execution time for the distributed steps 1 and 2 is measured. That is convenient, since the master process carries out almost none necessary computation for the simulation algorithm.
Figure 6 shows the very good efficiency obtained for our implementation using up to 130 transputers as workers, depending on the angle between adjacent rays generated by the transmitter. For good results in real rooms (see fig. 8) this angle must be lower than 0.3° according to our tests.

6 Results

The results of our test are encouraging. Many classes of typical sonar ranging errors (see chapter. 2) are repeatable. Hence, they are predictable features of the sonar sensor.
Figure 7 shows the results of simulation runs with different threshold levels in a specular room, consisting of planes, rectangular convex and rectangular concave corners. By convex, we denote a corner which relative to the incoming signal has an angle smaller than 180 degree.
Observe that a simulated sonar sector scan comprises of sequences of readings at which the range value measured is almost constant. LEONARD and DURRANT-WHITE [LDW92] called the sequences *regions of constant depth* (RCDs). It is well known that real scans also consist of RCDs. In the simulation plots there are also a lot of further characteristics of real scans.
So the main RCD of a wall is around the point at which the sonar waves intersect the wall perpendicularly. The RCDs of walls and of rectangular convex corners cannot be distinguished. Walls and at least almost rectangular convex corners have wider RCDs than concave corners. Walls and almost rectangular convex corners produce specular returns whereas concave corners diffract the pulse. The width and the number of RCDs increases as the threshold level is lowered. Echoes of diffracting objects and weak returns occur.
Furthermore, here the wall next to the sensor and the concave corner at its east end do not produce any echoes. The distance to the transducer is shorter than the minimal range of the sensor. The south-east corner is out of range. At more sensitive threshold levels there occur multiple reflection echoes far behind the wall next to the transducer. At medium threshold levels there

occur multiple reflection readings at the shorter south wall.

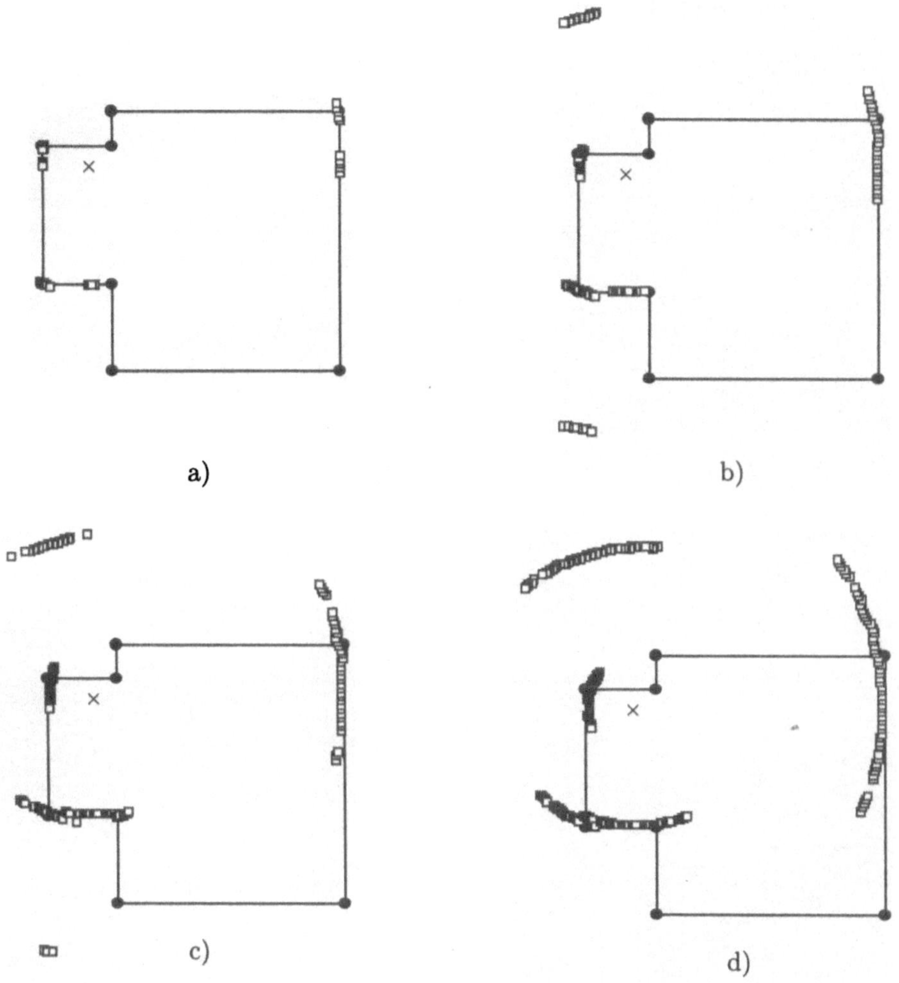

Figure 7: Simulated round about scans in a specular room with various threshold levels: a) -3 db, b) -20 db, c) -40 db and d) -60 db. The crossing shows the position of the sensor.

We also simulated more complex real rooms, for which there also exist measured sonar data. The data given here have been gathered in the floor of our institute.

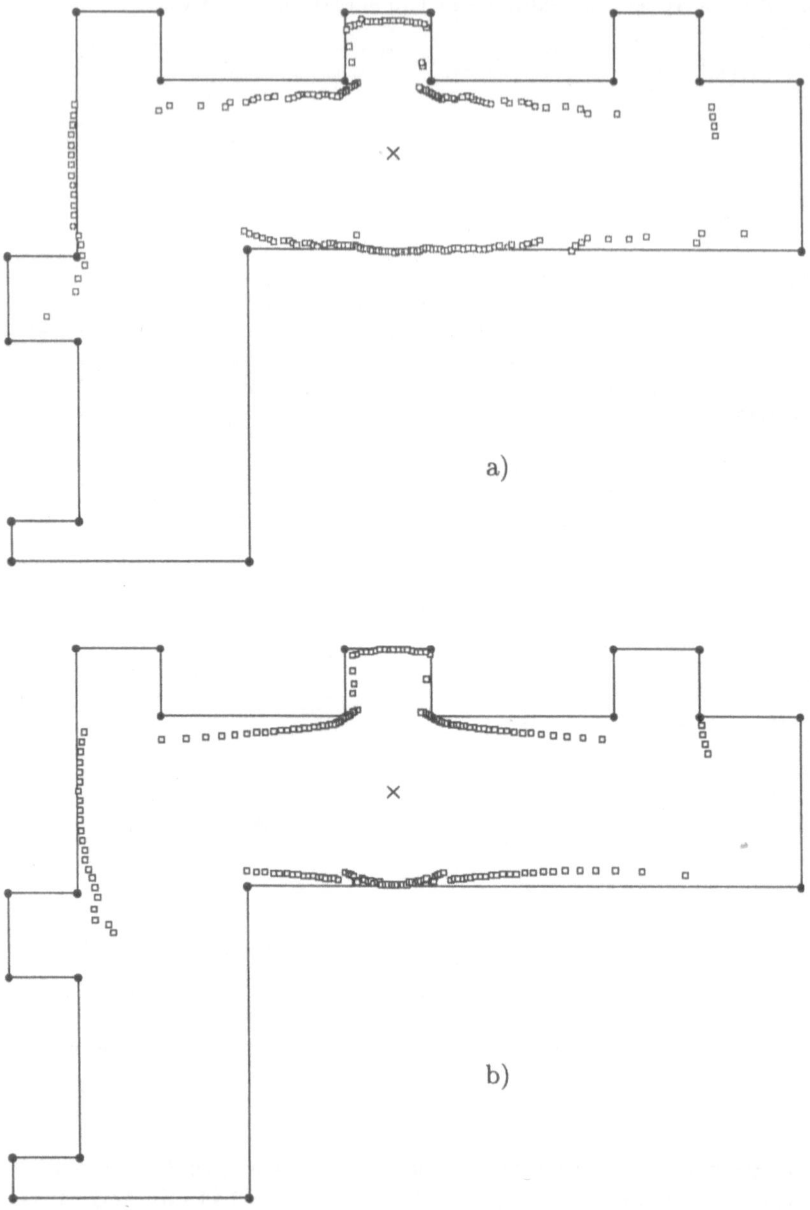

Figure 8: Round about scans in a floor at the institute: a) measured data and b) simulated data.

In its upper part figure 8 shows the original measurement, and in its lower part, a simulated sonar scan from the same position. The result of the simulation is obviously very close to the gathered data. Similar results we obtained for quite different rooms.

References

[KF62] L. E. Kinsler and A. R. Frey. *Fundamentals of Acoustics.* John Wiley & Sons, Inc., New York, 1962.

[KS87] R. Kuc and M. W. Siegel. Physically based simulation model for acoustic sensor robot navigation. *IEEE Transaction on Pattern Analysis and Machine Intelligence*, PAMI–9(6), 1987.

[LDW92] J. J. Leonard and H. F. Durrant-Whyte. *Directed Sonar Sensing for Mobile Robot Navigation.* Kluwer Academic Publishers, Boston, 1992.

[MI68] P. M. Morse and K. U. Ingard. *Theoretical Acoustics.* MacGraw-Hill, New York, 1968.

[OS75] A. V. Oppenheim and R. W. Schafer. *Digital Signal Processing.* Prentice-Hall, Englewood Cliffs, 1975.

[Rei95] A. Reich. *Simulation der Entfernungsmessung anhand des Laufzeitverfahrens mit einem Ultraschallsensor.* Fortgeschrittenenpraktikum, Institut für Angewandte Mathematik und Statistik, TU München, 1995.

[RT95] M. Rempter and M. Troll. *Interactive Simulation and Visualization of Sonar Data.* Information Processing for Active CIM Subsystems 8, TU München, 1995.

[SS96] R. Schöne and W. Schultz. Remote access to a transputer workstation. In H. Fischer, B. Riedmüller, and S. Schäffler, editors, *Applied Mathematics and Parallel Computing.* Physica, 1996.

[W+92] D. Wilkes et al. Modelling sonar range sensors. In C. Archiblad and E. Petriu, editors, *Advances in Machine Vision Strategies and Applications*, pages 361–370. World Scientific, Singapore, 1992.

[ZD63] L. A. Zadeh and C. A. Desoer. *Linear System Theory.* McGraw-Hill, New York, 1963.

A Modular Architecture for Optimization Tutorials

Bruno Riedmüller and Guillermina Schröder

Institut für Angewandte Mathematik und Statistik
Technische Universität München

Abstract

Treatment of algorithms in teaching requires practical experience of the students. This is true in particular when working with optimization algorithms. Here Computer-Based Instruction can be established by means of appropriate software. We propose a system with modular architecture for such software. This approach leads to tools which make it relatively easy to set up teaching software.

Keywords. Computer-Based Instruction, Tutorials, Optimization.

1. Introduction

Algorithms are a central part in teaching applied mathematics, especially optimization. Teaching algorithms in a course, for example in linear, quadratic or nonlinear programming, involve considerable effort. First of all, the teacher must pose the problem and give the theoretical basis for its solution by an algorithm, and finally he has to let his students work with the algorithm to obtain experience with it and to become familiar with it, so that they are then in a position to assess the algorithm themselves.

To this end the students need to work thoroughly with the algorithm. But there is a problem: Working with an algorithm requires a lot of time, above all for number crunching in real problems. It is certainly not desirable for the students to spend their "valuable time" doing calculations instead of learning the basic principles.

So the computer is a suitable device to do calculations. As a second task, it can be used to produce diagrams, graphs, figures and so on, and, finally, it checks and controls the learning process of the student.

This is an educational use of the computer. In the field of Computer-Based Instruction Alessi & Trollip [1] deal with five major types of programs: tutorials, drills, simulations, games, and tests. Our use of the computer is associated with tutorials, but there are special requests: Our programs, especially in the field of optimization, have to do a lot of calculations, and the learning process of the student should go on by solving problems with the algorithms developed in the course.

This needs special software which we will call *problem-oriented tutorials*. The aim of this paper is to describe a suitable architecture for such problem-oriented tutorials. Based on this architecture we then will be able to generate large parts of the necessary code automatically by special convenient tools. The initializing incentive to establish problem-oriented tutorials came from Michael J. Best and Klaus Ritter.

2. Structure and characteristic properties of problem-oriented tutorials

Creating educational software with an authoring system available on the market provides an easy way to implement lessons, but there is an important handicap: Such software often keeps the mathematical contents in the background, the mathematics is treated as mere appendage. Furthermore, control in such software in general is maintained by the teaching program: It is difficult or even impossible to involve results of complex mathematical processes in the teaching program control. For this reason we choose another concept for problem-oriented tutorials.

We split a problem-oriented tutorial into two parts, a calculating program and a teaching program. The calculating program covers the mathematical contents of the algorithm, it is the non-interactive knowlegde base. The didactical design leads to the interactive teaching program. The two modules together form the problem-oriented tutorial (see Figure 1).

During run time the non-interactive knowledge base (i.e. the calculating program) has the control and calls parts of the interactive teaching program. At the points of intersection the teaching program receives data and control information from the knowledge base and uses it in its next part autonomously. At the end of each part the teaching program returns choices of the user and information about his learning process to the base.

Thus, the structure of the knowledge base (i.e. of the main program) is characterized by two types of sections. The first type includes calculations and assignments to the data base as in a pure calculating

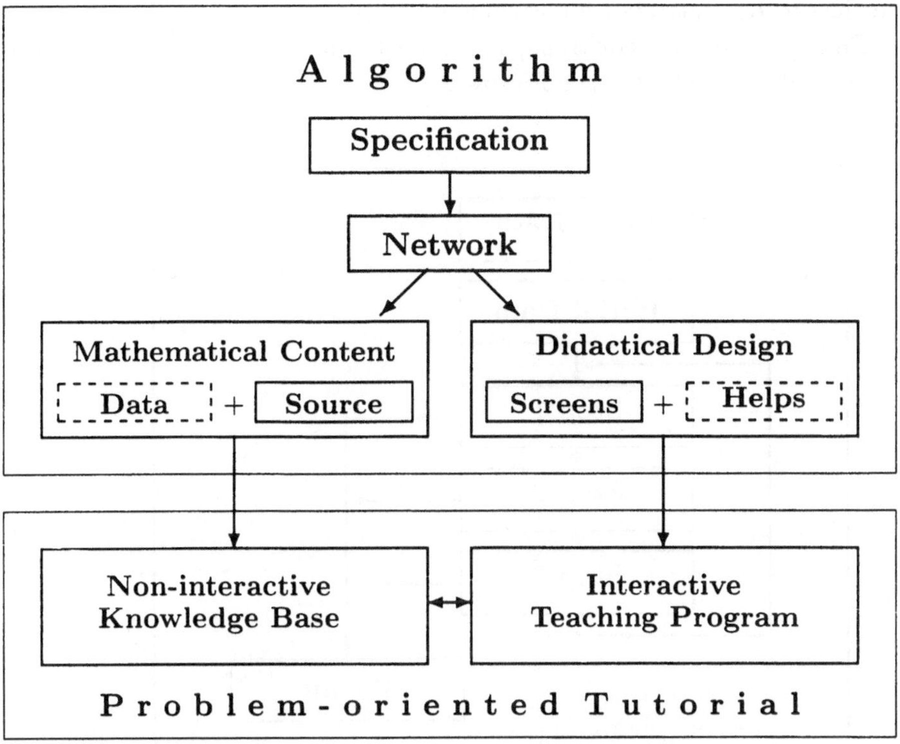

Figure 1. From the algorithm to the problem-oriented tutorial.

program. The other includes assignments to the control base and calls parts of the interactive teaching program, which we call *chapters*. Each section of the first type is always followed by a section of the second type.

The structure of the teaching program is defined by a directed graph. It is a network of learning units, called *pages* or *screens*. They are the vertices of the graph, and they are connected by directed edges (arrows) in such a manner that an arrow leads from page i to page k if page k succeeds page i immediately. This means that there is a direct path in the interactive program from page i to page k. There is just one source und just one sink in the network. From the source, each page is reachable, and from each page the sink is reachable.

Principally, the user of an interactive program should be allowed to move from here to there, to jump forward and backward as he/she pleases. However, in a problem-oriented tutorial the free play of the user has to be restricted for various reasons.

Firstly, iterative algorithms, and in particular optimization algorithms, are typically structured, see Figure 2.

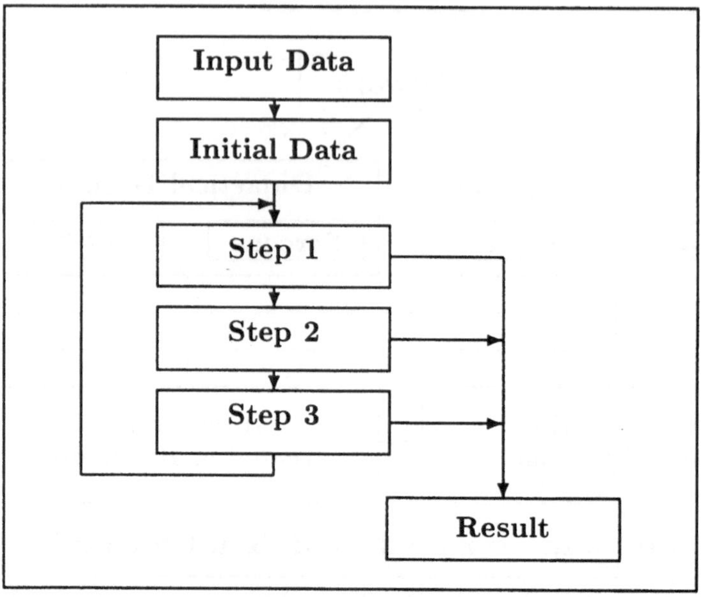

Figure 2. Typical structure of an iterative algorithm.

Since problem-oriented tutorials are based on explicit numerical examples, forward jumps are not possible. For instance, step 3 of an iteration needs numerical results of step 2, and in turn step 2 requires results of step 1. Thus, it can not be allowed to jump from step 1 to step 3.

The second reason for restrictions is the organization of transferring data between the non-interactive base program and the interactive teaching program. Here one can conceive several possibilities. For instance, one could execute the complete base program and store all intermediate results and relevant control information. We decided to choose another model: Guided by the structure of the algorithm we arrange the teaching program pages in equivalence classes, the *chapters* mentioned above. The base program forwards data and relevant control information to the teaching program chapter-wise. For this reason actual data are available only within a chapter, and hence repetitions are possible only within this chapter.

In order to structure the teaching program network transparent, further assessments have to be established. It is understood that from each page of a chapter a *repetition arrow* points to the beginning of that chapter, and a *termination arrow* points to the end of the program. Such repetition and termination arrows are intrinsic and need not be inserted explicitely, see Figure 3.

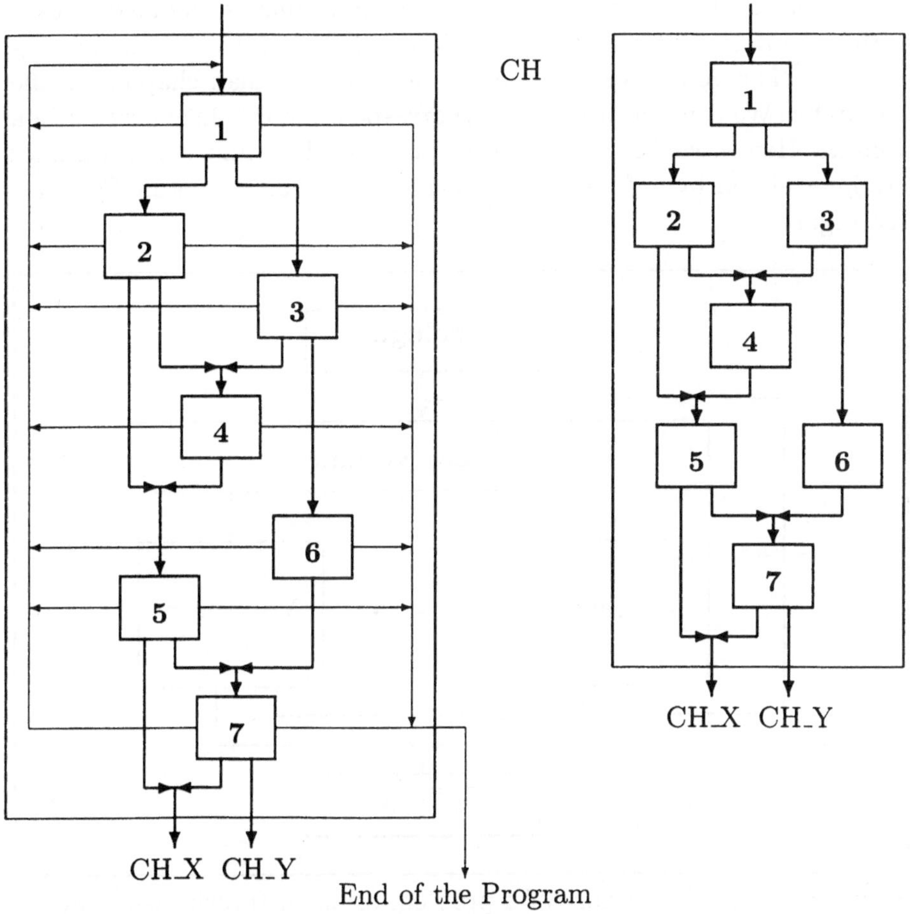

1,2,...,7: Pages of the Chapter CH
CH_X, CH_Y: Exits from Chapter CH

Figure 3. Network of a chapter with/without the arrows for repetition and termination.

Within a chapter we equip the collection of pages with a partial order, taken aside the repetition arrows. This order must not contain cycles. And we require that the order has exact one source. Then the chapter, provided with data and control information from the base program, can run autonomously.

Finally, we associate with each page of the teaching program at most one help-page. Thus, the structure of all the original pages carries over to the help-pages.

Let us consider a problem-oriented tutorial that is intended to deal with *several* algorithms. This situation can be handled by introducing *volumes*. The volumes shall rank one level higher than chapters in the hierarchy. We combine all chapters corresponding to one algorithm in a volume. Hence, we have as many volumes as algorithms. The volumes are interrelated by a *Prologue*, the *Choice of Volume*, and an *Epilogue*, see Figure 4.

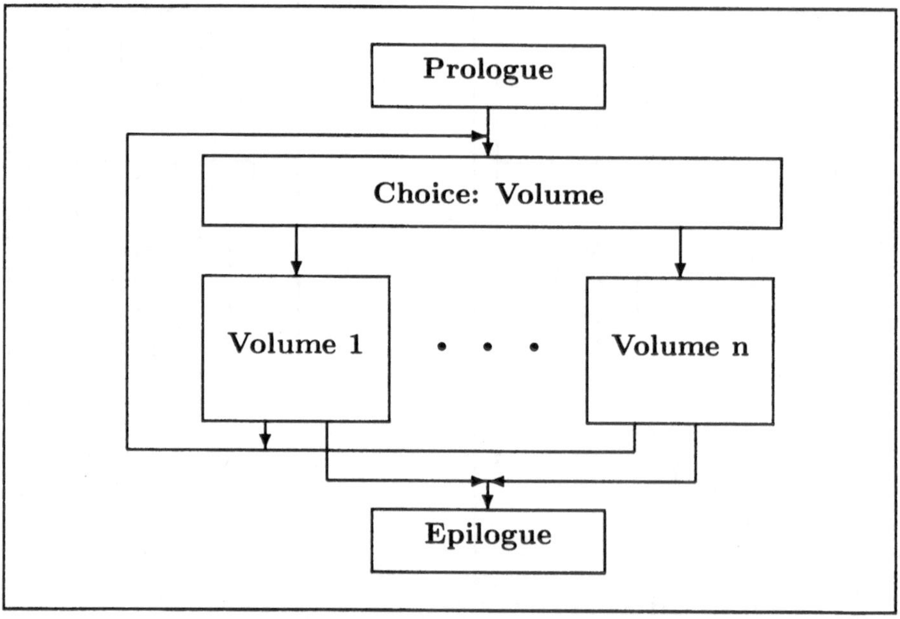

Figure 4. Typical structure of a problem-oriented tutorial with several volumes.

Best & Ritter [2] describe several algorithms for linear optimization problems. They deal with Algorithm 1, starting with an extreme point of the feasible region, and with Algorithm 2, starting with a feasible point. Both algorithms are applicable to problems with inequality constraints. The book also contains an Algorithm 3 for problems with inequality and equality constraints, the revised simplex method, and dual algorithms. All these algorithms are very similar in their basic structure.

In a problem-oriented tutorial for these algorithms, each is represented in a separate volume (see LINPRO; Riedmüller & Schröder [5]). The structure of the algorithms and the requirements of a flexible data input induces the structure of the respective volumes, and within each volume the structure of the chapters, see Figure 5. Many of the chapters deal with data input. The student can use the default problems or he can specify his own problems. He may not always be able to find suitable initial data and in this case the program has to support him.

Now we can see a decisive advantage of the architecture we introduced. Certain chapters of one volume can be reused unaltered in other volumes. This fact considerably reduces the effort in establishing the problem-oriented tutorial.

The initial point problems and the algorithm for parametric linear programming described in Best & Ritter [2] require a slightly different structure of the respective volumes, see Riedmüller & Schröder [5] and Riedmüller [4].

3. Internal structure of the pages of the teaching program

In addition to the modular structure of the problem-oriented tutorial network one has to design the internal structure of the pages. Hereby diverse tutorial concepts such as information, question, reasoning, choice, input, and others influence the particular design. In order to enhance a fairly automatic code generation for the teaching program we introduce four types of pages: input-page, choice-page, question-page, and information-page. Each of these types carries its characteristic internal structure.

Input-pages serve as means for interactive data input including confirmation. Choice-pages offer a menu of alternatives, for instance which algorithm is to be used, or which default problem should be selected. A question-page poses a question concerning the algorithm, more precisely, it checks whether the user is familiar with the algorithm, or it asks for specific details of the algorithm. Of special importance are questions concerning continuation or termination criteria. The answers to such questions often depend on actual numerical data, they may differ from iteration to iteration. Very often it suffices to pose multiple-choice questions. An information-page presents the latest information about

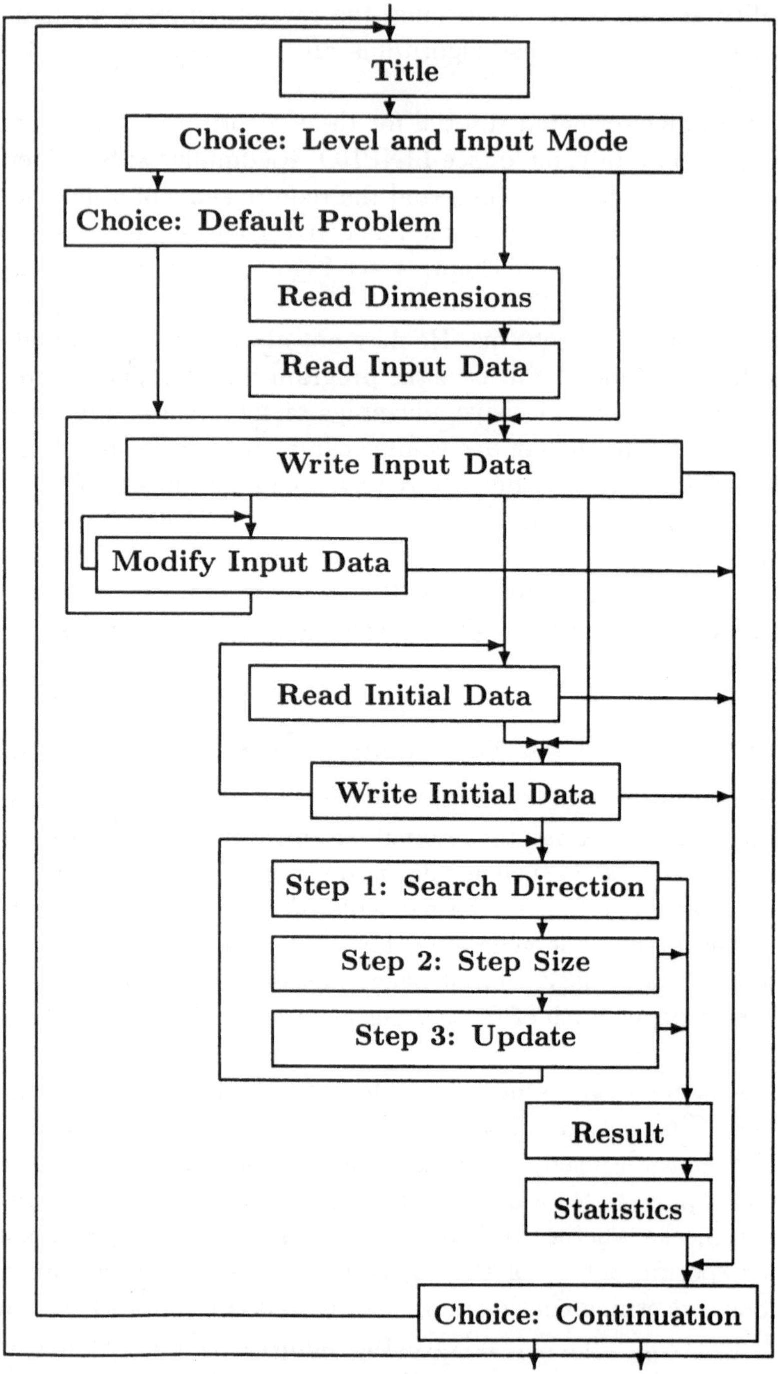

Figure 5. Typical structure of a volume in a problem-oriented optimization tutorial (Detail of LINPRO).

the stage of the algorithm. An information-page is easy to design, though it has to incorporate intermediate numerical results and its consequences.

An application program in general hides intermediate numerical results. This is not so with a problem-oriented tutorial. Here all relevant intermediate results have to be exhibited to let the student see what is going on. Therefore, the design of the pages has to incorporate the possibility of *displaying varying actual data* and resulting graphics.

This flexibility of a page can be achieved using a specific simple syntax. The author of a problem-oriented tutorial uses this syntax for designing pages following the WYSIWYG-principle (What You See Is What You Get). Details can be found in Riedmüller [3] and in Riedmüller & Schröder [7]. The syntax covers the consideration of control information, provided by the base program, to govern the teaching program, in particular in the case of branching. The syntax also permits calling help-pages and controlling the "speed" of the tutorial. According to his/her progress the user can skip pages which would tell him/her nothing new.

4. Conclusion

A modular network architecture for a problem-oriented tutorial and a specific internal structure of pages are the means for an automatic generation of teaching program code (Riedmüller & Schröder [6]). Of course, this can not cover the implementation of the underlaying non-interactive base program.

In this way teaching programs for algorithmic processes, in particular for optimization algorithms, can be designed and developed. Such tutorials result in an effective teaching and learning of algorithms. They enhance the familiarity with algorithms and the comprehension and judgement of algorithmic processes in general.

References

[1] Alessi, St. M.; Trollip, St. R. *Computer-Based Instruction. Methods and Development*. Englewood Cliffs, New Jersey: Prentice-Hall, 1985

[2] Best, M. J.; Ritter, K. *Linear Programming. Active Set Analysis and Computer Programs.* Englewood Cliffs, New Jersey: Prentice-Hall, 1985

[3] Riedmüller, B. *Interaktive Übungsprogramme zur Behandlung von Algorithmen - Entwurf, Entwicklung und Einsatz.* Diss. Univ. Augsburg, 1990

[4] Riedmüller, B. *LINPRO/QUAPRO. Interaktive Übungsprogramme zur linearen und quadratischen Optimierung.* In: PC-Einsatz in der Hochschulausbildung. Hrsg. K. Dette. Berlin, Springer, 1992, 601-609

[5] Riedmüller, B.; Schröder, G. *LINPRO - LINear PROgramming. Ein interaktives Übungsprogramm zur linearen Optimierung. Dokumentation.* München: Institut für Angewandte Mathematik und Statistik, Technische Universität München, 1991

[6] Riedmüller, B.; Schröder, G. *TUTLAB - TUTorials LABoratory. Ein Werkzeug zur Erstellung tutorieller Lernprogramme und interaktiver Simulationsprogramme. User's Guide.* München: Institut für Angewandte Mathematik und Statistik, Technische Universität München, 1992

[7] Riedmüller, B.; Schröder, G. *TUTLAB (TUTorials LABoratory) - Ein Werkzeug zur Erstellung problemorientierter Tutorials für Algorithmen.* - In: *Multimedia, Vernetzung und Software für die Lehre.* Hrsg. K. Dette. Berlin: Springer, 1992, 247-254

[8] Winter, H. *Allgemeine Ziele für den Mathematikunterricht.* Zentralblatt für Didaktik der Mathematik 7. 1995, 106-116

Differential Stability Conditions for Saddle Problems on Products of Convex Polyhedra

Stephen M. Robinson *

Abstract

In this paper we consider variational inequalities expressing necessary conditions for saddle problems over products of convex polyhedra. We show that a certain algebraic property of bisymmetric matrices leads to a condition that can greatly simplify the verification of stability conditions for such problems. We show how to apply this property in connection with recent advances in stability analysis, and we demonstrate that in a particular case derived from nonlinear programming the condition is satisfied without any additional requirements.

AMS (MOS) Subject Classifications: 49B50, 90D05, 90C25
IAOR 1973 Subject Classification: Games; Programming, convex
OR/MS Index Subject Classifications: Primary: Games/group decisions, noncooperative
Key Words: Saddle problems, saddle functions, stability, strong regularity, normal maps

---

*Address: Department of Industrial Engineering, University of Wisconsin–Madison, 1513 University Avenue, Madison, WI 53706-1572. Email smr@cs.wisc.edu; Fax 608-262-8454; Phone 608-263-6862.

The research reported here was sponsored by the U. S. Army Research Office under Grant DAAH04-95-1-0149. The U. S. Government has certain rights in this material, and is authorized to reproduce and distribute reprints for Governmental purposes notwithstanding any copyright notation thereon. The views and conclusions contained herein are those of the author and should not be interpreted as necessarily representing the official policies or endorsements, either expressed or implied, of the sponsoring agency or the U. S. Government.

1 Introduction

This paper deals with stability properties of necessary conditions associated with saddle problems on products of convex polyhedra. For motivation, we begin with the fundamental problem of finding, for a fixed value of p, a saddle point of the real-valued function $K(x, y, p)$ where the variables x and y are constrained to lie in convex polyhedral sets $X \subset \mathbf{R}^n$ and $Y \subset \mathbf{R}^m$ respectively. The variable x represents the choices available to the minimizing player and y represents those available to the maximizing player. K itself is defined on a product $Q \times P$, where Q is some open set that meets $X \times Y$ and P is a normed linear space representing possible perturbations of K. In practice P is often either $\mathbf{R}^k$ or a function space.

This basic saddle problem can represent many different practical situations. One well known situation for which it is a model is that of two-person zero-sum noncooperative games that are not finite: that is, in which the strategies of the players can be any elements $(x, y) \in X \times Y$. Here K represents the payoff function. A great many problems of competitive decision making can be cast in this form.

Another well known and widely used situation is that in which the saddle problem represents the Lagrangian optimality condition associated with a constrained nonlinear programming problem in which the nonlinear constraints have been removed by introducing multipliers (the linear constraints remain, in the form of X). In that case K has a special structure and, in particular, if the standard Lagrangian has been used then K is linear in the second variable y. This Lagrangian saddle condition is necessary and sufficient for optimality if the problem is of convex type and a constraint qualification holds; if the problem is not convex but the constraint qualification, still holds, the variational inequality representing the first-order condition for the Lagrangian problem is necessary for optimality, and in fact it is the primary vehicle used for computing solutions of such problems.

However, it should be noted that the standard Lagrangian is certainly not the only functional form available, and in many cases one may wish to use different kinds of Lagrangians to reflect appropriate ways of expressing constraints, penalties, or other aspects of the problem [12, 13]. Regardless of the form used, however, the Lagrangian condition has the properties described above, though with other forms one does not always have such properties as linearity in the second variable.

Accordingly, the saddle problem – and particularly its first-order necessary condition – has great importance in applications. In this work, the main problem that we consider is that of the stability of those first-order conditions. We shall make, for the remainder of the paper, the blanket assumption that K has second partial derivatives in (x, y) that are continuous at (x_0, y_0, p_0).

Let us now suppose temporarily that $Q \subset X \times Y$, that for fixed p_0 $K(\cdot, \cdot, p_0)$ is convex-concave, and that we have found a saddle point (x_0, y_0) of $K(\cdot, \cdot, p_0)$ on $X \times Y$. A necessary and sufficient condition for (x_0, y_0)

to be a saddle point of $K(\cdot,\cdot,p_0)$ on $X \times Y$ is then that (x_0, y_0, p_0) satisfy a variational inequality that can conveniently be written as the generalized equation

$$\begin{bmatrix} 0 \\ 0 \end{bmatrix} \in \begin{bmatrix} K_x(x_0, y_0, p_0) \\ -K_y(x_0, y_0, p_0) \end{bmatrix} + N_{X\times Y}(x_0, y_0). \tag{1.1}$$

Moreover, even if we drop the assumption that $K(\cdot,\cdot/p_0)$ is convex-concave, the conditions (1.1) can still have fundamental importance; for example, as we illustrate in Section 3, they formulate the basic first-order necessary optimality conditions of (nonconvex) nonlinear programming in a very general form.

In this paper we exhibit differential conditions that are sufficient and, if the perturbation structure is rich enough, also necessary for Lipschitzian stability of (1.1) under small perturbations in p. Specifically, these conditions are to guarantee that there are neighborhoods Φ of p_0 and Θ of (x_0, y_0), and a Lipschitzian function $w(p) = (x(p), y(p))$ defined from Φ into Θ, such that for each $p \in \Phi$, $w(p)$ is the unique solution in Θ of the perturbed generalized equation

$$\begin{bmatrix} 0 \\ 0 \end{bmatrix} \in \begin{bmatrix} K_x(x_0, y_0, p) \\ -K_y(x_0, y_0, p) \end{bmatrix} + N_{X\times Y}(x_0, y_0). \tag{1.2}$$

Thus, in the convex-concave case we have a Lipschitzian family of parametrized saddle points of K, whereas in the general case we have a similar family of solutions of the first-order conditions (1.2), which could be local saddle points if additional conditions are imposed and which may be of considerable interest even otherwise.

Further, we shall show that if certain geometric conditions hold then the nature of these conditions can be drastically simplified, so that they become easily verifiable in practice. We shall see that these geometric conditions often do hold, and in fact in certain common applications (such as the nonlinear programming problem with the standard Lagrangian) they *always* hold.

The rest of this paper is organized in two sections. Section 2 develops certain properties of matrices that will be important in the stability analysis. Section 3 then applies these properties, together with known results from the literature, to establish the stability properties that we are after.

2 Some properties of bisymmetric matrices

This section deals with properties of a class of matrices closely involved in the analysis of (1.1). To see what this class is and why it is related to (1.1), suppose that we differentiate the nonlinear function shown in (1.1). We obtain a linear transformation represented by the matrix

$$\begin{bmatrix} K_{xx}(x_0, y_0, p_0) & K_{xy}(x_0, y_0, p_0) \\ -K_{yx}(x_0, y_0, p_0) & -K_{yy}(x_0, y_0, p_0) \end{bmatrix}.$$

Recalling that K_{xx} and K_{yy} are symmetric and that $K_{xy} = K_{yx}^T$, we see that although this matrix is not itself symmetric, it does have a property known as *bisymmetry* (*cf.* [3], p. 4). To define this property, let us agree that if M is a matrix and if σ and τ are subsets of the row and the column indices of M respectively, we shall denote by $M_{\sigma\tau}$ the submatrix of M obtained by selecting those elements whose row indices are in σ and whose column indices are in τ. With this convention, we define an $n \times n$ square matrix M to be *bisymmetric* if the integers $\{1,\ldots,n\}$ can be partitioned into two disjoint subsets α and β such that $M_{\alpha\alpha}$ and $M_{\beta\beta}$ are symmetric, while $M_{\alpha\beta} = -M_{\beta\alpha}^T$. This bisymmetry property underlies some of the structure discussed by Cottle [1] in his early studies of symmetric dual quadratic programs, and it appears to be related also to the symmetric nonlinear duality structure noted by Dantzig, Eisenberg, and Cottle [4]. It is clear that the saddle point problems we are considering will lead to bisymmetric matrices when we differentiate as we did above.

When referring to a bisymmetric matrix M in this paper, we may add the words "with respect to (α,β)" when necessary, to indicate a complementary partition of the row (and column) indices having the properties just specified. The matrix M need not determine α and β uniquely, as may be seen by considering the identity matrix of order 2 or higher.

Now suppose M is a square matrix of order n and partition the indices $\{1,\ldots,n\}$ into two disjoint sets α and β. Suppose $M_{\beta\beta}$ is nonsingular; then the **Schur complement of $M_{\beta\beta}$ in M**, written $(M/M_{\beta\beta})$, is the matrix $M_{\alpha\alpha} - M_{\alpha\beta}M_{\beta\beta}^{-1}M_{\beta\alpha}$. For more information about Schur complements and their applications, see [2].

The following theorem shows how the bisymmetry of a matrix can be inherited by Schur complements derived from it.

Theorem 2.1 *Let M be a matrix that is bisymmetric with respect to (μ,ν). Partition the indices in μ into disjoint subsets μ_1 and μ_2, and those in ν into disjoint subsets ν_1 and ν_2. Let $\gamma = \mu_2 \cup \nu_2$. If $M_{\gamma\gamma}$ is nonsingular, then the Schur complement $(M/M_{\gamma\gamma})$ is bisymmetric with respect to (μ_1,ν_1).*

Proof. For brevity we shall use the integers 1 through 4 to denote the index sets μ_1, μ_2, ν_1, and ν_2. By assumption the principal submatrix

$$S = \begin{bmatrix} M_{22} & M_{24} \\ M_{42} & M_{44} \end{bmatrix}$$

is nonsingular; write its inverse as

$$S^{-1} = \begin{bmatrix} M^{22} & M^{24} \\ M^{42} & M^{44} \end{bmatrix}.$$

The hypotheses imply that M_{22} and M_{44} are symmetric, and that $M_{24} = -M_{42}^T$; therefore S is bisymmetric with respect to (μ_2,ν_2). However, from this we can also conclude that S^{-1} is bisymmetric with respect to the same index sets. Indeed, if we premultiply S by a diagonal matrix J whose ith diagonal element is $+1$ if $i \in \mu_2$ and -1 if $i \in \nu_2$, we find that JS is symmetric; hence so is its inverse $S^{-1}J^{-1}$, and this implies that M^{22} and M^{44} are symmetric and that $M^{24} = -M^{42^T}$.

The Schur complement $(M/M_{\gamma\gamma})$ that we need is given by

$$\begin{bmatrix} M_{11} & M_{13} \\ M_{31} & M_{33} \end{bmatrix} - \begin{bmatrix} M_{12} & M_{14} \\ M_{32} & M_{34} \end{bmatrix} \begin{bmatrix} M^{22} & M^{24} \\ M^{42} & M^{44} \end{bmatrix} \begin{bmatrix} M_{21} & M_{23} \\ M_{41} & M_{43} \end{bmatrix}.$$

If we carry out the algebra we find that this matrix is of the form

$$\begin{bmatrix} A & B \\ C & D \end{bmatrix},$$

with

$$\begin{aligned} A &= M_{11} - M_{12}M^{22}M_{21} - (M_{12}M^{24}M_{41} + M_{14}M^{42}M_{21}) \\ &\quad -M_{14}M^{44}M_{41}, \\ B &= M_{13} - M_{12}M^{22}M_{23} - M_{12}M^{24}M_{43} - M_{14}M^{42}M_{23} \\ &\quad -M_{14}M^{44}M_{43} \\ C &= M_{31} - M_{32}M^{22}M_{21} - M_{32}M^{24}M_{41} - M_{34}M^{42}M_{21} \\ &\quad -M_{34}M^{44}M_{41} \\ D &= M_{33} - M_{32}M^{22}M_{23} - (M_{32}M^{24}M_{43} + M_{34}M^{42}M_{23}) \\ &\quad -M_{34}M^{44}M_{43}. \end{aligned}$$

The matrices M_{IJ} are symmetric if the index sets I and J are subsets of the same one of the two sets μ and ν, and skew-symmetric otherwise. Therefore we have

$$\begin{array}{lll} M_{12} = M_{21}^T, & M_{13} = -M_{31}^T, & M_{14} = -M_{41}^T, \\ M_{23} = -M_{32}^T, & M_{24} = -M_{42}^T, & M_{34} = M_{34}^T. \end{array}$$

By making these substitutions in the formulas for A, B, C, and D we find that the matrices in parentheses in the formulas for A and D are symmetric, and therefore that A and D are themselves symmetric. Further, $B^T = -C$. Therefore the Schur complement is bisymmetric with respect to (μ_1, ν_1), as claimed. □

A particular consequence of Theorem 2.1 is that if either $\mu_2 = \mu$ or $\nu_2 = \nu$, then the resulting Schur complement is symmetric (because the complementary index set is empty). We shall use this observation in the following section as part of the stability analysis for the generalized equation (1.1).

3 Stability analysis of the first-order conditions

At this point we turn from matrix properties to the main concern of the paper, namely conditions that will ensure that the perturbed generalized equation (1.2) will be solvable, and that the solutions will be well behaved as functions of the parameter p. The theoretical basis for such stability results is by now fairly well understood, and we give here only a summary of the most important points. For further information, see *e.g.* the expository paper [10].

To analyze (1.2) we first consider a generalized equation over a polyhedral convex set C in $\mathbf{R}^k$, of the form

$$0 \in f(x,p) + N_C(x), \tag{3.1}$$

where f is a function from the product of an open set Ω in $\mathbf{R}^k$ and a normed linear space P, to the space $\mathbf{R}^k$. Suppose that $x_0 \in \Omega \cap C$ and $p_0 \in P$, and that the pair (x_0, p_0) satisfies (3.1). Write Π_C for the Euclidean projector on C. It is well known (and easy to show) that the generalized equation $0 \in f(x,p) + N_C(x)$ is equivalent to the single-valued *normal-map equation*

$$f(\cdot,p)_C(z) = 0, \tag{3.2}$$

where for any $z \in \Pi_C^{-1}(\Omega \cap C)$ we define

$$f(\cdot,p)_C(z) = f(\Pi_C(z),p) + z - \Pi_C(z).$$

One can go back and forth between solutions of the generalized equation and the normal-map equation by using the formulas $z = x - f(x,p)$, $x = \Pi_C(z)$. However, it is sometimes more convenient to employ the normal-map formalism and we shall do so here to the extent that it is helpful.

The construction of the normal map involves the set C. However, if we are interested only in local analysis near the point z_0, then another polyhedral convex set becomes relevant, namely the *critical cone* of C at the point z_0. This is a polyhedral convex cone, which we write $K(z_0)$, defined by

$$K(z_0) = T_C(\Pi_C(z_0)) \cap [z_0 - \Pi_C(z_0)]^{\perp},$$

where for an element y of $\mathbf{R}^k$ we define $[y]$ to be the subspace spanned by y, and denote by the superscript $\perp$ the orthogonal complement. The symbol $T_C(w)$ denotes the tangent cone of C at the point w in the sense of convex analysis [11]. In words, we obtain the critical cone of C at z_0 by projecting z_0 onto C to obtain a point x_0, then evaluating the tangent cone $T_C(x_0)$. The portion of $T_C(x_0)$ that is orthogonal to the vector $z_0 - x_0$ then constitutes the critical cone $K(z_0)$.

Now suppose we make the assumption that f has a strong partial derivative $f_x(x_0, p_0)$ with respect to x at (x_0, p_0), which for brevity we shall denote by D. Write K for the critical cone $K(z_0)$. It is known that, under the conditions noted above, in order for (3.2) to have a locally unique, Lipschitzian solution $z(p)$ defined on some neighborhood of p_0, it suffices for the normal-map equation

$$D_K(w(y)) = y \tag{3.3}$$

to define a Lipschitzian function $w(y)$ on $\mathbf{R}^k$. Further, Dontchev and Rockafellar have recently shown that if the perturbation structure is rich enough this condition is also necessary (see [5], Theorem 1).

Moreover, if such a $w(y)$ exists it is a piecewise linear function of y, and its existence is in turn equivalent to the *coherent orientation* of the normal map D_K. By that term we mean the following: if one considers the maximal subsets of $\mathbf{R}^k$ on which the Euclidean projector Π_K onto K is linear, it turns out that these subsets are of dimension n, and they constitute the *n-cells* of a certain subdivided piecewise-linear manifold, called the *normal manifold* of K. Coherent orientation of D_K simply means that on each of these n-cells the linear transformation representing the normal map D_K has a determinant with the same nonzero sign. Further discussion of these topics can be found in [10].

Now let us return to the variational inequality (1.2), which for convenience we reproduce here:

$$\begin{bmatrix} 0 \\ 0 \end{bmatrix} \in \begin{bmatrix} K_x(x_0, y_0, p) \\ -K_y(x_0, y_0, p) \end{bmatrix} + N_{X \times Y}(x_0, y_0). \tag{3.4}$$

To apply the stability theory we have discussed here to this problem, we let

$$f(x_0, y_0) = \begin{bmatrix} K_x(x_0, y_0, p) \\ -K_y(x_0, y_0, p) \end{bmatrix},$$

and $z_0 = (x_0, y_0) - f(x_0, y_0)$. As we did above, we let D be a strong derivative of f (with respect to (x, y)) at (x_0, y_0). By combining Theorems 1 and 3 of [10] we find that in order to have a locally unique, Lipschitzian solution $z(p) = (x(p), y(p))$ of (3.4) defined for all p near p_0, it suffices that D_K be coherently oriented, where K is the critical cone of $X \times Y$ at the point z_0. That means that we need to look at the matrix

$$D = \begin{bmatrix} K_{xx} & K_{xy} \\ -K_{yx} & K_{yy} \end{bmatrix},$$

where from here on we suppress the arguments (x_0, y_0, p_0), and at the critical cone K, which will be the product of the critical cones X_0 (of X at $x_0 - K_x$) and Y_0 (of Y at $y_0 - K_y$). Further, by using some results of [10] we can reduce the problem somewhat.

To carry out this reduction, we select mutually orthogonal subspaces N_1, N_2, and N_3 of $\mathbf{R}^n$ and M_1, M_2, and M_3 of $\mathbf{R}^m$, in such a way that N_3 (M_3) is the orthogonal complement of the affine hull of X_0 (Y_0) and N_2 (M_2) is the lineality space of X_0 (Y_0). Then $X_0 = U_0 + N_2$, where $U_0 = X_0 \cap N_1$, and $Y_0 = V_0 + M_2$, where $V_0 = Y_0 \cap N_1$. Further, U_0 and V_0 are polyhedral convex cones containing no lines and having the dimensions of N_1 and M_1 respectively. Now let us suppose that D has been written in accordance with this set of bases, and partition the principal submatrix of D corresponding to the subspaces N_1, N_2, M_1, and M_2 as

$$\begin{bmatrix} D_{\alpha\alpha} & D_{\alpha\beta} & D_{\alpha\gamma} & D_{\alpha\delta} \\ D_{\beta\alpha} & D_{\beta\beta} & D_{\beta\gamma} & D_{\beta\delta} \\ D_{\gamma\alpha} & D_{\gamma\beta} & D_{\gamma\gamma} & D_{\gamma\delta} \\ D_{\delta\alpha} & D_{\delta\beta} & D_{\delta\gamma} & D_{\delta\delta} \end{bmatrix}.$$

Applying Proposition 1 of [10] and the preceding discussion in this section, we see that in order for D_K to be coherently oriented it is necessary and sufficient that (a) the principal submatrix

$$E = \begin{bmatrix} D_{\beta\beta} & D_{\beta\delta} \\ D_{\delta\beta} & D_{\delta\delta} \end{bmatrix}$$

be nonsingular, and (b) the normal map $(D/E)_{U_0 \times V_0}$ be coherently oriented, where (D/E) is the Schur complement discussed in Section 2. So we have established the general conditions that we were after; however, although theoretically meaningful these conditions are frequently very difficult to check in practice. Verifying coherent orientation of a normal map, particularly one taken with respect to polyhedral convex cones that are not products of orthants and subspaces, can be extremely difficult. However, as we will show, in a certain special case this work becomes much easier.

To understand what this case is and how it arises, note that the matrix D is bisymmetric with respect to the index sets $(\alpha \cup \beta, \gamma \cup \delta)$. Applying Theorem 2.1 to D, we see that if E is nonsingular then the Schur complement (D/E) will be symmetric if either α or γ is empty. We will explore the geometric meaning of this condition presently, but for now let us see why it might be significant.

Recall that U_0 and V_0 were polyhedral convex cones of full dimension in N_1 and M_1 respectively, and containing no lines. Therefore $U_0 \times V_0$ is also of full dimension and contains no lines. Theorem 3.1 of [9] shows that for such a cone, if the matrix (D/E) is symmetric then coherent orientation of $(D/E)_{U_0 \times V_0}$ is *equivalent* to positive definiteness of (D/E). Therefore, in this case the coherent orientation condition reduces to a simple positive definiteness condition that can fairly easily be checked computationally. For this reason it is of interest to interpret geometrically the conditions under which the symmetry condition holds. We do this for V_0; a similar analysis applies to U_0.

For α to be empty it is necessary and sufficient that the critical cone Y_0 of Y at $y_0 + K_y$ be a subspace. In turn, if we recall that Y_0 was the face of the tangent cone $T_Y(y_0)$ orthogonal to K_y, and that K_y belonged to the normal cone $N_Y(y_0)$, then we find that a necessary and sufficient condition for Y_0 to be a subspace is that K_y belong not only to $N_Y(y_0)$ but in fact to the relative interior $\operatorname{ri} N_Y(y_0)$; this is proved, for example, in Lemma 2.1 of [8], where it is also shown that in that case the cone V_0 will actually be the lineality space of Y (as we would expect).

An example might make clearer the meaning of this relative interiority condition. Suppose we consider the case in which Y is the product of an orthant $\mathbf{R}^j_+$ and a subspace $\mathbf{R}_i$. Then K_y will have to be in the product of $\mathbf{R}^j_-$ and $\{0\}^i$, and will have to satisfy $\langle K_y, y_0\rangle = 0$. The critical cone Y_0 then consists of all $w \in \mathbf{R}^{j+i}$ whose first j components are nonnegative and for which $w_r = 0$ whenever $(K_y)_r < 0$ (note that any such r must be among the first j indices). After taking the appropriate bases as explained above, and removing the orthogonal complement of the affine hull of Y_0, we are left with the cone $V_0 \times M_2$ where $V_0 = \mathbf{R}^s_+$ and $M_2 = \mathbf{R}^i$, with the orthant V_0 representing those indices r for which both $(y_0)_r = 0$ and $(K_y)_r = 0$. These indices therefore make up the index set γ, and the condition that γ should be empty is therefore the condition that there should be no such indices: that is, that y_0 and K_y should exhibit the property of *strict complementary slackness,* well known in the literature.

Finally, we remark that in a particular special case frequently found in applications, this condition will *always* hold. The case in question is that in which the first-order conditions (1.1) come from a nonlinear programming problem by way of the standard Lagrangian. The most general perturbed smooth nonlinear programming problem involving polyhedral convex sets can be written, after some reformulation if needed, as

$$\begin{array}{rrcl} \text{minimize}_x & f(x,p) & & \\ \text{subject to} & h(x,p) & = & 0, \\ & x & \in & C, \end{array}$$

where C is a polyhedral convex set. The standard Lagrangian is given by the function $K(x, y*) = f(x) - \langle y^*, h(x)\rangle$ on the product $C \times \mathbf{R}^j$, where j is the number of components in h. First-order necessary optimality conditions of the form (1.2) then follow in the presence of standard constraint qualifications [6, 7]. In this case the set Y is the subspace $\mathbf{R}^j$, and this will also be the critical cone. Accordingly, the index set γ will necessarily be empty, and our positive definiteness condition applies.

We have thus demonstrated that a particular property of bisymmetric matrices leads to a condition which, when satisfied, greatly simplifies the verification of stability conditions for variational inequalities derived from saddle functions. We have shown how to apply this property in connection with recent advances in stability analysis for variational problems, and we have demonstrated that in special cases it leads to simple criteria such as strict complementary slackness. Further, in a very common case derived from nonlinear programming, the property applies in all cases, with no additional requirements whatever.

References

[1] R. W. Cottle, Symmetric dual quadratic programs, *Quarterly of Applied Mathematics* **21** (1963) 237–243.

[2] R. W. Cottle, Manifestations of the Schur complement, *Linear Algebra and Its Applications* **8** (1974) 189–211.

[3] R. W. Cottle, J.–S. Pang, and R. E. Stone, *The Linear Complementarity Problem*, (Academic Press Inc., San Diego, CA, 1992).

[4] G. B. Dantzig, E. Eisenberg, and R. W. Cottle, Symmetric dual nonlinear programs, *Pacific Journal of Mathematics* **15** (1965) 809–812.

[5] A. L. Dontchev and R. T. Rockafellar, Characterizations of strong regularity for variational inequalities over polyhedral convex sets, Preprint, August 1995.

[6] S. M. Robinson, Local structure of feasible sets in nonlinear programming, Part I: Regularity, in: V. Pereyra and A. Reinoza, eds., *Numerical Methods*, Lecture Notes in Mathematics No. 1005 (Springer-Verlag, Berlin, 1983) 240–251.

[7] S. M. Robinson, Local structure of feasible sets in nonlinear programming, Part II: Nondegeneracy, *Mathematical Programming Study* **22** (1984) 217–230.

[8] S. M. Robinson, Local structure of feasible sets in nonlinear programming, Part III: Stability and sensitivity, *Mathematical Programming Study* **30** (1987) 45–66.

[9] S. M. Robinson, Nonsingularity and symmetry for linear normal maps, *Mathematical Programming* **62** (1993) 415–425.

[10] S. M. Robinson, Sensitivity analysis of variational inequalities by normal-map techniques, in: F. Giannessi and A. Maugeri, eds., *Variational Inequalities and Network Equilibrium Problems* (Plenum Press, New York, 1995) 257–269.

[11] R. T. Rockafellar, *Convex Analysis* (Princeton University Press, Princeton, NJ, 1970).

[12] R. T. Rockafellar, Lagrange multipliers and optimality, *SIAM Review* **35** (1993) 183-238.

[13] R. T. Rockafellar, Nonsmooth optimization, in: J. R. Birge and K. G. Murty, eds., *Mathematical Programming: State of the Art 1994* (University of Michigan, Ann Arbor, MI, 1994) 248–258.

Large-Scale Global Optimization on Transputer Networks

Stefan Schäffler

Institut für Angewandte Mathematik und Statistik
Technische Universität München
Arcisstr. 21, 80290 München, Germany

Abstract

This paper presents a stochastic method for solving large-scale global optimization problems. An efficient algorithm is developed and investigated. Since this algorithm is very suitable for parallel implementation, an implementation of the method in a parallel version of C on a transputer network is briefly described. Numerical results for optimization problems in up to 60,000 variables are considered.

Keywords

global optimization, stochastic differential equations, parallel computing, transputer networks, simulated annealing.

1 Introduction

A large number of questions in science, technology, and economics leads to the following global optimization problem (GOP): Compute a point in $\mathbb{R}^n$ at which a given twice continuously differentiable objective function takes its global minimum. In nonlinear optimization one tries to compute a minimizer of f iteratively starting at a chosen starting point. For that, a search direction and a stepsize are determined in each iteration. Since the search directions are descent directions, only local minimizers are computed in general. In order to be able to compute global minimizers for a wide class of objective functions, two approaches are proposed in the literature quite often (an overview is given in [6], [7], [18]): A number of starting points is computed randomly (with Monte Carlo methods). Each starting point is used for a local minimization. A great disadvantage of this approach is obvious: A very large number of starting points is necessary for problems in higher dimensions ($n \geq 10$) in order to obtain a realistic possibility for the computation of a global minimizer. The second approach consists in the transformation of the global optimization problem to a sequence of local minimization problems using special auxiliary objective functions. If minimizers of f are computed, one tries to generate – based on penalty functions – a new objective function which has approximately the same minimizers as f except those computed so far. The aim is to compute all minimizers of f by the solution of a sequence of local minimization problems. This approach is useful only for a very restricted class of objective functions and the iterative application of penalty functions may lead to numerical difficulties.

The idea of our approach goes back to 1953, when Metropolis et al. proposed an algorithm for the simulation of the evolution of a solid to thermal equilibrium(cf. [13]). In 1970, Pincus noticed the analogy of statistical mechanics and optimization in a paper on optimization of continuous functions (cf. [14]). This paper and the paper of Khachaturyan et al. (cf. [11]) on the same subject have remained overlooked. In 1982, Kirkpatrick et al. (cf. [12]) and, independently, Cerny (cf. [4]) applied the analogy of minimizing the cost function of a combinatorial optimization problem and the slow cooling of a solid until it reaches its low energy ground state using the paper of Metropolis et al. and substituting cost for energy. They obtained a combinatorial optimization algorithm, which they called "simulated annealing". Details about simulated annealing including historical background are given in [1] and [19].

Since 1985 one tries to use the idea of simulated annealing for unconstrained global optimization problems with twice continuously differentiable objective functions (e.g. in [2], [9], and [5]). Unfortunately, the whole concept of simulated annealing is taken over ignoring the special properties of (GOP) such as the existence of efficient local minimizing procedures for example. As a result, cooling strategies, which firstly restrict the class of objective functions and secondly lead to unsatisfactory results, are used for the treatment of (GOP).

In this paper, we introduce and analyse a stochastic method for solving (GOP), which is based on the numerical integration of a specific stochastic differential equation. This stochastic differential equation combines local minimization using the curve of steepest descent and random search by the choice of a special parameter. Hence, good theoretical and practical properties can be proven for a wide class of objective functions independent of the dimension of the problem. The necessary calculus for our theoretical considerations is summarized in [10]. Since the numerical analysis of our approach shows that a lot of computations can be done simultaneously at different levels, we decided on a parallel implementation of our developed algorithm using a transputer network.

2 Probability Theory

Let Ω be the set of all continuous functions $v: [0,\infty[\to \mathbb{R}^n$, $n \in \mathbb{N}$. We define a metric d on Ω by:

$$d(\omega_1,\omega_2) := \sum_{m=1}^{\infty} \frac{1}{2^m} \min\left(\max_{0\leq t\leq m} \|\omega_1(t)-\omega_2(t)\|_2; 1\right),$$

where $\|\cdot\|_2$ denotes the Euclidean norm. By $\mathcal{B}(\Omega)$ we mean the smallest σ-field containing all open sets of Ω in the topology defined by the metric d. The open sets in $\mathbb{R}^p$, $p \in \mathbb{N}$, are given by the Euclidean norm. Throughout this paper, we use only the probability spaces $(\Omega, \mathcal{B}(\Omega), \mathcal{W})$ where $\mathcal{W}$ denotes a n-dimensional Wiener measure, which we consider later on. Let $\bar{\mathbb{R}} := \mathbb{R} \cup \{\pm\infty\}$ be the compactification of $\mathbb{R}$ with $0\cdot(\infty) = 0\cdot(-\infty) = (\infty)\cdot 0 = (-\infty)\cdot 0 := 0$ and let $\mathcal{B}(\bar{\mathbb{R}})$ be the Borel σ-field of $\bar{\mathbb{R}}$ given by:

$$B \in \mathcal{B}(\bar{\mathbb{R}}) \iff (B \cap \mathbb{R}) \in \mathcal{B}(\mathbb{R}).$$

Definition 2.1: *A function $g : \Omega \to \bar{\mathbb{R}}$ is called a numerical function.*

Definition 2.2: *A collection $\{X_t\}$ of n-dimensional real random variables X_t, $t \geq 0$, defined on Ω is called a stochastic process.*

Definition 2.3: *Let $\{X_t\}$ be a stochastic process. For each fixed $\omega \in \Omega$ the function $X_\omega : [0,\infty[\to \mathbb{R}^n, t \mapsto X_t(\omega)$ is called a path of $\{X_t\}$.*

Definition 2.4: *Let $\{X_t\}$ be a stochastic process. $\{X_t\}$ is called continuous, if each path of $\{X_t\}$ is a continuous function.*

For each $n \in \mathbb{N}$ our Wiener measure $\mathcal{W}$ is uniquely determined by the following conditions using the stochastic process $\{B_t\}$ defined by $B_t(\omega) := \omega(t)$ for all $t \geq 0$ (a proof is given in [3]):

(i) $B_0(\omega) = 0$ $\mathcal{W}$-almost surely.

(ii) For every $0 = t_0 < t_1 < \ldots < t_m$, $m \in \mathbb{N}$, the random variables $B_0, B_{t_1} - B_{t_0}, \ldots, B_{t_m} - B_{t_{m-1}}$ are stochastically independent.

(iii) For every $0 \leq s < t$ the random variable $B_t - B_s$ is $N(0,(t-s)I_n)$ Gaussian distributed, where I_n denotes the n-dimensional identity matrix.

Now, we are able to define the Brownian motion process.

Definition 2.5: *The stochastic process $\{B_t\}$ defined by $B_t(\omega) := \omega(t)$ is called n-dimensional Brownian motion.*

The following definition is very important for the investigation of some properties of stochastic processes.

Definition 2.6: *A $\mathcal{B}(\Omega)$-$\mathcal{B}(\bar{\mathbb{R}})$-measurable numerical function g is called a random time.*

An important class of random times is given by the following theorem, which we restate without proof (see, e.g. [15]).

Theorem 2.7: *Let $\{X_t\}$ be a continuous stochastic process, then the function $g: \Omega \to \bar{\mathbb{R}}$,*

$$\omega \mapsto \begin{cases} \inf\{t \geq 0 \mid X_\omega(t) \in A\} & \textit{if } \bigcup_{t\geq 0}(X_\omega(t) \cap A) \neq \emptyset \\ \infty & \textit{else} \end{cases}$$

is a random time for each open or closed set $A \in \mathcal{B}(\mathbb{R}^n)$, $n \in \mathbb{N}$.

The function value $g(\omega)$ of g at ω indicates the shortest time at which the path X_ω of $\{X_t\}$ hits the Borel set A. The following result, which we restate without proof (see, e.g. [3]), is used in the proof of a theorem formulated in Section 3.

Theorem 2.8: *Let $\{B_t\}$ be a one-dimensional Brownian motion, then*

$$\liminf_{\substack{t\to 0\\ t>0}} \frac{B_t}{\sqrt{2t\ln(\ln(1/t))}} = -1 \quad \mathcal{W}\text{-almost surely,}$$

$$\limsup_{\substack{t\to 0\\ t>0}} \frac{B_t}{\sqrt{2t\ln(\ln(1/t))}} = 1 \quad \mathcal{W}\text{-almost surely.}$$

3 Large-Scale Global Optimization

Now, we consider a stochastic method for the solution of the unconstrained problem:

(GOP) Compute $x^* \in \mathbb{R}^n$ with $f(x^*) \leq f(x)$ for all $x \in \mathbb{R}^n$, where $f: \mathbb{R}^n \to \mathbb{R}$, $f \in C^2$ is a given objective function.

The following results are restated from [17] without proofs. The following assumption concerning the objective function is fundamental for further investigations.

Assumption A: There exists an $\varepsilon > 0$ such that

$$x^\top \nabla f(x) \geq \frac{1 + n\varepsilon^2}{2} \max(1, \|\nabla f(x)\|_2)$$

for all $x \in \mathbb{R}^n \backslash \{x \in \mathbb{R}^n \mid \|x\|_2 \leq r\}$ with some $r > 0$, where $\nabla f(x)$ denotes the gradient of f at x.

Assumption A describes a special behaviour of f outside a ball with radius r for which only the existence is postulated. Starting at the origin, the objective function f must increase sufficiently fast along each direction $s \in \mathbb{R}^n$, $\|s\|_2 = 1$ outside the mentioned ball. It is obvious that each function for which Assumption A is fulfilled has a global minimizer. The opposite is not true as the function $\sin(x)$ shows.

For the solution of (GOP) we investigate the following class of stochastic Itô-differential equations (SDE), where $\{B_t\}$ is a n-dimensional Brownian motion, $\omega \in \Omega$, and f is the given objective function.

(SDE) $$dX_t = -\nabla f(X_t)\, dt + \varepsilon\, dB_t, \quad X_0 = x_0,$$

with $\varepsilon > 0$, $x_0 \in \mathbb{R}^n$. We study existence, uniqueness, and regularity of the solution of (SDE).

Theorem 3.1: *Consider (SDE). For all $x_0 \in \mathbb{R}^n$ and for all ε, for which Assumption A is fulfilled, we obtain the following:*

(i) *there exists a unique stochastic process $\{X_t^{\varepsilon,x_0}\}$ that solves (SDE);*

(ii) *all paths of $\{X_t^{\varepsilon,x_0}\}$ are continuous;*

(iii) $X_0^{\varepsilon,x_0} \equiv x_0$.

Let $\bar{x} \in \mathbb{R}^n$ and consider $S(\bar{x},\rho) := \{x \in \mathbb{R}^n \mid \|x-\bar{x}\|_2 \leq \rho\}, \rho > 0$. For the investigation of the relations between solutions of (SDE) and (GOP) we need the random times $s_{\bar{x},\rho} : \Omega \to \mathbb{R}^n$,

$$\omega \mapsto \begin{cases} \inf\{t \geq 0 \mid \|X_t^{\varepsilon,x_0}(\omega) - \bar{x}\|_2 \leq \rho\} \\ \qquad\qquad \text{if } \{t \geq 0 \mid \|X_t^{\varepsilon,x_0}(\omega) - \bar{x}\|_2 \leq \rho\} \neq \emptyset, \\ \infty \qquad\qquad \text{else.} \end{cases}$$

Theorem 3.1 allows us to apply some important results from [10] that we formulate in Theorem 3.2. The proof of this theorem leads to Lyapunov's stability calculus for stochastic differential equations and to the theory of parabolic partial differential equations.

Theorem 3.2: *Consider (SDE) with $\varepsilon > 0$ such that Assumption A is fulfilled. Then we obtain the following for each $x_0, \bar{x} \in \mathbb{R}^n$, $\rho > 0$:*

(i) $\mathcal{W}(\{\omega \in \Omega \mid s_{\bar{x},\rho}(\omega) < \infty\}) = 1$.

(ii) $E(s_{\bar{x},\rho}) < \infty$ ($E(\cdot)$ denotes expectation).

(iii) Each random variable X_t^{ε,x_0}, $t > 0$, has a density $p_{\varepsilon,x_0,t}$ with

$$\lim_{t\to\infty} p_{\varepsilon,x_0,t}(x) = p_\varepsilon(x),$$

where p_ε is the unique density function satisfying

$$\frac{1}{2}\varepsilon^2 \sum_{i=1}^{n} \frac{\partial^2 p_\varepsilon(x)}{\partial x_i^2} + \sum_{i=1}^{n} \frac{\partial}{\partial x_i}\left(\frac{\partial f(x)}{\partial x_i} p_\varepsilon(x)\right) \equiv 0.$$

It is important to recognize that the limit p_ε is independent of the starting point x_0. Now, we are interested in the density p_ε and its relationship to (GOP). For the explicit formulation of p_ε we need the following theorem.

Theorem 3.3: *Let f_{gl} be the global minimum of f, then*

$$C_\varepsilon := \int_{\mathbb{R}^n} \exp\left(-\frac{2(f(x) - f_{gl})}{\varepsilon^2}\right) dx < \infty$$

for all ε satisfying the Assumption A.

With Theorem 3.3 we are able to give p_ε explicitly:

$$p_\varepsilon(x) = \frac{1}{C_\varepsilon} \exp\left(-\frac{2(f(x) - f_{gl})}{\varepsilon^2}\right)$$

Each global minimizer of f is a global maximizer of p_ε. This is the most important relationship between the solution of (SDE), where the densities of X_t^{ε,x_0} converge pointwise to p_ε, and (GOP). From Theorem 3.2 we know that for each local minimizer $\bar{x}$ $\mathcal{W}$-almost all paths of $\{X_t^{\varepsilon,x_0}\}$ hit any ball $S(\bar{x}, \xi)$ centered at $\bar{x}$ for arbitrary chosen $\xi > 0$ after a finite time for all $x_0 \in \mathbb{R}^n$. Moreover, the expectation of $s_{\bar{x},\rho}$ is finite.

Now, we investigate the question of how to choose ε. For that, we consider (SDE):

$$dX_t = -\nabla f(X_t)\, dt + \varepsilon\, dB_t, \quad X_0 = x_0.$$

The parameter ε is a measure for the balance between local minimization along the curve of steepest descent

$$X_t^{x_0} = x_0 - \int_0^t \nabla f(X_\tau^{x_0})\, d\tau,$$

and a random search with realizations of Gaussian distributed random vectors with increasing variance

$$X_t^{x_0}(\omega) = x_0 + B_t(\omega) - B_0(\omega).$$

If we choose ε for a fixed ω such that local minimization dominates, then the chosen path of $\{X_t^{\varepsilon,x_0}\}$ spends a very long time close to (local) minimizers of f. If we choose ε such that random search dominates, then the minimizers of f play no significant role along this path of $\{X_t^{\varepsilon,x_0}\}$.

The optimal balance between random search and local minimization and therefore, the choice of ε depends on the objective function f and the scale used. If one observes, during the numerical computations of a path of $\{X_t^{\varepsilon,x_0}\}$ that this path spends a very long time close to local minimizers of f, then ε is too small. If the minimizers of f play no significant role along this path, then ε is too large.

Now, we discuss the use of cooling strategies (functions $\varepsilon : \mathbb{R} \to \mathbb{R}$). Consider the density

$$p_\varepsilon(x) = \frac{1}{C_\varepsilon} \exp\left(-\frac{2(f(x) - f_{gl})}{\varepsilon^2}\right).$$

For decreasing $\varepsilon > 0$ the probability measure given by p_ε is more concentrated on the global minimizers of f. This is the idea of cooling strategies: Find a strictly monoton decreasing function $\varepsilon : \mathbb{R} \to \mathbb{R}$ that converges to 0, for which the probability measures given by p_ε are more and more concentrated on the global minimizer of f. Theoretical investigations (see [8]) show that under special assumptions on the objective function f (comparable to our Assumption A) cooling strategies must be of the following type:

(i) $\varepsilon(0) \geq C$, where C is an unknown constant,

(ii) $\lim_{t\to\infty} \varepsilon(t) = 0$,

(iii) $\lim_{t\to\infty} \varepsilon^2(t)\ln(t) > 0$.

If $\varepsilon(0) < C$ or $\lim_{t\to\infty} \varepsilon^2(t)\ln(t) = 0$, the probability measure given by the limit distribution is only concentrated on local minimizers. The convergence to the limit distribution is very slow, as (ii) shows. Therefore, it is inefficient to implement cooling strategies.

For the numerical computation of a path of $\{X_t^{\varepsilon,x_0}\}$ we consider the following iteration scheme, which results from a standard approach in the numerical analysis of ordinary differential equations (semi-implicit Euler method with incomplete Jacobi matrix). For a fixed stepsize σ let $H(x)$ be the Hessian of f at x and set

$$
\begin{aligned}
x_{j+1}^1 &:= x_j \\
&\quad -[I_n + \sigma \operatorname{diag} H(x_j)]^{-1}[\sigma \nabla f(x_j) - \varepsilon n_3 (\sigma/2)^{1/2}], \quad (1) \\
x(\sigma/2) &:= x_j - [I_n + (\sigma/2) \operatorname{diag} H(x_j)]^{-1} \\
&\quad \cdot [(\sigma/2)\nabla f(x_j) - \varepsilon n_1 (\sigma/2)^{1/2}], \quad (2) \\
x_{j+1}^2 &:= x(\sigma/2) - [I_n + (\sigma/2) \operatorname{diag} H(x(\sigma/2))]^{-1} \\
&\quad \cdot [(\sigma/2)\nabla f(x(\sigma/2)) - \varepsilon n_2 (\sigma/2)^{1/2}], \quad (3)
\end{aligned}
$$

where n_1 and n_2 are realizations of independent $N(0, I_n)$ normally distributed random vectors, which are computed by pseudo-random numbers, and $n_3 = n_1 + n_2$.

For a fixed positive δ we take $\bar{x}_{j+1} = x_{j+1}^2$ if $\|x_{j+1}^1 - x_{j+1}^2\|_2 \leq \delta$; otherwise, steps (1) and (3) have to be repeated with $\sigma/2$ instead of σ. If one of the matrices $I_n + \sigma \operatorname{diag} H(x_j)$ or $I_n + (\sigma/2) \operatorname{diag} H(x(\sigma/2))$ is not positive definite, $\sigma/2$ has to be used instead of σ. After the choice of a maximal number J of iterations we accept the point $\bar{x}_j \in \{x_0, \bar{x}_1 \ldots, \bar{x}_J\}$ with the smallest function value f($\bar{x}_j$) as an approximate optimal solution, and this point is chosen as a starting point for local minimization.

The use of the diagonal matrix $\operatorname{diag} H(x_j)$ instead of $H(x_j)$ permits the application of this method to large-scale global optimization problems.

For the computation of the integration error of the above numerical scheme we denote the solution $\{X_{t+\sigma}^{\varepsilon,\xi}\}$ of the stochastic Itô-differential equation

$$dX_{t+\sigma} = -\nabla f(X_{t+\sigma})\, d\sigma + \varepsilon\, dB_\sigma, \quad X_t = \xi$$

for a path $\omega \in \Omega$ by $X^\varepsilon(t+\sigma; \xi, \sigma, \omega)$. With regard to the numerical scheme considered above we have to compare $X^\varepsilon(t+\sigma; \xi, \sigma, \omega)$ with $\bar{X}^\varepsilon(t+\sigma; \xi, \sigma, \omega) := \xi - (I_n + \sigma \operatorname{diag} H(\xi))^{-1}(\sigma \nabla f(\xi) - \varepsilon(B_{t+\sigma}(\omega) - B_t(\omega)))$.

Theorem 3.4: *For each ε such that Assumption A is fulfilled and for each $t \geq 0$, $\xi \in \mathbb{R}^n$, we obtain the existence of $c > 0$ with*

$$\lim_{\sigma \to 0} \frac{\|\bar{X}^\varepsilon(t+\sigma; \xi, \sigma, \omega) - X^\varepsilon(t+\sigma; \xi, \sigma, \omega)\|_2}{\sqrt{\sigma^3 \ln(\ln(1/\sigma))}} \leq c \quad \mathcal{W}\text{-almost surely.}$$

Proof: From the law of iterated logarithm for Brownian motions we obtain for each $t \geq 0, \xi \in \mathbb{R}^n$, the existence of $c^* > 0$ with

$$\lim_{\sigma \to 0} \frac{\|X^\varepsilon(t+\sigma;\xi,\sigma,\omega) - \xi\|_2}{\sqrt{\sigma \ln(\ln(1/\sigma))}} \leq c^* \quad \mathcal{W}\text{-almost surely, because}$$

$$\lim_{\sigma \to 0} \frac{\|X^\varepsilon(t+\sigma;\xi,\sigma,\omega) - \xi\|_2}{\sqrt{\sigma \ln(\ln(1/\sigma))}} = \lim_{\sigma \to 0} \frac{\|\varepsilon(B_{t+\sigma}(\omega) - B_t(\omega))\|_2}{\sqrt{\sigma \ln(\ln(1/\sigma))}} \quad \text{and}$$

$\{(B_{t+\sigma} - B_t)_i\}$ is a one-dimensional Brownian motion in $\sigma \geq t$ for all $1 \leq i \leq n$. The Taylor expansion of ∇f yields:

$$\begin{aligned} X^\varepsilon(t+\sigma;\xi,\sigma,\omega) &= \xi + \varepsilon(B_{t+\sigma}(\omega) - B_t(\omega)) - \sigma \nabla f(\xi) \\ &\quad - \int_t^{t+\sigma} \int_0^1 \nabla^2 f(\xi + \theta(X^\varepsilon(\tau;\xi,\sigma,\omega) \\ &\quad - \xi))\, d\theta (X^\varepsilon(\tau;\xi,\sigma,\omega) - \xi)\, d\tau. \end{aligned}$$

Let $\operatorname{diag}(\lambda_1, \ldots, \lambda_n) := \operatorname{diag} H(\xi)$, then we obtain

$$\begin{aligned} \bar{X}^\varepsilon(t+\sigma;\xi,\sigma,\omega) &= \xi - (\operatorname{diag}(1+\sigma\lambda_1, \ldots, 1+\sigma\lambda_n))^{-1} \\ &\quad \cdot(\sigma \nabla f(\xi) - \varepsilon(B_{t+\sigma}(\omega) - B_t(\omega))) \\ &= \xi - \left(\operatorname{diag}\left(\frac{1}{1+\sigma\lambda_1}, \ldots, \frac{1}{1+\sigma\lambda_n}\right)\right) \\ &\quad \cdot(\sigma \nabla f(\xi) - \varepsilon(B_{t+\sigma}(\omega) - B_t(\omega))) \\ &= \xi - \left(I_n - \sigma \operatorname{diag}\left(\frac{\lambda_1}{1+\sigma\lambda_1}, \ldots, \frac{\lambda_n}{1+\sigma\lambda_n}\right)\right) \\ &\quad \cdot(\sigma \nabla f(\xi) - \varepsilon(B_{t+\sigma}(\omega) - B_t(\omega))) \\ &= \xi + \sigma^2 \operatorname{diag}\left(\frac{\lambda_1}{1+\sigma\lambda_1}, \ldots, \frac{\lambda_n}{1+\sigma\lambda_n}\right) \nabla f(\xi) \\ &\quad + \varepsilon(B_{t+\sigma}(\omega) - B_t(\omega)) - \sigma \nabla f(\xi) \\ &\quad - \sigma \operatorname{diag}\left(\frac{\lambda_1}{1+\sigma\lambda_1}, \ldots, \frac{\lambda_n}{1+\sigma\lambda_n}\right) \\ &\quad \cdot \varepsilon(B_{t+\sigma}(\omega) - B_t(\omega)). \end{aligned}$$

Hence,

$$\begin{aligned}
X^\varepsilon(t+\sigma;\xi,\sigma,\omega) \quad - \quad & \bar{X}^\varepsilon(t+\sigma;\xi,\sigma,\omega) = \\
= \quad & - \int_t^{t+\sigma} \int_0^1 \nabla^2 f(\xi + \theta(X^\varepsilon(\tau;\xi,\sigma,\omega) \\
& -\xi))\, d\theta (X^\varepsilon(\tau;\xi,\sigma,\omega) - \xi)\, d\tau \\
& -\sigma^2 \operatorname{diag}\left(\frac{\lambda_1}{1+\sigma\lambda_1},\ldots,\frac{\lambda_n}{1+\sigma\lambda_n}\right) \nabla f(\xi) \\
& +\sigma \operatorname{diag}\left(\frac{\lambda_1}{1+\sigma\lambda_1},\ldots,\frac{\lambda_n}{1+\sigma\lambda_n}\right) \\
& \cdot\varepsilon(B_{t+\sigma}(\omega) - B_t(\omega)).
\end{aligned}$$

Since there exist $c_1 > 0$ and $c_2 > 0$ with

$$\lim_{\sigma\to 0} \frac{\sigma\|\varepsilon(B_{t+\sigma}(\omega) - B_t(\omega))\|_2}{\sqrt{\sigma^3 \ln(\ln(1/\sigma))}} \le c_1 \quad \mathcal{W}\text{-almost surely, and}$$

$$\lim_{\sigma\to 0} \frac{\|\int_t^{t+\sigma} (X^\varepsilon(\tau;\xi,\sigma,\omega) - \xi)\, d\tau\|_2}{\sqrt{\sigma^3 \ln(\ln(1/\sigma))}} \le c_2 \quad \mathcal{W}\text{-almost surely,}$$

we obtain the existence of $c > 0$ with

$$\lim_{\sigma\to 0} \frac{\|X^\varepsilon(t+\sigma;\xi,\sigma,\omega) - \bar{X}^\varepsilon(t+\sigma;\xi,\sigma,\omega)\|_2}{\sqrt{\sigma^3 \ln(\ln(1/\sigma))}} \le c \quad \mathcal{W}\text{-almost surely.} \quad \square$$

4 Implementation on Transputer Networks

In this section we discuss the application of transputer networks for the implementation of the mentioned procedure in a parallel version of C (see [16]).

The transputer T805-25 from INMOS is a 32-bit microprocessor with a 64-bit floating point unit (FPU), graphics support, and 4 Kbytes on-chip RAM. This processor is able to perform 1.875 Mflops at a processor speed of 25MHz. A transputer has a configurable memory interface and four bidirectional communication links, which allow networks of transputers constructed by direct point-to-point connections without external logic. Each link can transfer data at rates up to 2.35 Mbytes/sec. The processor has access to a maximum of 4Gbytes of memory via the external memory interface. In a transputer, the

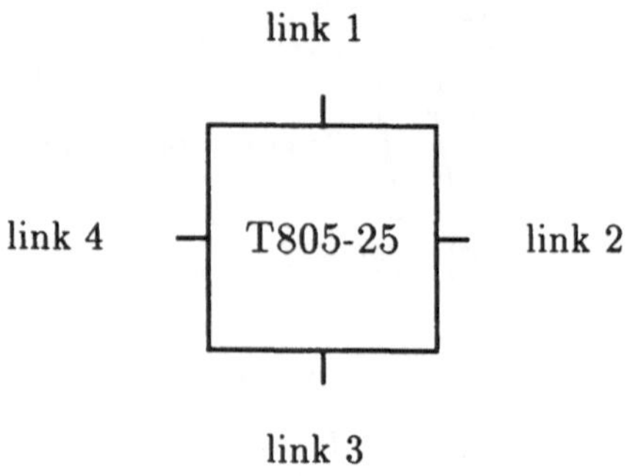

Figure 1: Transputer T805-25

FPU, the central processing unit (CPU), and the data transfer work physically in parallel.

For the transputer architecture just described it is advantageous to interpret the software as a set of processes that can be performed in serial or in parallel. After starting execution, a process performs a number of actions and then either terminates or pauses. Each action may be an assignment, an input, or an output. The communication between two processes is achieved by means of channels, where a channel between processes executing on the same transputer is implemented by a single word in the memory and a channel between different transputers is implemented by point-to-point links.

Transputers can be programmed in most high level languages. Using a parallel version of C it is possible to exploit the advantages of the transputer architecture. This version can be used to program an individual transputer or a network of transputers. When parallel C is used to program an individual transputer, the transputer shares its time between the concurrent processes and channel communication is implemented by moving data within the memory. Communication between processes on different transputers is implemented directly by transputer links. Since each transputer has four bidirectional links, one can conveniently use trinary trees as topologies for transputer networks.

Using the network of figure 2 we outline how one can parallelize the procedure given above. First of all, the number n of variables has to be divided into 13 parts (corresponding to the 13 transputers), where each part gets nearly the same number of variables. Thus, each

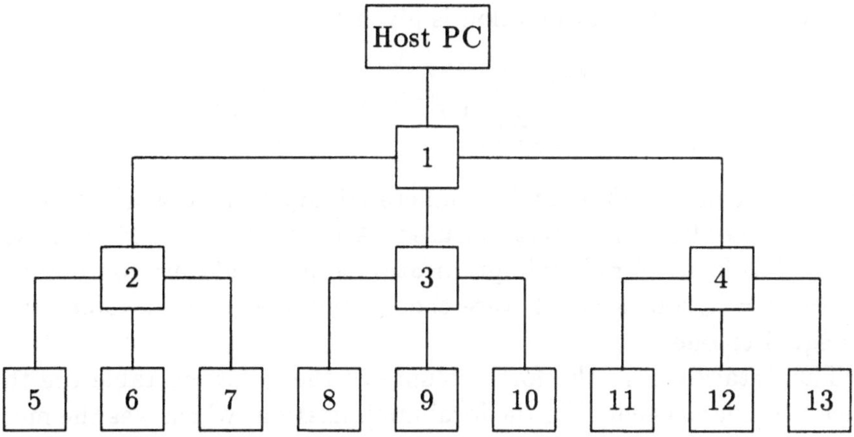

Figure 2: Trinary tree of depth 3 with 13 transputers

transputer is assigned a particular set of variables. Furthermore, each transputer has the formulas for the corresponding components of the gradient and the diagonal elements of the Hessian matrix in its local memory. After sending the current iterate x_j to all transputers, each transputer computes the corresponding components of the gradient and the Hessian matrix and sends the data back to the first transputer. These computations are done in parallel. The first transputer checks the stopping condition, computes the new iterate, and sends the results to the Host PC.

5 Numerical Results

In this section we summarize the numerical results for the global minimization of the pure Lennard-Jones potential function $f : \mathbb{R}^{3N} \to \mathbb{R}$, for N atoms given in [16]. Besides the application of a special heuristic for the computation of a starting point for global optimization and besides a procedure for local minimization, the global optimization procedure proposed in this paper is used in [16]. Let $\eta_i \in \mathbb{R}^3$ denote the cartesian position of the i-th atom in the cluster and let $m_{i,j}(x) := \|\eta_i - \eta_j\|_2^2$ with $x := (\eta_1, \ldots, \eta_N)^\top \in \mathbb{R}^{3N}$, then the

Lennard-Jones potential function is given by

$$f(x) = \sum_{i=1}^{N-1} \sum_{j=i+1}^{N} \left(m_{i,j}^{-6}(x) - 2m_{i,j}^{-3}(x) \right).$$

For all the computations an incomplete trinary tree network of depth 5 consisting of 95 transputers was used. A PC with an 80486 processor was used as host. For global optimization a maximal iteration number $J = 50, \varepsilon = 2$, and $\delta = 0.1$ were chosen and the starting value for σ was equal to one.

The data given in the forth column of the following table are the times in seconds needed for the local minimization, which was the most time consuming part of the method.

Size N	Transputers	L-J potential	Time
1000	95	-7.11780e+03	939
2000	95	-1.48269e+04	2135
3000	95	-2.26451e+04	4876
4000	95	-3.05479e+04	8000
5000	95	-3.85245e+04	11100
6000	95	-4.64918e+04	22109
7000	95	-5.45212e+04	26866
8000	95	-6.25840e+04	39409
9000	95	-7.06315e+04	48733
10000	95	-7.87336e+04	63324
20000	95	-1.60152e+05	304190

References

[1] Aarts, E.H.L., Korst, J., *Simulated Annealing and Boltzmann Machines.* Chichester, New York 1990.

[2] Aluffi-Pentini, F., Parisi, V., Zirilli, F., *Global Optimization and Stochastic Differential Equations.* JOTA 47, 1985, 1–16.

[3] Bauer, H., *Wahrscheinlichkeitstheorie.* Berlin, New York, 1991[4].

[4] Cerny, V., *Thermodynamic Approach to the Traveling Salesman Problem: An Efficient Simulation Algorithm.* JOTA 45, 1985, 41–51.

[5] Chiang, T., Hwang, C., Sheu, S., *Diffusion for Global Optimization in* $\mathbb{R}^n$. SIAM J. Control and Optimization 25, 1987, 737–753.

[6] Dixon, L.C.W., Szegö, G.P., *Towards Global Optimization.* Amsterdam, 1975.

[7] Dixon, L.C.W., Szegö, G.P., *Towards Global Optimization 2.* Amsterdam, 1978.

[8] Gelfand, S.B., Mitter, S.K., *Metropolis-type annealing algorithms for global optimization in* $\mathbb{R}^n$. SIAM J., Control Optim., 31, 1993, 111–131.

[9] Geman, S., Hwang, C., *Diffusions for Global Optimization.* SIAM J. Control and Optimization 24, 1986, 1031–1043.

[10] Hasminskij, R.Z., Stochastic Stability of Differential Equations. Amsterdam, 1980.

[11] Khachaturyan, A., Semenovsskaja, S., Vainshtain, B., *The Thermodynamical Approach to the Structure Analysis of Crystals.* Acta. Cryst. A37, 1981, 174–754.

[12] Kirkpatrick, S., Gelatt, C.D., Vecchi, M.P., *Optimization by simulated annealing.* Science 220, 1983, 671–680.

[13] Metropolis, N., Rosenbluth, A.W., Rosenbluth, M.N., Teller, A.H., Teller, E., *Equation of state calculation by fast computing machines.* J. Chem. Phys. 21, 1953, 1087–1092.

[14] Pincus, M., *A Monte Carlo Method for the Approximate Solution of Certain Types of Constrained Optimization Problems.* Oper. Res. 18, 1970, 1225–1228.

[15] Protter, P., *Stochastic Integration and Differential Equations.* Berlin, New York, 1990.

[16] Ritter, K., Robinson, S.M., Schäffler, S., *Global Minimization of Lennard-Jones Functions on Transputer Networks.* submitted.

[17] Schäffler, S., *Unconstrained Global Optimization Using Stochastic Integral Equations.* Optimization 35, 1995, 43–60.

[18] Törn, A., Zilinskas, A., *Global Optimization.* Berlin, Heidelberg, New York, 1990.

[19] van Laarhoven, P.J.M., Aarts, E.H.L., *Simulated Annealing: Theory and Applications.* Dordrecht, Boston, 1987.

The Statistical Art of Maximizing the Likelihood

Walter Schlee
Technische Universität München
Institut für Angewandte Mathematik und Statistik

Abstract

The maximum likelihood method is the most popular method in the statistical community for obtaining reasonable estimators for the unknown parameters of a family of distributions. In simple cases the maximum likelihood estimators may be given explicitly, in most cases they can be computed only numerically. This article reviews only methods, which use the special structure of the maximum likelihood function coming from the underlying statistical model, and which are formulated in statistical terms.

Keywords: maximum likelihood; globally convergent algorithms; missing information samples.

1 The Likelihood Function

Maximizing the likelihood function is the most popular method in the statistical community for obtaining estimators for the unknown parameters of a family of distributions (see for instance [9]). The maximum of the likelihood function is also needed for the likelihood ratio test, a general method of testing hypotheses. The term Maximum Likelihood was used for the first time (see e.g. [2]) in 1922 by R.A.Fisher on page 323 in his article "On the mathematical foundation of theoretical statistics", Philosophical Transactions of the Royal Society of London, Ser. A Vol. 222(1922), 309-368. The term Likelihood Ratio was used later (see e.g. [2]) in 1931 by Neyman, J. and Pearson E.S. on page 480 in their article "On the problem of k-samples",Bulletin de l'Académie Polonaise des Sciences et Lettres, Ser. A,460-481. The popularity stems from the fact, that the intuitively constructed estimators for various simple cases can be regained by the maximum likelihood method. Under appropriate assumptions the maximum likelihood estimators (MLEs) have all the properties desireable for estimators.
In simple cases the MLE may be explicitly given, in most cases it can only be determined numerically. Forgetting the origin, the likelihood function (LF) is simply a function where all the elaborate optimization methods can be applied. In this article we would like to give a review on methods, which use

the special structure of the LF coming from the underlying statistical model, and which are formulated in statistical terms.

2 Consistency results for Maximum Likelihood Estimators

Let $X = (X_1, \ldots, X_n)$ be a random vector of observations with the joint distribution $f_n(x|\theta)$. The parameter space is assumed to be $\theta \in \Theta \subset R^s$. For given x the function $L(\theta|x) = f_n(x|\theta) : \Theta \to R$ is called the Likelihood function (LF).
Any $\hat{\theta} = \hat{\theta}(x)$ which maximizes $L(\theta|x)$ is called a maximum likelihood estimate (MLE) for the unknown true parameter θ_0.
Assuming differentiability of the LF we may consider the likelihood equation (or system of equations) (LEQ):

$$\frac{\partial log(L(\theta|x))}{\partial\theta} = 0 \ .$$

Under a lengthy list of assumptions (see e.g. [11]) a result of Cramér asserts that one of the solutions $\hat{\theta}_n(x)$ is consistent for the unknown θ_0.
However the theorem gives no constructive way how to find this consistent solution in the set of solutions. There is also a counterexample by Kraft and LeCam [6] which shows that the MLE does not necessarily coincide with the consistent root of the LEQ. Clearly in the case of a unique solution of the LEQ, the theorem of Cramér shows that this solution must be consistent and must be the MLE. In the case of a regular exponential familiy of probability distributions the logarithm of the likelihood function is strictly concave and therefore the LEQ has at most one solution (see e.g [1]).

3 Exponential Families of Probability Distributions

We summarize some results for this family of distribution functions.
The probability distributions of an exponential family with parameter $\zeta \in \Upsilon$ have probability densities of the following type with respect to some σ-finite measure μ:

$$g(x|\zeta) = a(\zeta)exp\{\theta(\zeta)^T t(x)\}b(x) \ .$$

This family of probability distributions is called s-parametric , if 1, $\theta_1(\zeta)$, ..., $\theta_s(\zeta)$ are linearly independent functions and 1, $t_1(x)$, ..., $t_s(x)$ are linearly independent in the complement of every set with μ-measure zero.
Besides this curved version there is also a normalized version, called regular representation of an exponential family with a density with respect to some

σ-finite measure ν:

$$g(x|\theta) = \frac{1}{c(\theta)} exp\{\theta^T t(x)\} .$$

The set $\Theta = \{\theta | 0 < \int exp\{\theta^T t(x)\} d\nu < \infty\}$ is called the natural parameter space which is convex and contains an s-dimensional interval with interior points. If the family of distribution functions is indexed with this set of parameter values, it is called full. The statistic $T(X)$ is sufficient for θ. All moments of the random vector $T(X)$ exist. If θ_0 is an interior point of the natural parameter space, the moments can be computed as

$$c(\theta_0)\mathbf{E}_{\theta_0}\left(\prod_{i=1}^{s}(T_i(X))^{m_i}\right) = \frac{\partial^{m_1+\ldots+m_s}}{\partial\theta_1^{m_1}\cdots\partial\theta_s^{m_s}} c(\theta)|_{\theta_0} \ .$$

If there is an observation t(x) of the sufficient statistic T(X) which lies in the interior of the convex support of the measure $\kappa = \nu \circ t^{-1}$ then the MLE exists and is unique and the LEQ may be written as

$$\mathbf{E}_\theta T(X) = t(x)$$

4 Full Information Samples

In this section we consider random experiments where all involved variables can be observed. This is in contrary to the next section, where we assume that some variables are not observable. We use also the terms "full information samples" and "missing information samples". For samples with full information the interesting results of Jensen, Johannson and Lauritzen [4] are applicable. They prove in the cited article the global convergence of an iterative procedure for the computation of MLE's in exponential families. Their most general result for exponential families is the following theorem on an iterative partial maximization procedure:

Theorem 1 *If the observation of the sufficient statistic from a regular s-dimensional exponential family is contained in the interior of its convex support, then the maximum likelihood estimate can be calculated by successively maximizing over each canonical parameter, keeping the others fixed.*

They prove this theorem as a special case of a more general one. Essential part of their results is the maximization in the case of one-dimensional exponential families.

They start with a modified Newton algorithm for solving the equation $h(x) = 0$ where $h :]a, b[\rightarrow R$.

$$g(x) = x - h(x)/h'(x) \quad x_{n+1} = \begin{cases} \frac{1}{2}(x_n + a) \text{ if } & g(x_n) \leq a \\ g(x_n) \text{ if } a < & g(x_n) < b \\ \frac{1}{2}(x_n + b) \text{ if } & g(x_n \geq b \end{cases}$$

Specializing this algorithm to the computation of the maximum likelihood estimator they found that the reciprocal likelihood function fulfills the convergence condition of this Newton algorithm.

Theorem 2 *Consider n independent and identically distributed observations from a regular one-dimensional exponential family. Assume that the value of the sufficient statistic $t(x)$ is contained in the interior of its convex support. Then the modified Newton algorithm, applied to the derivative of the nth root of the reciprocal likelihood function, converges to the maximum likelihood estimate for any starting value in the full parameter space.*

Using the definitions

$$\mu(\theta) = \mathbf{E}_\theta(T) \quad \sigma^2(\theta) = \mathbf{V}_\theta(T)$$

and the following representation of the nth root of the reciprocal likelihood function

$$\frac{1}{L(\theta|t)} = c(\theta)exp(-\theta\bar{t}) = \int exp\{\theta(t-\bar{t})\}d\kappa$$

where $\bar{t}$ is the average of the observed sufficient statistics.
The resulting modified Newton algorithm is described by the function

$$g(\theta) = \theta + \frac{\bar{T} - \mu(\theta)}{\sigma^2(\theta) + [\bar{T} - \mu(\theta)]^2} \,. \tag{1}$$

Examples:
(1) In the case of the one-dimensional exponential family with κ as the Borel-Lebesgue measure on R_+ the MLE can be computed explicitly. Nevertheless using the algorithm we see that it succeeds immediately if we choose as initial value $\theta = 1/\bar{t}$, and taking into account that

$$\mu(\theta) = \frac{1}{\theta} \qquad \sigma^2(\theta) = \frac{1}{\theta^2} \,.$$

(2) A more elaborate example is the two-dimensional exponential family of a gamma distribution with scale parameter $1/\lambda$ and shape parameter β. The density is

$$\begin{aligned} g(x|\lambda,\beta) &= \Gamma(\beta)^{-1}\lambda^\beta x^{\beta-1}\exp(-x\lambda) \\ &= (\Gamma(\beta)^{-1}\lambda^\beta)\exp(\beta\ln(x) - \lambda x)x \,. \end{aligned}$$

In the case of a sample $X_1, X_2, \ldots, X_n$ of size n we have the sufficient statistics

$$T(X) = \frac{1}{n}\sum_{i=1}^{n}\ln(X_i) \text{ for } \beta \text{ and } T_\lambda(X) = \frac{1}{n}\sum_{i=1}^{n}X_i = \bar{X} \text{ for } \lambda \,.$$

We need only $T(X)$ for the Newton iteration on β. Given β the MLE of λ can be computed explicitely

$$\hat{\lambda} = \frac{\hat{\beta}}{\bar{X}} .$$

Using $\psi(\beta) = \Gamma'(\beta)/\Gamma(\beta)$ then

$$\mathbf{E}(T(X)) = \psi(\beta) - \ln(\lambda) \quad \mathbf{V}(T(X)) = \psi'(\beta)$$

Inserting these quantities into formula(1) the Newton iteration for β is

$$\beta_{p+1} = \beta_p + \frac{T(x) + (\ln(\lambda) - \psi(\beta_p))}{\psi'(\beta_p) + (T(x) + \ln(\lambda) - \psi(\beta_p))^2} .$$

We assume now a sample with $T(x) = 0.209$ and $\bar{x} = 1.679$. Each line of the following table gives the iterates β_{pi} $i = 0, \ldots, 5$ for the specified λ_p in the first column. We choose $\lambda_p = \beta_{p0}/\bar{x}$ and $\beta_{p+1,0} = \beta_{p5}$. The starting value is $\beta_{00} = 3$. The following table shows the progress of the iteration.

p	λ_p	β_{p0}	β_{p1}	β_{p2}	β_{p3}	β_{p4}	β_{p5}
0	1.79	3.00	2.68	2.68	2.68	2.68	2.68
1	1.60	2.68	2.44	2.45	2.45	2.45	2.45
2	1.46	2.45	2.27	2.27	2.27	2.27	2.27
3	1.35	2.27	2.14	2.15	2.15	2.15	2.15
4	1.27	2.15	2.05	2.05	2.05	2.05	2.05
5	1.22	2.05	1.98	1.98	1.98	1.98	1.98
6	1.18	1.98	1.92	1.92	1.92	1.92	1.92
7	1.15	1.92	1.88	1.88	1.88	1.88	1.88
8	1.12	1.88	1.85	1.85	1.85	1.85	1.85
9	1.10	1.85	1.83	1.83	1.83	1.83	1.83
10	1.09	1.83	1.81	1.81	1.81	1.81	1.81
11	1.08	1.81	1.80	1.80	1.80	1.80	1.80
12	1.07	1.80	1.79	1.79	1.79	1.79	1.79
13	1.07	1.79	1.79	1.79	1.79	1.79	1.79

The iteration is getting stable up to two decimal digits with the iteration number 13 of λ. Therefore the result is

$$\hat{\lambda} = 1.07 \text{ and } \hat{\beta} = 1.79$$

A further theorem of Jensen, Johannson and Lauritzen [4] asserts, that we may take instead of the reciprocal likelihood function the likelihood function itself, if the sufficient statistic $T(X)$ has an infinitely divisible distribution.

Theorem 3 *Given a regular one-dimensional exponential family with the sufficient statistic having infinitely divisible distribution. If the observation of the sufficient statistic is contained in the interior of its convex support, then the modified Newton algorithm applied to the derivative of the log likelihood function converges to the maximum likelihood estimate for any starting value in the full parameter space.*

As example we may take here also the gamma distribution. Shanbhag and Sreehari [12] proved, that the logarithm of a gamma variable is infinitely divisible.
The Newton algorithm derived from the likelihood function itself would be

$$g(\theta) = \theta + \frac{\bar{T} - \mu(\theta)}{\sigma^2(\theta)} .$$

5 Missing Information Samples

We assume that the observed data vector y lies in the sample space $\mathcal{Y}$, and that the complete data vector is $x \in \mathcal{X}$. The complete data vector x is not observed due to particular circumstances, e.g. missing values, censoring or latent parameters of the statistical model. In this case the estimation of unknown parameters use a modified log-likelihood function. The modification is called the missing information principle, described for instance in Laird [7]. Also in this case we restrict our review to regular s-dimensional exponential families.
Let again θ denote the s-dimensional unknown parameter vector. To each observed data vector y corresponds a set $\mathcal{X}(y)$ of complete data vectors. Let $g(y|\theta)$ be the density of y and $f(x|\theta)$ the density of x. Then

$$g(y|\theta) = \int_{\mathcal{X}(y)} f(x|\theta) d\mu(x) . \tag{2}$$

The conditional density of the complete data given the observed data is

$$k(x|y,\theta) = f(x|\theta)/g(y|\theta) .$$

Denoting the log-likelihood of the complete resp. the observed data by $LL(\theta|x)$ resp. $LL(\theta|y)$ and that based on the conditional density of x on y by $LL(\theta|x,y)$ we obtain

$$LL(\theta|x) = LL(\theta|y) + LL(\theta|x,y) .$$

To get rid of the unknown complete data vector x the expectation is taken on both sides of this equation with respect to the conditional density $k(x|y,\theta_a)$ where θ_a is some arbitrary value. Then

$$Q(\theta|\theta_a) = \mathbf{E}(LL(\theta|x)|y,\theta_a) ,$$

$$H(\theta|\theta_a) = \mathbf{E}(LL(\theta|x,y)|y,\theta_a) ,$$

$$Q(\theta|\theta_a) = LL(\theta|y) + H(\theta|\theta_a) .$$

The function $H(\theta|\theta_a)$ is maximized by θ_a. Let be θ_L the maximum likelihood estimate, that is the value which maximizes $LL(\theta|y)$. If we now set $\theta_a = \theta_L$ then

$$Q(\theta|\theta_L) = LL(\theta|y) + H(\theta|\theta_L)$$

is maximzed by $\theta = \theta_L$. The following theorem results:

Theorem 4 *Assuming uniqueness, the value θ^* which maximizes $Q(\theta|\theta_a)$ is some function $M(\theta_a)$: $\theta^* = M(\theta_a)$. The maximum likelihood estimate θ_L based on the observed data y satifies the fixed point equation*

$$\theta_L = M(\theta_L) \ .$$

This theorem suggests an iterative algorithm with

$$\theta^{(p+1)} = M(\theta^{(p)}) \ . \tag{3}$$

Theorem 5 *For the sequence in (3)*

$$L(\theta^{(p+1)}|y) \geq L(\theta^{(p)}|y)$$

is valid with equality if and only if

$$Q(\theta^{(p+1)}|\theta^{(p)}) = Q(\theta^{(p)}|\theta^{(p)}) \text{ and } k(x|y,\theta^{(p+1)}) = k(x|y,\theta^{(p)}) \ .$$

In 1977 Dempster et al. [3] summarized and extended previous results concerning this concept and gave a proof for the convergence of this iteration. But only in 1983 C.F.J.Wu [13] has given a correct proof. The iteration algorithm is called the EM-algorithm (expectation-maximization-algorithm):

$$\begin{array}{ll} \text{E-step: determine} & Q(\theta|\theta^{(p)}) \\ \text{M-step: determine} & \theta^{(p+1)} = \arg\max_{\theta} Q(\theta|\theta^{(p)}) \end{array} \tag{4}$$

In the case of a regular exponential family the LEQ is

$$\mathbf{E}(T(X)|\theta^*) = \mathbf{E}(T(X)|Y,\theta^*)$$

Hereof it follows that the maximization of $Q(\theta|\theta^{(p)})$ as a function of θ is equivalent to the following two steps
E-step: $t^{(p)} = \mathbf{E}(T(X)|y,\theta^{(p)})$
M-step: define $\theta^{(p+1)}$ as the solution of $t^{(p)} = \mathbf{E}(T(X)|\theta)$.
More precise the above mentioned function M has to be regarded as a point-to-set map. $M(\theta^{(p)})$ is the set of θ-values which maximize $Q(\theta|\theta^{(p)})$. In most applications the underlying family of probability distributions involve curved exponential families. In these cases the E-step and M-step take special forms. On the other hand there are also cases where the full performance of an M-step cannot be recommended due to numerical properties, e.g. very lengthy

computations. Therefore in [3] also a GEM-algorithm is defined. The map M is then point-to-set, such that

$$Q(\theta'|\theta) \geq Q(\theta|\theta) \quad \forall \theta' \in M(\theta) .$$

The EM algorithm is a special GEM algorithm. The GEM algorithm is only an algorithmic scheme to be filled with further detailed instructions. Recently there are two proposals for such instructions. The first one was published 1993 by Meng and Rubin [10], the second one is the EM gradient algorithm by Lange [8] published in 1995. Both refer to the general convergence results of Wu [13], for which we will now give an overview. The following assumptions are made

$$\Theta \subseteq R^s \tag{5}$$

$$\Theta' = \{\theta \in \Theta | LL(\theta|y) \geq LL(\theta'|y)\} \tag{6}$$

$$\text{is compact for any } LL(\theta'|y) > -\infty$$

$$LL \text{ is continuous in } \Theta \text{ and differentiable in the interior of } \Theta \tag{7}$$

$$\Theta' \text{ is in the interior of } \Theta \text{ for } \theta' \in \Theta \tag{8}$$

A map M from points of Θ to subsets of Θ is called a point-to-set map on Θ. It is said to be closed at θ if $\theta^{(p)} \to \theta$, $\theta^{(p)} \in \Theta$ and $\theta^{*(p)} \to \theta^*$, $\theta^{*(p)} \in M(\theta^{(p)})$ imply $\theta^* \in M(\theta)$.

Let be $\mathcal{M}$ the set of local maxima of $LL(\theta|y)$ in the interior of Θ and $(\mathcal{S})$ the set of stationary points in the interior of Ω.

We call a sequence $\{\theta^{(p)}\}$ an instance of a GEM algorithm, if $\theta^{(p+1)} \in M(\theta^{(p)})$. If $\theta^{(p+1)}$ maximizes $Q(\theta|\theta^{(p)})$ we call the sequence $\{\theta^{(p)}\}$ an instance of an EM algorithm.

Theorem 6 *Let $\{\theta^{(p)}\}$ be an instance of a GEM algorithm and suppose additionally to the above assumptions (5) - (8), that*
(i) M is a closed point-to-set map over the complement of $\mathcal{S}$ (resp. $\mathcal{M}$),
(ii) $LL(\theta^{(p+1)}|y) > LL(\theta^{(p)}|y)$ $\forall \theta^{(p)} \notin \mathcal{S}$ (resp. $(\mathcal{M})$.
Then all the limit points of $\{\theta^{(p)}\}$ are stationary points (local maxima) of LL, and $LL(\theta^{(p)}|y)$ converges monotonically to $LL(\theta^*|y)$ for some $\theta^* \in \mathcal{S}$ (resp. $\mathcal{M}$).

In the case of the EM algorithm a sufficient condition for the closedness of M is

$$Q(\theta'|\theta) \text{ is continuous in both } \theta' \text{ and } \theta . \tag{9}$$

Theorem 7 *Suppose the assumptions (5) - (8), (9) are fulfilled. Then all limit points of any instance $\{\theta^{(p)}\}$ of an EM algorithm are stationary points of LL and $LL(\theta^{(p)}|y)$ converges monotonically to $LL(\theta^*|y)$ for some stationary point θ^*.*

We get some theorems which are more useful in applications, if we further assume

$$\frac{\partial}{\partial \theta'} Q(\theta'|\theta) \text{ is continuous in both } \theta' \text{ and } \theta \, . \tag{10}$$

Theorem 8 *Suppose the assumptions (5) - (8), (10) are valid. Let $\{\theta^{(p)}\}$ be an instance of a GEM algorithm with the additional property*

$$\left.\frac{\partial}{\partial \theta'} Q(\theta'|\theta^{(p)})\right|_{\theta^{(p+1)}} = 0 \, .$$

Then $\theta^{(p)}$ converges to a stationary point θ^* with $LL(\theta^*|y) = LL^*$, the limit of $LL(\theta^{(p)}|y)$, if either
(a)$LL^{-1}(LL^*) = \{\theta^*\}$ or
(b)$||\theta^{(p+1)} - \theta^{(p)}|| \to 0$ as $p \to \infty$ and $LL^{-1}(LL^*)$ is discrete.

Theorem 9 *Let the assumptions and notations of the previous theorem (8) be valid. Suppose that $LL(\theta|y)$ is unimodal in Θ with θ^* being the only stationary point. Then for any such instance $\{\theta^{(p)}\}$ of an EM algorithm $\theta^{(p)}$ converges to the unique maximizer θ^* of $LL(\theta|y)$.*

Example: This is an example by Milton Weinstein, described in [7], which we recompute here with two different initial values. This example deals with a medical problem. It is assumed, that two different screening tests exist for some disease. Each test gives a dichotomous result, the true disease status is unknown. We assume, that the two test results are independent, given the disease status, and further let be the false rate zero for a positive result for each test. Let be the test sensitivities S_1 resp. S_2 and let be the disease prevalence rate π.
The sample results may be summarized as

		Test 2	
		+	−
Test 1	+	y_{11}	y_{12}
	−	y_{21}	y_{22}

The complete data are x_{11}, x_{12}, x_{21}, x_{221}, x_{222}. Then $y_{ij} = x_{ij}$ for $(i,j) \neq (2,2)$ and $y_{22} = x_{221} + x_{222}$. x_{221} is the number of persons falsely classified as nondiseased by both tests, x_{222} is the number of persons correctly classified as nondiseased by both tests.
We denote $\sum_{ij} y_{ij}$ by N, and $(N - x_{222})$ by N_D. As usual $x_{\bullet 1} = \sum_i x_{ij}$ and $x_{1\bullet} = \sum_j x_{ij}$.
These assumptions yield the following representation of the likelihood function for the complete sample values:

$$
\begin{aligned}
L(\pi, S_1, S_2|x) &= (S_1S_2)^{x_{11}}\{S_1(1-S_2)\}^{x_{12}}\{S_2(1-S_1)\}^{x_{21}} \\
&\times \{(1-S_1)(1-S_2)\}^{x_{221}}\pi^{N-x_{222}}(1-\pi)^{x_{222}}
\end{aligned}
$$

Collecting the parameters gives

$$
\begin{aligned}
L(\pi, S_1, S_2|x) &= S_1^{x_{1\bullet}}(1-S_1)^{N_D-x_{1\bullet}}S_2^{x_{\bullet 1}} \\
&\times (1-S_2)^{N_D-x_{\bullet 1}}\pi^{N_D}(1-\pi)^{N-N_D} .
\end{aligned}
$$

Hereof it follows that $x_{1\bullet}$, $x_{\bullet 1}$, and N_D are jointly sufficient for S_1, S_2, and π.
A further rearrangement of the complete data and the application of equation(2) help us to compute the observed data likelihood $L(\pi, S_1, S_2|y)$.

$$
\begin{aligned}
L(\pi, S_1, S_2|x) &= (S_1S_2)^{x_{11}}\{S_1(1-S_2)\}^{x_{12}}\{S_2(1-S_1)\}^{x_{21}} \\
&\times \pi^{N-y_{22}}\{(1-S_1)(1-S_2)\pi\}^{y_{22}-x_{222}}(1-\pi)^{x_{222}} \\
&= (S_1S_2)^{x_{11}}\{S_1(1-S_2)\}^{x_{12}}\{S_2(1-S_1)\}^{x_{21}} \\
&\times \pi^{N-y_{22}}u^{y_{22}} \\
&\times \left\{\frac{\pi(1-S_1)(1-S_2)}{u}\right\}^{y_{22}-x_{222}}\left(\frac{1-\pi}{u}\right)^{x_{222}} ,
\end{aligned}
$$

where $u = (1-S_1)(1-S_2)\pi + (1-\pi)$.

$$
\begin{aligned}
L(\pi, S_1, S_2|y) &= \int_{\mathcal{X}(y)} L(\pi, S_1, S_2|x)d\mu(x) \\
&= (S_1S_2)^{y_{11}}\{S_1(1-S_2)\}^{y_{12}}\{S_2(1-S_1)\}^{y_{21}} \\
&\times \pi^{N-y_{22}}u^{y_{22}} .
\end{aligned}
$$

This result ist obtained, recognizing that x_{222} is binomial distributed with parameters $(y_{22}, \frac{1-\pi}{u})$ and $\mathcal{X}(y)$ is the set of all possible outcomes of the N experiments with the y-values given, especially given y_{22}.
Using this result and formula(4) we obtain for the E-step, superscripting with the iteration number p:

$$
\begin{aligned}
\mathbf{E}(x_{1\bullet}|y, \pi^{(p)}, S_1^{(p)}, S_2^{(p)}) &= y_{1\bullet} \\
\mathbf{E}(x_{\bullet 1}|y, \pi^{(p)}, S_1^{(p)}, S_2^{(p)}) &= y_{\bullet 1} \\
\mathbf{E}(N_D|y, \pi^{(p)}, S_1^{(p)}, S_2^{(p)}) &= N - y_{22}(1-\pi^{(p)})/u^{(p)} =: N_D^{(p+1)} .
\end{aligned}
$$

For the M-Step, see formula (4), we solve the following equations for the $(p+1)$th iterated value:

$$\begin{aligned}
\mathbf{E}(x_{1\bullet}|\pi^{(p+1)}, S_1^{(p+1)}, S_2^{(p+1)}) &= N_D^{(p+1)} S_1^{(p+1)} = y_{1\bullet} \\
\mathbf{E}(x_{\bullet 1}|\pi^{(p+1)}, S_1^{(p+1)}, S_2^{(p+1)}) &= N_D^{(p+1)} S_2^{(p+1)} = y_{\bullet 1} \\
\mathbf{E}(N_D|\pi^{(p+1)}, S_1^{(p+1)}, S_2^{(p+1)}) &= N\pi^{(p+1)} = N_D^{(p+1)} .
\end{aligned}$$

N_D is according to its definition in the range $[N - y_{22}, N]$. From a starting point of N_D the starting values of π, S_1 and S_2 are obtained. The results of the iteration for two different starting values N_D are the following:

	Starting point 1				Starting point 2			
p	$\pi^{(p)}$	$S_1^{(p)}$	$S_2^{(p)}$	$LL^{(p)}$	$\pi^{(p)}$	$S_1^{(p)}$	$S_2^{(p)}$	$LL^{(p)}$
0	0.4000	0.625	08750	-106.9882	0.8000	0.3125	0.4375	-115.9097
1	0.4182	0.5978	0.8730	-106.4288	0.7644	0.3270	0.4579	-114.9082
2	0.4270	0.5855	0.8197	-106.3158	0.7252	0.3447	0.4826	-113.7720
3	0.4317	0.5792	0.8108	-106.2862	0.6834	0.3658	0.5122	-112.5223
4	0.4342	0.5758	0.8061	-106.2776	0.6402	0.3905	0.5467	-111.2070
5	0.4356	0.5739	0.8034	-106.2750	0.5978	0.4182	0.5855	-109.9054
6	0.4364	0.5728	0.8019	-106.2750	0.5583	0.4478	0.6269	-109.9054
7	0.4369	0.5722	0.8011	-106.2742	0.5240	0.4771	0.6680	-108.7217
8	0.4372	0.5719	0.8006	-106.2739	0.4963	0.5038	0.7053	-107.7579
9	0.4373	0.5717	0.8004	-106.2738	0.4756	0.5257	0.7360	-107.0728
10	0.4374	0.5716	0.8002	-106.2738	0.4612	0.5421	0.7589	-106.6549
⋮								
∞	0.4375	0.5714	0.8000	-106.2738	0.4375	0.5714	0.8000	-106.2738

The $LL^{(p)}$ quantities are the values of the log-likelihood function of the incomplete data y. These values show the monotone convergence asserted in theorems (5) and (6) above.

6 Closing Remarks

In [10] the easy implementation is pointed out as the main advantage of an EM/GEM type algorithm. Furthermore the maximization can often be done in closed form similiar to the example in the preceding section. Nevertheless the maximization step may also be in many interesting cases a crucial point of the algorithm. As an example we may think of the gamma distribution in the section(4) assuming now censored data. For these kind of problems and more complicated ones Meng and Rubin proposed and proved in [10] the ECM algorithm which is a combination of the EM algorithm and the method of successive maximization over the various paramters of [4]. Lange [8] stresses the fact that an iteration within an iteration is in most cases a time consuming procedure. Therefore Lange proposes to solve the M-step only approximately with one gradient step.This may be considered as a special evaluation of a GEM algorithm.

References

[1] O. Barndorff-Nielsen. *Information and Exponential Families in Statistical Theory.* John Wiley & Sons, New York, 1978.

[2] H.A. David. First (?) occurence of common terms in mathematical statistics. *The Amer. Statistician*, 49(2):121–133, 1995.

[3] A. P. Dempster, N.M. Laird, and D.B. Rubin. Maximum likelihood from incomplete data via the EM algorithm(with discussion). *J.R.Statist.Soc. Ser. A*, 39:1–38, 1977.

[4] Søren Tolver Jensen, Johannsen Søren, and Steffen L. Lauritzen. Globally convergent algorithms for maximizimg a likelihood function. *Biometrika*, 78(4):867–877, 1991.

[5] Samuel Kotz, L. Norman Johnson, and B. Campbell Read, editors. *Encyclopedia of Statistical Sciences Vol. 5.* John Wiley & Sons, New York, 1985.

[6] C. Kraft and L. LeCam. *Ann. Math. Statistics*, 27:1174–1177, 1956.

[7] Nan Laird. Missing information principle. In Kotz et al. [5], pages 548–552.

[8] Kenneth Lange. A gradient algorithm locally equivalent to the EM algorithm. *J.R.Statist.Soc. Ser. B*, 57(2):425–437, 1995.

[9] E. L. Lehmann. *Theory of Point Estimation.* John Wiley & Sons, New York, 1983.

[10] Xiao-Li Meng and Donald B. Rubin. Maximum likelihood estimation via the ECM algorithm: A general framework. *Biometrika*, 1993(2):267–268, 80.

[11] F. W. Scholz. Maximum likelihood estimation. In Kotz et al. [5], pages 340–351.

[12] D.N. Shanbhag and M. Sreehari. On certain self-decomposable distributions. *Z. Wahr. verw. Geb.*, 38:217–222, 1977.

[13] C.F.Jeff Wu. On the convergence properties of the EM algorithm. *Annals of Statistics*, 11(1):95–103, 1983.

Remote Access to a Transputer Workstation

R. Schöne, W. Schultz

Abstract

The need for high computing performance in many simulation and optimization problems leads to the use of massively parallel system architectures which are rather expensive. Obviously it is desirable to have remote access to an existing system anywhere in the world via internet.

As at the chair of Prof. Ritter a parallel computer with presently 140 computing nodes was developed and built, this system has been connected to the internet to provide access for cooperating partners.

1 Introduction

The term *TransStation* (TS) is used to describe a parallel computing system. This name is derived from the name of the processor, which is used as computing node: the *Transputer* [1].

The front end host of a TS is an industrial standard personal computer (PC) with standard peripheral devices - such as Monitor, Keyboard, Mouse, Printer, Mass storage devices - attached.

As a special peripheral device, there is a so called *Link Interface* installed, which forms the gateway to a multiprocessor system consisting of a local Transputer network. Data flow through the network is provided by *Links* [1], which form point-to-point connections between the processors and the PC in turn. (Fig. 1)

Presently there are four TransStations which can be used stand alone or interconnected to form a multi TS. They can therefore also be considered as transputer clusters (TC) in a multi TS configuration.

One of the TCs in the configuration presented here contains a high performance PC and an Ethernet controller to provide access to the internet. It uses Linux as its operating system. This TC is used as master TC.

The other three TCs use MS-DOS for local configuration and local debugging purposes.

2 Configuration

Each transputer has four physical links. To establish data paths between the individual transputers these links have to be connected to each other to build up the topology for the transputer network. This can be done dynamically with programmable crossbar switches.

The basic construction module of the TS is the so called *Transputer Board* (TB). There are two versions of TBs : the *Root* board (Bd0) and the *Worker* board (Bd1..Bd4).

On Bd0 three computing nodes are implemented, named M, A, and G. Node M can be considered as master. Node A has access to the industrial VMEbus [7]. Node G has graphics capabilities and additional memory. Furthermore, on Bd0 the primary data link handles the data transfer from and to the front end host computer.

The master host has one fixed link (X0) to the cluster. The cluster is subdivided into one master cluster and three slave clusters.

There is one fixed link connection from each slave cluster to the master cluster (links Q0, Q1 and Q2).

Each of the TCs contains a host PC, one root board and four to six worker boards. Presently the four TCs are equally equipped with four worker boards, so there are 35 Transputers within each of the TCs (Fig. 2).

The topology of the TCs must be configured locally. For this purpose three standard RS232 serial communication lines COM2 .. COM4 are used. Thus there is no need for a physical access to the slave clusters and the configuration can be done as shown in chapter 3 (Software Considerations, step 3). However, it is planned to implement an additional configuration link to have a single file for the description of the the configuration for all of the TCs.

As an aid for the overall configuration and the distribution of processes, the Virtual Channel Router [4] can be used.

Software development and debugging can completely be done without the use of the TransStation. This task is accomplished by CfC [5], which emulates the transputer-specific links and channels by an extension of the widely used programming language C.

If the application produces graphics output on node G, it can be saved in a bitmapped file and downloaded to the remote user via internet.

3 Remote Access to the TransStation

Hardware Considerations

The high performance transputer workstation at the TUM consists of four TransStations as shown in Fig. 2.

Only the master TS however is connected to the WAN (internet) and hence only the master's host can be remotely accessed. Configuration commands to the three slave TSs are passed via three local serial lines. Although this command passing mechanism is transparent to the user, it has to be kept in mind during the topology definition phase that there are four rather independent workstations (Fig. 3).

Fig. 4 shows the hardware configuration with prewired channels X0..7 / Y0..7.

Software Considerations

To make a parallel program run on a TS several steps have to be performed:

1. Execute

   ```
   rlogin sun0.statistik.tu-muenchen.de -l ACCOUNT
   ```

   from your local Unix workstation to the host of the TransStation. Replace `ACCOUNT` by your home directory name.

2. Edit the configuration file(s) to describe the desired topology of the TS cluster, e.g. the files

   ```
   top-master.cfg, top-slave1.cfg, top-slave2.cfg ..
   ```

   See Fig. 5 and 6 for a configuration example. The software configuration for master and slave clusters is shown together with the corresponding configuration file listings.

3. Execute a configuration script with the appropriate parameters [3], e.g.

```
#!/bin/sh
echo "cfg top-slave1.cfg" >/dev/cua1
echo "cfg top-slave2.cfg" >/dev/cua2
echo "cfg top-slave3.cfg" >/dev/cua3
sleep 2
cfg top-master.cfg
```

4. Edit the source file of your parallel application program [2], e.g.

```
first-job.c
```

5. Execute the make script with the appropriate parameters, e.g.

```
make first-job.c
```

6. Load the TS cluster with the executable object and launch the program by executing the running script with the appropriate parameters, e.g.

```
run first-job
```

7. Interprete the results.

8. Return to step 4, if you are not satisfied with the results.

If something goes wrong, there are some utility tools to help debugging:

- To verify the net topology after step 2 you can use the 'check' utility [6].
- To abort a running (or hanging) application program after step 6 or 7 you can execute

```
res_net
```

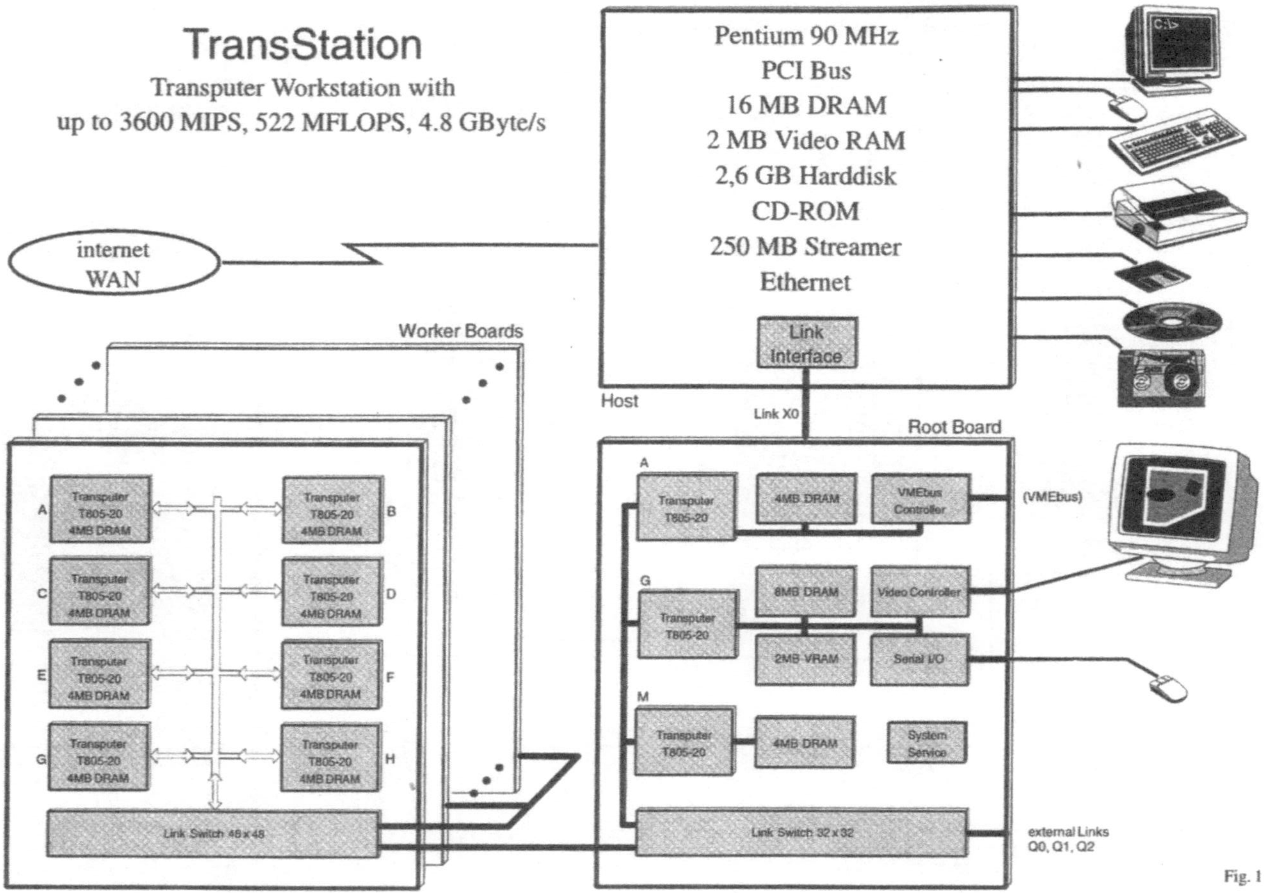
TransStation
Transputer Workstation with
up to 3600 MIPS, 522 MFLOPS, 4.8 GByte/s
internet
WAN
Pentium 90 MHz
PCI Bus
16 MB DRAM
2 MB Video RAM
2,6 GB Harddisk
CD-ROM
250 MB Streamer
Ethernet
Link Interface
Host
Link X0
Root Board
Worker Boards
A
Transputer T805-20
4MB DRAM
VMEbus Controller
(VMEbus)
G
8MB DRAM
Video Controller
2MB VRAM
Serial I/O
M
System Service
Link Switch 32 x 32
external Links
Q0, Q1, Q2
Transputer T805-20 4MB DRAM
A
B
C
D
E
F
G
H
Link Switch 48 x 48

Fig. 1

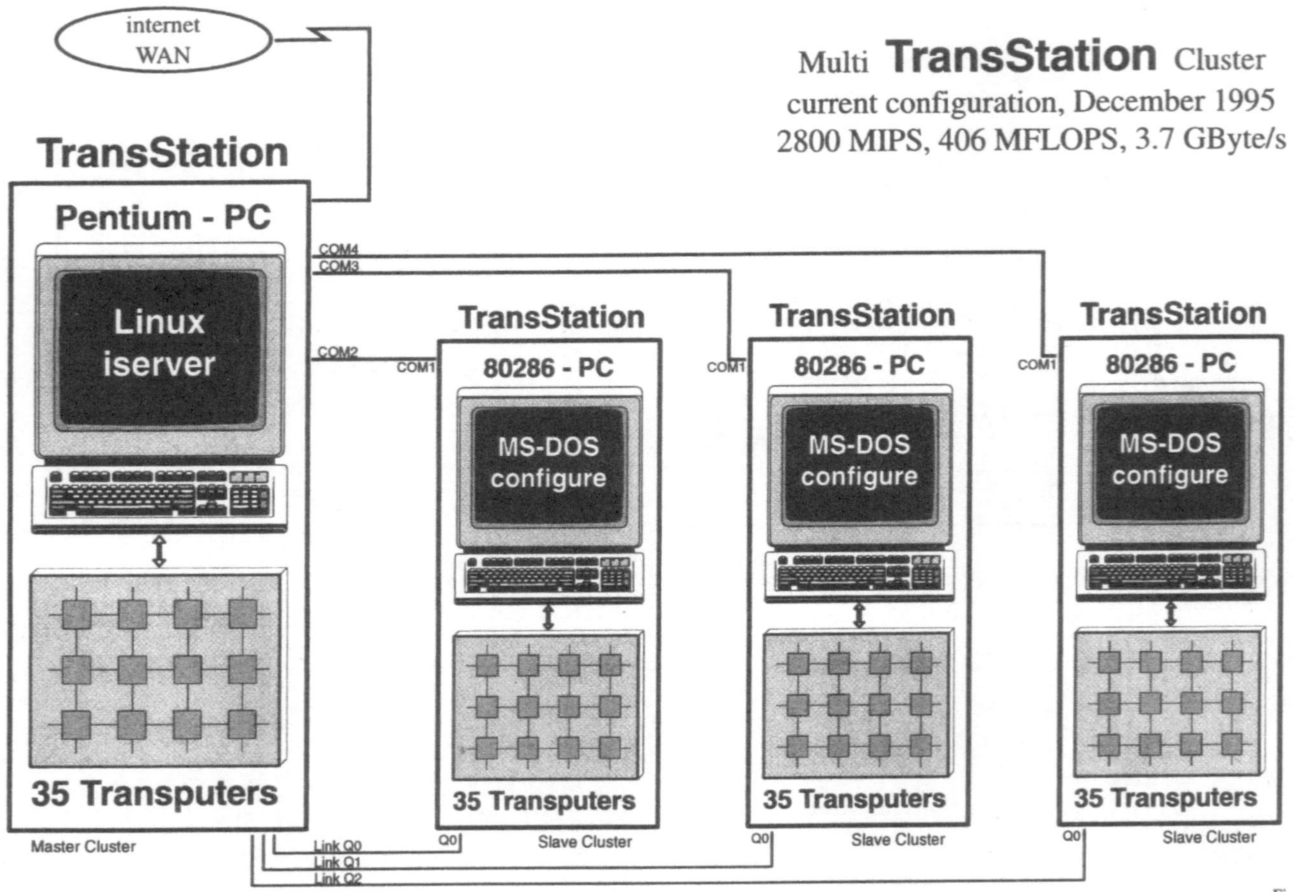
Multi TransStation Cluster
current configuration, December 1995
2800 MIPS, 406 MFLOPS, 3.7 GByte/s
internet
WAN
TransStation
Pentium - PC
Linux
iserver
35 Transputers
Master Cluster
COM4
COM3
COM2
Link Q0
Link Q1
Link Q2
TransStation
80286 - PC
MS-DOS
configure
35 Transputers
Slave Cluster
COM1
Q0
TransStation
80286 - PC
MS-DOS
configure
35 Transputers
Slave Cluster
COM1
Q0
TransStation
80286 - PC
MS-DOS
configure
35 Transputers
Slave Cluster
COM1
Q0

Fig. 2

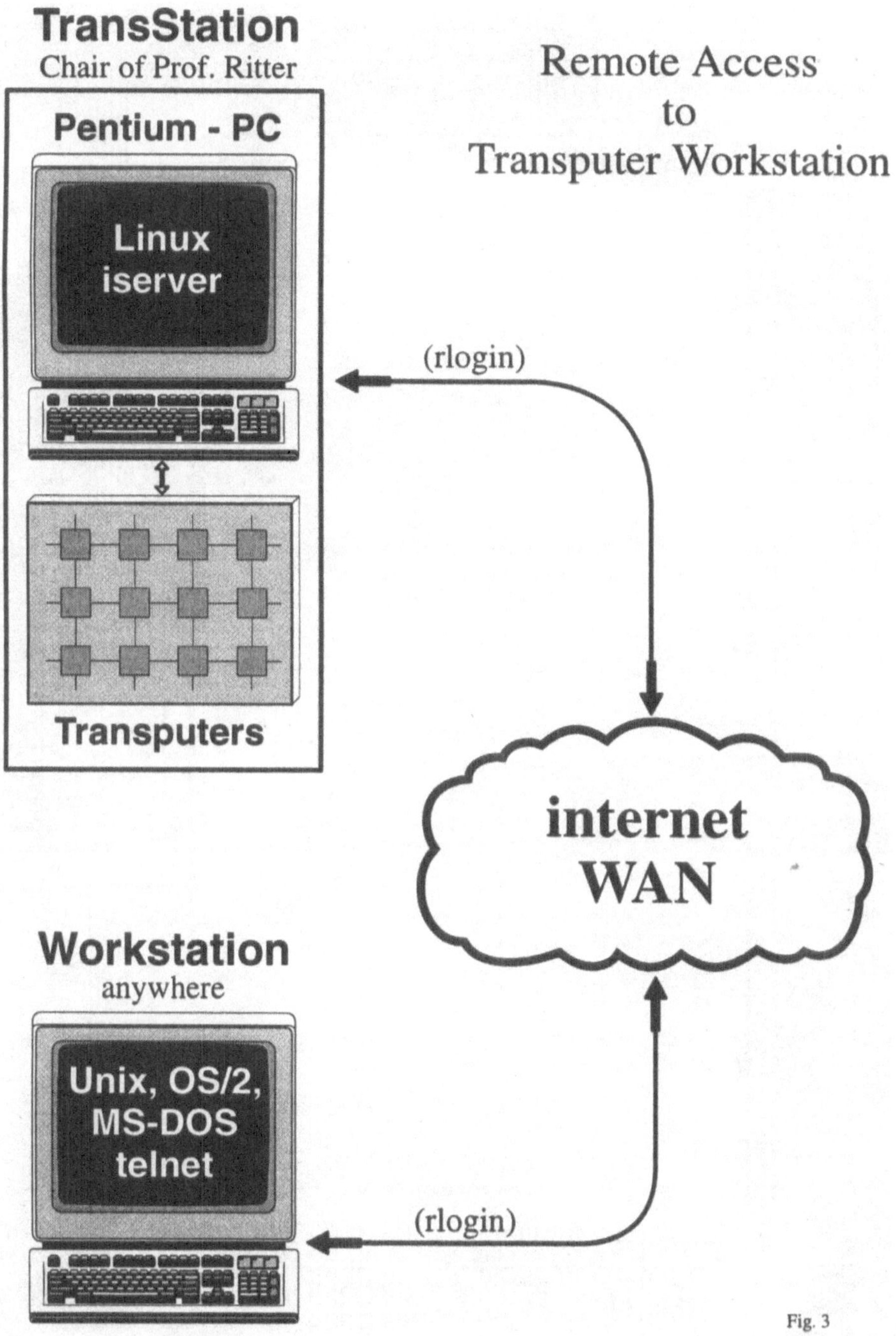
TransStation
Chair of Prof. Ritter
Pentium - PC
Linux
iserver
Transputers
Remote Access
to
Transputer Workstation
(rlogin)
internet
WAN
Workstation
anywhere
Unix, OS/2,
MS-DOS
telnet
(rlogin)

Fig. 3

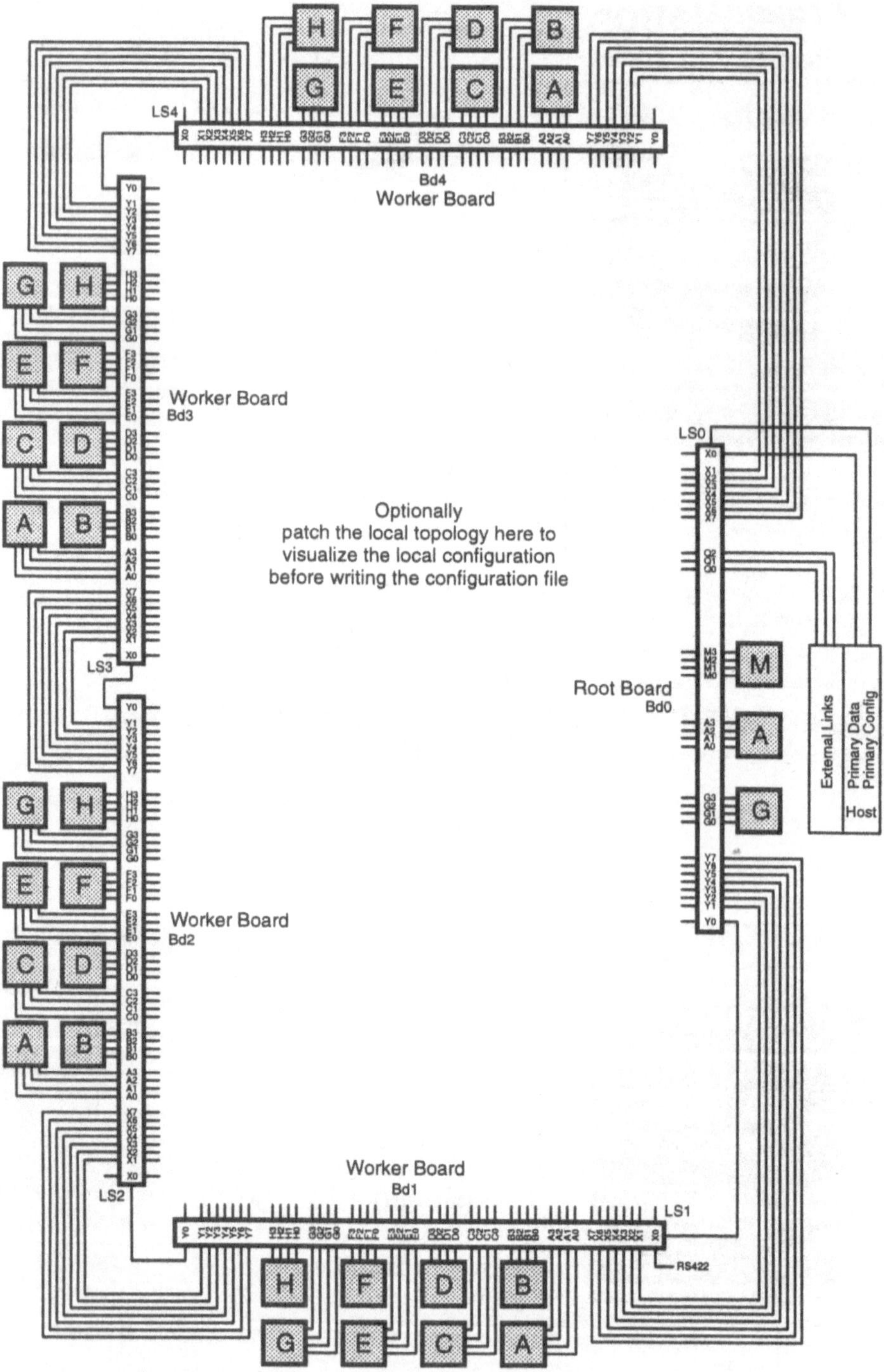
Bd4
Worker Board
LS4
Worker Board
Bd3
LS3
Optionally
patch the local topology here to
visualize the local configuration
before writing the configuration file
LS0
Root Board
Bd0
External Links
Primary Data
Primary Config
Host
Worker Board
Bd2
LS2
Worker Board
Bd1
LS1
RS422

Fig. 4

Configuration Example "Tree"

Configuration File Listing

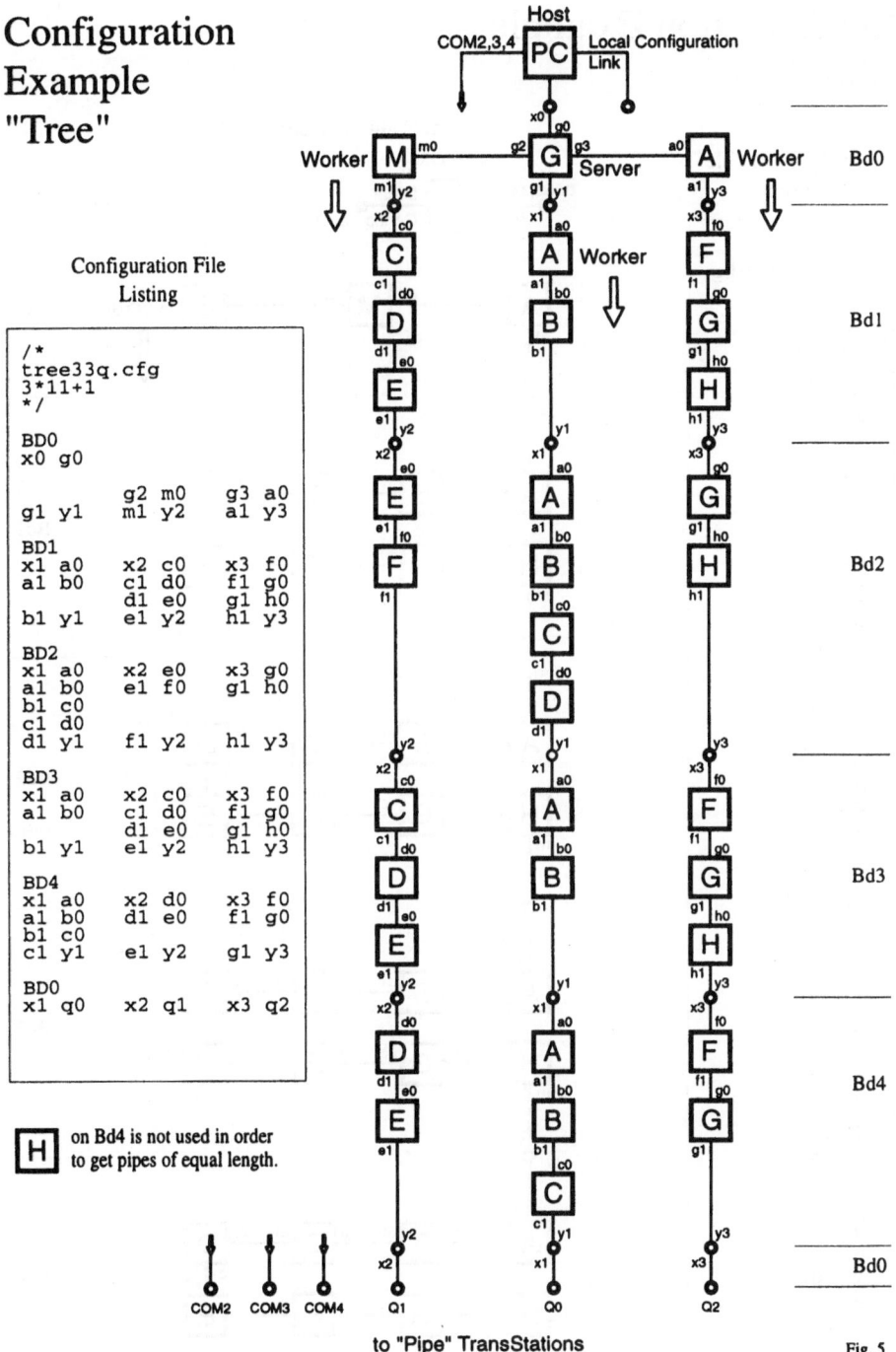

```
/*
tree33q.cfg
3*11+1
*/

BD0
x0 g0

         g2 m0    g3 a0
g1 y1    m1 y2    a1 y3

BD1
x1 a0    x2 c0    x3 f0
a1 b0    c1 d0    f1 g0
         d1 e0    g1 h0
b1 y1    e1 y2    h1 y3

BD2
x1 a0    x2 e0    x3 g0
a1 b0    e1 f0    g1 h0
b1 c0
c1 d0
d1 y1    f1 y2    h1 y3

BD3
x1 a0    x2 c0    x3 f0
a1 b0    c1 d0    f1 g0
         d1 e0    g1 h0
b1 y1    e1 y2    h1 y3

BD4
x1 a0    x2 d0    x3 f0
a1 b0    d1 e0    f1 g0
b1 c0
c1 y1    e1 y2    g1 y3

BD0
x1 q0    x2 q1    x3 q2
```

H on Bd4 is not used in order to get pipes of equal length.

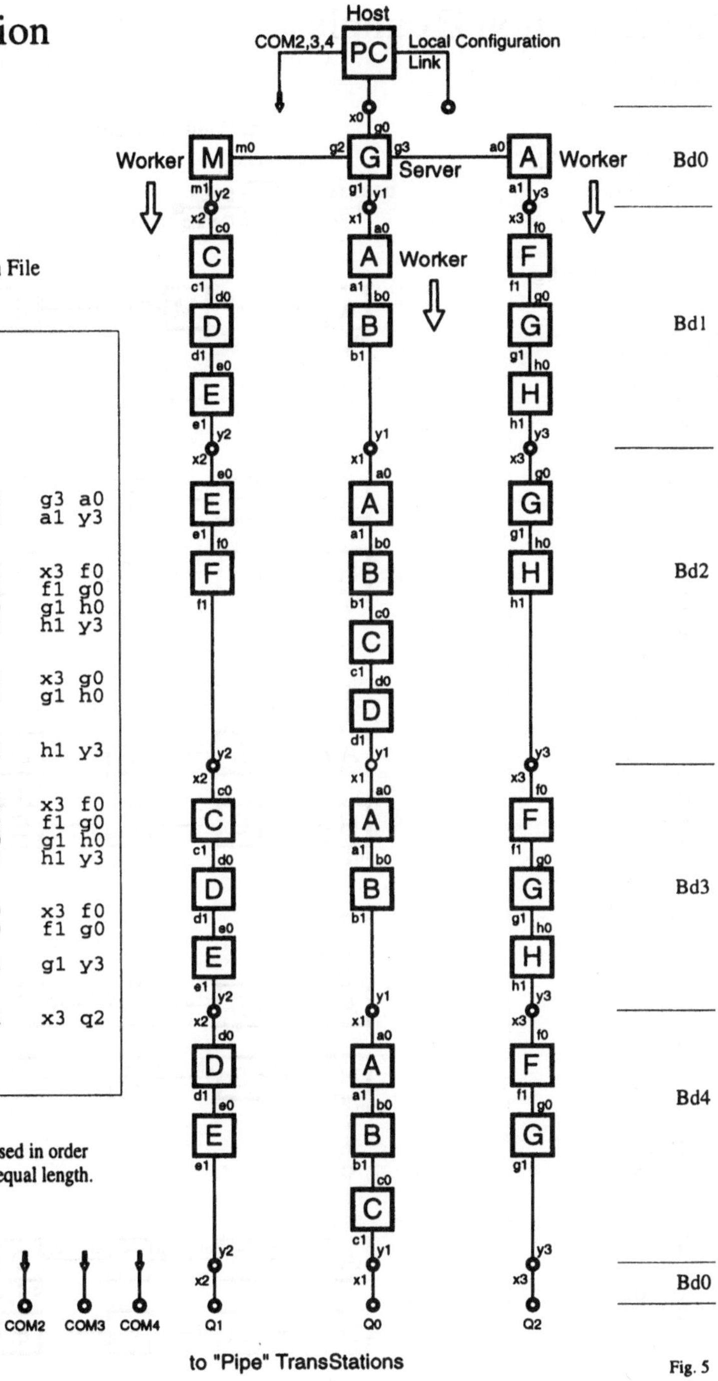

Fig. 5

Configuration Example "Pipe"

Configuration File Listing

```
/*
qpipe35.cfg
*/

BD0
q0 g0
g1 m0
m1 a0
a1 y1

BD1
x1 a0
a1 b0
b1 c0
c1 d0
d1 e0
e1 f0
f1 g0
g1 h0
h1 y1

BD2
x1 a0
a1 b0
b1 c0
c1 d0
d1 e0
e1 f0
f1 g0
g1 h0
h1 y1

BD3
x1 a0
a1 b0
b1 c0
c1 d0
d1 e0
e1 f0
f1 g0
g1 h0
h1 y1

BD4
x1 a0
a1 b0
b1 c0
c1 d0
d1 e0
e1 f0
f1 g0
g1 h0
```

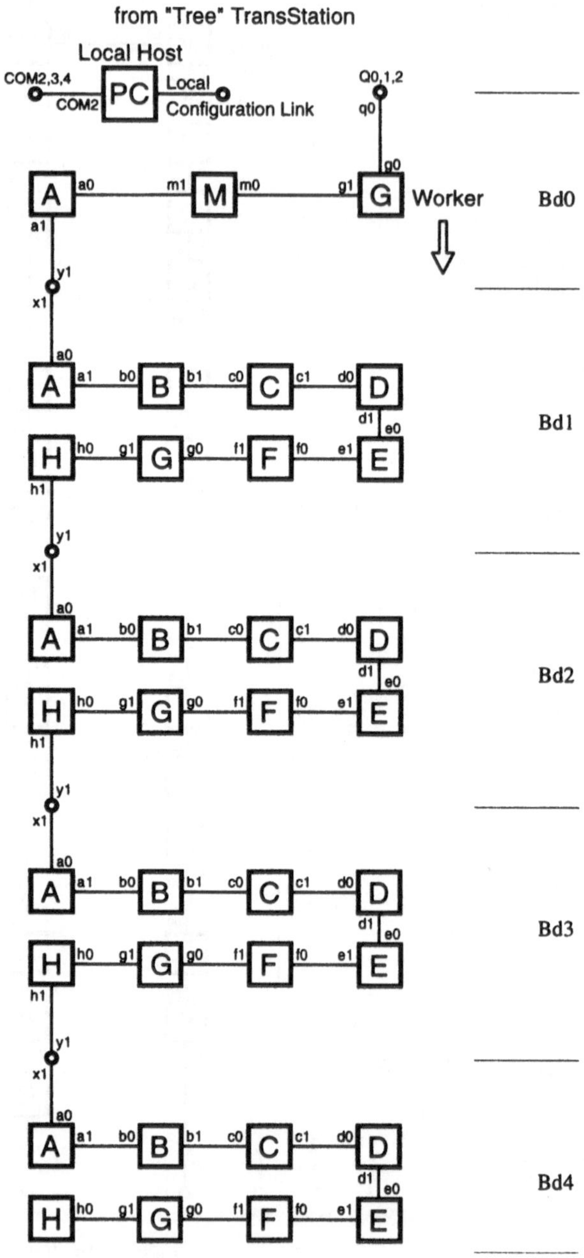

Fig. 6

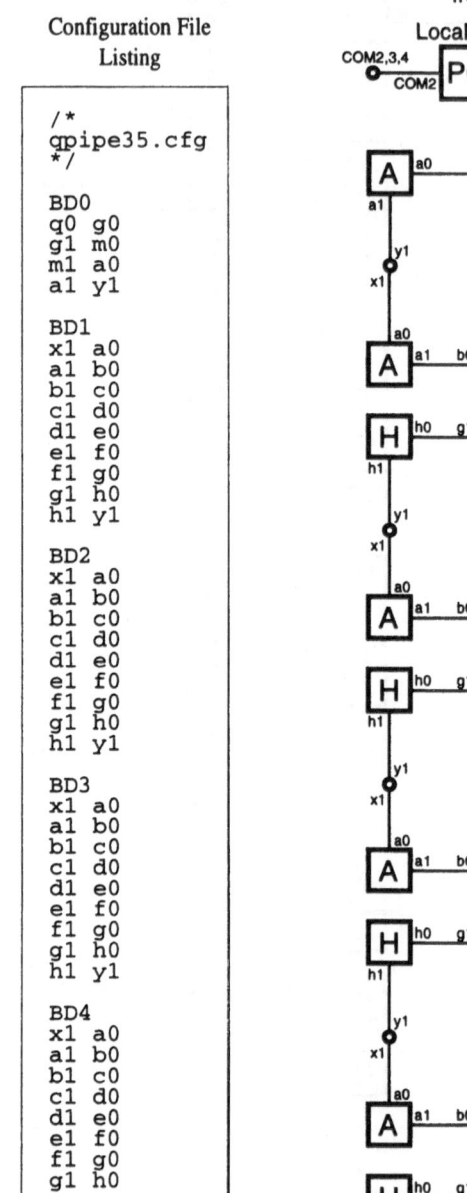

References

[1] inmos: *The Transputer Data Book* (INMOS document number: 72 TRN 203 02), 1992

[2] inmos: *ANSI C Toolset User Manual* (INMOS document number: 72 TDS 225 00), 1990

[3] Schultz: Program *CFG*, Configuration Program for MS-DOS and Linux, 1991

[4] Debbage, Hill, Nicole: *Virtual Channel Router Version 2.0*
- User Guide, 1991
- RPC Support and Run-Time Library, 1991
- VCR 2.0k Restrictions, 1992

[5] Ulbrich, Ulbrich: *Development of Parallel C-Programs with 'Channels for C' (CfC)*, Copernicus Research Report 9506, 1995

[6] Rabagliati: Program *CHECK*, Network Test Program, 1990

[7] Motorola: *VMEBus Specification Manual,* 1982

[8] Schöne, Schultz: *TransStation*, Hardware and Firmware Documentation, 1991

An Extension of Multivariate Reliability Systems

Alfred Schöttl

Institute of Applied Mathematics and Statistics SCA

Technical University of Munich

Abstract

In reliability theory the behaviour of a system with respect to the eventual failure of parts of the system is investigated. Classic reliability systems consist of a finite number of parts, so-called components, with only two states, working or failed. We extend this concept to continuous failure positions and continuous failure degrees. The component states of these systems are modeled by a discrete measure on the component set. We introduce system functions based on an appropriate ordering and classification method for discrete measures. Random component states and the evaluation of the system reliabilities are investigated.

1 Basic Notions

A classic reliability system (see e.g. [1]) consists of a finite number of components $i \in \{1, \dots, m\} =: \mathcal{C}$. Each component i is associated with exactly one state x_i in the state set $\mathcal{S} := \{0, 1\}$. The *component state* $x = (x_i)_{i \in \mathcal{C}}$ indicates the state of all components in the system. The *component state space* $\mathcal{S}^m$ is denoted by $\mathcal{Z}$. A component i is called *working*, if $x_i = 0$, and failed otherwise.

A *system function* $\varphi : \mathcal{Z} \to S$ assigns to each component state $x \in \mathcal{Z}$ a *system state* $\varphi(x) \in \mathcal{S}$. Since it is in general very hard to specify a system function (there are 2^m entries in the function table of φ), we have to restrict ourselves to so–called *maximal path systems.*

A component state $v \in \mathcal{Z}$ is called a *path* if the state yields a working system, i.e. $\varphi(v) = 0$. A path $v \in \mathcal{Z}$ is called *maximal* if

$$\forall w \in \mathcal{Z} : w > v \Rightarrow \varphi(w) = 1 \quad .$$

The *maximal path set* $\mathcal{V}$ is the set of all maximal paths of the system.

Example 1 We consider a classic reliability system with $m = 2$ components. The system is assumed to be working only if both components are working. The according map is given for all component states $\boldsymbol{x} \in \{0,1\}^2$ by

$$\varphi(\boldsymbol{x}) = \begin{cases} 0 & : \quad \boldsymbol{x} = (0,0) \\ 1 & : \quad \text{else} \end{cases} .$$

The maximal path set consists of only one component state, $\boldsymbol{v} = (0,0)$.

If the system was working if one component is working (i.e. a system with a so-called hot reserve) we would have two maximal paths, namely $\boldsymbol{v}_1 = (0,1)$ and $\boldsymbol{v}_2 = (1,0)$.

Various authors (see e.g. [2]) consider components with multistate state sets, i.e. $S = \{0,1,\ldots,M\}$, $M \in \mathbb{N}$, to treat components that are neither completely working nor completely failed.

Example 2 Assume we have a pump station consisting of $M \in \mathbb{N}$ independently working pumps. All except one pump serve as hot reserves. A reasonable state set of the component "pump station" is $S = \{0,\ldots,M\}$. The pump station is assumed to be in state $x \in S$ if exactly x pumps failed.

Multistate reliability systems may be handled with methods equivalent to those of classic systems (see [5], [9]).

Recently, systems with "continuous structures" are considered (see [3], [4]). The state of a component in a continuous structure is supposed to be an element of the interval $S = [0,1]$. Hence, it is possible to model components with stageless utilities.

Example 3 The output of each of the M pumps in example 1 depends on its (continuous) state. Hence, a deteriorating pump yields the output to fall off.

All these systems have in common a finite number of components. Furthermore, the geometric configuration of the system is not considered. These facts cause problems in some applications.

Example 4 There are a lot of examples for reliability systems that are not treatable with the reliability models stated above, e. g.:

1. A pipeline that is possibly damaged or leaking.
2. A pipeline in which a fluid is flowing. A cooling around the pipeline guarantees the temperature not to exceed a certain threshold. The cooling of the pipeline may fail in some (random) intervals.
3. Dust particles on the surface of a wafer cause problems in the chip production especially if they accumulate in a region.

In our model the component set C is defined to be a connected, closed set. Since there are up to an uncountable number of elements in C it is reasonable to call the elements "failure positions" instead of "components". To ease the notation we consider the onedimensional case. The extension on higher–dimensional spaces is straight–forward.

Definition 1 Let $C := [a, b]$, $a, b \in \mathbb{R}$, be the component set and $\mathcal{S} := \mathbb{R}_+$ be the state set of a reliability system. A pair $(x, y) \in C \times \mathcal{S}$ of a *failure position* and a *failure degree* is called a *failure*. Let $(x_i, \alpha_i)_{i \in \mathbb{N}} \in (C \times \mathcal{S})^{\mathbb{N}}$ be a sequence of failures with pairwise different failure positions and let $n \in \mathbb{N}_0$ be a constant, the *number of failures*. The associated discrete measure

$$z = \sum_{i=1}^{n} \alpha_i \delta_{x_i}$$

is called a *component state* of the system. The set $\mathcal{Z}$ of all discrete measures on C with finite support is called the *component state space*.

The structure of the system is determined by a system function operating on the set $\mathcal{Z}$ of discrete measures on C.

Definition 2 A function $\varphi : \mathcal{Z} \to \mathcal{S}$ is called a *system function* iff $\varphi(0_C) = 0$. Hereby the zero measure 0_C on C denotes the *perfect component state*.

In the next section we prepare the representation of system functions.

2 An Ordering and Classification Method for Component States

In the first part of this section we introduce an (quasi–)ordering on the component set $\mathcal{Z}$.

[11] provides various definitions of orderings of discrete measures. They all are based on the behaviour of the distribution functions F_{μ_1}, F_{μ_2} of the underlying measures μ_1, μ_2. For example the relation $<_d$ is defined by

$$\mu_1 <_d \mu_2 \iff \forall t \in \mathbb{R} : \; F_{\mu_1}(t) \geq F_{\mu_2}(t) \quad . \tag{1}$$

Orderings of this type are very useful to compare sequences of random time points (such as repair times of a machine, see e.g. [8]). In our situation they are not appropriate due to the inherited polarity: Let C be the set of possible damage positions of a pipeline. A scenario $\mu_1 = \varepsilon\delta_{x-\alpha}$ with a small damage $\varepsilon > 0$ in the pipeline system at position $x - \alpha$ ($\alpha > 0$, small) should naturally be assumed to yield a better system than a scenario $\mu_2 = \gamma\delta_x$ with huge damage $\gamma \gg \varepsilon$ at a position x in close vicinity, but the corresponding distributions are not comparable with respect to $<_d$. On the other hand the slightly shifted component state $\mu_3 = \varepsilon\delta_{x+\alpha}$ is $<_d$–smaller than μ_2.

Due to these inconsistencies the introduced ordering is not based on distribution functions. We use functions that may be interpreted as realizations of kernel estimates.

Definition 3 A continuous, monotone decreasing function $u : \mathbb{R}_+ \to [0,1]$ with $u(0) = 1$ and $\lim_{t\to\infty} u(t) = 0$ is called a *valence function*. A valence function u is called *feasible*, iff

$$\forall a,b,c \in \mathbb{R}_+ : \quad \begin{aligned} & u(a)\,u(b) - u(a+b) \\ & + u(b+c)\,(u(a+b)\,u(b+c) - u(a+b+c)\,u(b)) \\ & - u(c)\,(u(a)u(b+c) - u(a+b+c)) \geq 0 \end{aligned} \tag{2}$$

holds true. The set of all feasible valence functions is denoted by U.

The purpose of the technical condition (2) will be obvious later. Examples of feasible valence functions are the functions $x \mapsto e^{-\alpha x^\beta}$, where $\beta \geq 1$ and $\alpha \geq 0$ (see [10]). All valence functions of this type with $\alpha > 0$ and $\beta = 1$ are called *standard valence functions*. In our model a valence function specifies how badly a failure (α, x) at position x affects neighboured parts of the system. At position $y \in C$ the failure (α, x) is supposed to have the effect $\alpha u(|x-y|)$.

The considered model is founded on two additional assumptions:

1. The valence function u does not depend on the failure position $x \in C$. Thus, given a valence function u the effect of the failure (α, x) at position $y \in C$ only depends on the failure degree $\alpha \geq 0$ and the distance $|x-y|$.

2. The effects of different failures are additive.

Definition 4 Let $z = \sum_{i=1}^{n} \alpha_i \delta_{x_i} \in \mathcal{Z}$ be a component state. Then the function $v : C \to \mathbb{R}$ with

$$v_z(y) := \sum_{i=1}^{n} \alpha_i u(|y - x_i|) \quad ,$$

is called *valuation function* of the state z. Valuation functions with respect to a standard valence function are called *standard valence functions*.

Example 5 We consider an isolated wire with eventual failures in the isolation. The failure positions $(X_i)_{i=1,\ldots,N}$ are assumed to be governed by a poisson process on C. Using the kernel $K_\beta : \mathbb{R}_+ \to \mathbb{R}_+$,

$$K_\beta(x) := c_\beta e^{-|x|^\beta}, \quad \text{with an appropriate constant } c_\beta > 0 \text{ and } \beta \geq 1 \quad ,$$

we get the kernel estimate v of the intensity function by

$$\forall y \in \mathcal{C}: \quad v(y) = c_\beta \sum_i e^{-|X_i - y|^\beta} \quad .$$

The estimate v/c_β of the intensity function is a valuation function. The special case $\beta = 1$ (K_1 is the so-called Picard-kernel with bandwidth ∞, see [12]) leads to a standard valuation function.

Example 6 Let $(\alpha_i, x_i)_{i=1,\dots,n}$ be a sequence of failures of the pipeline cooling in example 3.2. With appropriately scaled failure degrees $(\alpha_i)_{i=1,\dots,n}$ the inner temperature $T(y)$ of the pipeline at position y is approximated by (see [10])

$$T(y) = \sum_{i=1}^{n} \alpha_i e^{-\sigma(y-x_i)^*} \quad ,$$

where the $*$-function is defined by $x^* := x$ for all $x \geq 0$ and $x^* := \infty$ for $x < 0$. $\sigma > 0$ is a constant of physics. To get rid of the dependency on the direction of the flow we regard the sum v of the inner temperature function for both directions as a criterion for the quality of the pipeline,

$$\forall y \in \mathcal{C}: \quad v(y) = \sum_{i=1}^{n} \alpha_i e^{-\sigma|y-x_i|} \quad ,$$

(where the jumps of the temperature function at the positions $(x_i)_{i=1,\dots,n}$ have only been considered once.) Observe that v is a standard valuation function.

Valuation functions induce an ordering on $\mathcal{Z}$.

Lemma 5 Let $u \in U$ be a valence function and let $z_1, z_2 \in \mathcal{Z}$ be two states. Then the definition

$$z_1 \preceq_u z_2 \quad :\Longleftrightarrow \quad \forall x \in \mathcal{C}: \; v_{z_1}(x) \leq v_{z_2}(x)$$

provides an (quasi-)ordering $\preceq_u$ on $\mathcal{Z}$.

Hence, a state $z_1 \in \mathcal{Z}$ is $\preceq_u$-smaller than another state z_2 iff the estimated failure intensity function is pointwise smaller. In terms of example 5 a state z_1 is said to be $\preceq_u$-smaller than another state z_2 iff the "symmetrized" temperature function is pointwise smaller. The ordering relation can easily be extended to a classification method.

Definition 6 Let $f : \mathcal{C} \to \mathbb{R}_+$ be a function and $z \in \mathcal{Z}$ a state. Then

$$z \preceq_u f \quad :\Longleftrightarrow \quad \forall x \in \mathcal{C}: \; v_z(x) \leq f(x)$$

defines a classification method.

Since the condition $\mu_1 \preceq_u \mu_2$ (where $\preceq_u$ is used in the sense of definition 2) is equivalent to $\mu_1 \preceq v_{\mu_2}$ (where $\preceq_u$ is used in the sense of definition 6) the classification method actually is an extension of the ordering. If $z \preceq_u f$ holds true we call the state z *$\preceq_u$-small with respect to f*.

Figure 1 depicts the valuation function v_z of component state z and the graph of a u–subconcave function f. Observe that z is not $\preceq_u$-small with respect to f.

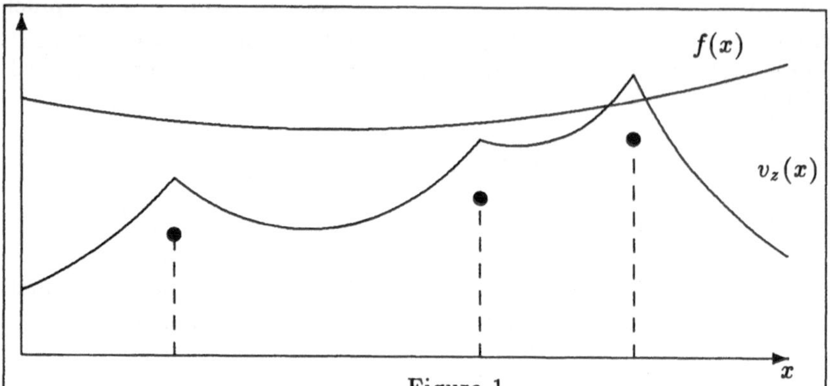

Figure 1.

3 Properties of $\preceq_u$

Let $z \in Z$ be a component state and let $f : C \to \mathbb{R}_+$ be a function. In general, the decision whether or not $z \preceq_u f$ holds true requires the evaluation of $f(x)$ and $v_z(x)$ for an infinite amount of failure positions $x \in C$. For a restricted class of functions a finite number of function evaluations is sufficient.

Definition 7 Let $u \in U$ be a valence function. A function $f : C \to \mathbb{R}_+$ is called *u–subconcave* iff there is a nonnegative, continuous extension $\bar{f}$ of f on $\mathbb{R}$ such that for all $t, t_0, t_1 \in \mathbb{R}$ with $t_0 < t < t_1$

$$(1 - \Delta^2)\bar{f}(t) \geq (\Delta_0 - \Delta\Delta_1)\bar{f}(t_0) + (\Delta_1 - \Delta\Delta_0)\bar{f}(t_1) \quad , \qquad (3)$$

where

$$\Delta := u(t_1 - t_0), \quad \Delta_0 := u(t - t_0) \text{ and } \Delta_1 := u(t_1 - t) \quad .$$

The set of all u–subconcave functions is denoted by F_u.

Examples of u–subconcave functions are given below.

- Any constant function $f : \mathcal{C} \to \mathbb{R}_+$, $f(t) = a$ with $a \geq 0$ is u–subconcave with respect to any valence function $u \in U$.

- The function $f : \mathcal{C} \to \mathbb{R}_+$, $f(t) = \sin(t + a) + 4$, $a \in \mathbb{R}$, turns out to be u–subconcave with respect to the standard valence function $u \in U$, $u(t) = e^{-2t}$, but is not concave.

The examples show that u–subconcavity does not imply concavity (whereas the converse is true). The following lemma gives an interpretation of the u–subconcavity.

Lemma 8 Let $u \in U$ be a valence function, f a u–subconcave function and let $x_0, x_1 \in \mathcal{C}$, $x_0 < x_1$ be two failure positions. Let $z = \alpha\delta_{x_0} + \beta\delta_{x_1} \in \mathcal{Z}$ be the (unique) component state such that $v_z(x_0) = f(x_0)$ and $v_z(x_1) = f(x_1)$. Then it holds

$$\forall x \in [x_0, x_1] : \quad v_z(x) \leq f(x) \quad .$$

In other words: Let z be a component state with two failures. If the graph of the valuation function v_z touches the graph of a u–subconcave function f at both of the failure positions x_0, x_1 then the graph of v_z does not exceed the graph of f at any position $x \in [x_0, x_1]$.

According to the next theorem, the number of evaluations of v_z and f necessary to decide whether or not $z \preceq_u f$ holds true remains finite for u–subconcave functions. As already mentioned this is not true for arbitrary (even not for analytical) functions $f : \mathcal{C} \to \mathbb{R}_+$.

Theorem 9 Let $u \in U$ be a feasible valence function, $f \in F_u$ a u–subconcave function and $z \in \mathcal{Z}$ a component state. Then

$$z \preceq_u f \iff \forall t \in \operatorname{supp} z : \; v_z(t) \leq f(t)$$

holds true.

Observe that in the situation of figure 1 we only have to check the value of f and v_z at 3 failure positions. We see that component state z is not $\preceq_u$–small with respect to f since v_z exceeds f at the third failure position.

Proof
"$\Rightarrow$" This implication follows immediately from definition 6.
"$\Leftarrow$" Let $z = \sum_i y_i \delta_{t_i} \in \mathcal{Z}$ be a discrete measure with disjoint support locations $t_i \in \mathbb{R}_+$ in its representation. We also assume $|\operatorname{supp} z| \geq 2$ (the case $|\operatorname{supp} z| \leq 1$ can be proved similarly). Let for each $t \in \mathbb{R}_+$

$$\overleftarrow{t} := \sup\{t_i : \; t_i \leq t, \; i = 1, \ldots, n\}$$
$$\overrightarrow{t} := \inf\{t_i : \; t_i \geq t, \; i = 1, \ldots, n\}$$

be the nearest element of the support that is not greater (smaller) than t and assume $\inf \operatorname{supp} z < t < \sup \operatorname{supp} z$ (the other cases may be treated similarly). Then there exists a discrete measure $\bar{z} := y_0 \delta_{\underleftarrow{t}} + y_1 \delta_{\underrightarrow{t}} \in \mathcal{Z}$ with

$$
\begin{aligned}
y_0 &:= \frac{v_z(\underleftarrow{t}) - u(\underrightarrow{t} - \underleftarrow{t})\, v_z(\underrightarrow{t})}{1 - u(\underrightarrow{t} - \underleftarrow{t})} \quad , \\
y_1 &:= \frac{v_z(\underrightarrow{t}) - u(\underrightarrow{t} - \underleftarrow{t})\, v_z(\underleftarrow{t})}{1 - u(\underrightarrow{t} - \underleftarrow{t})} \quad .
\end{aligned}
$$

The coefficients y_0, y_1 are chosen in such a way that

$$
\begin{aligned}
v_z(\underleftarrow{t}) = v_{\bar{z}}(\underleftarrow{t}) \leq f(\underleftarrow{t}) \quad \text{and} \\
v_z(\underrightarrow{t}) = v_{\bar{z}}(\underrightarrow{t}) \leq f(\underrightarrow{t}) \quad .
\end{aligned}
$$

Now we can expand the term $\Delta v := v_{\bar{z}}(t) - v_z(t)$ and abbreviate the "distances" by $a := t - \underleftarrow{t}$, $b := \underrightarrow{t} - t$ and $c := \underrightarrow{t} - \underleftarrow{t}$. Using inequality (2) it can be shown by elementary but lengthy calculations that

$$
\forall t \in [\underleftarrow{t}, \underrightarrow{t}] : v_{\bar{z}}(t) \geq v_z(t) \quad .
$$

Therefore the investigation can be restricted to discrete measures z satisfying $|\operatorname{supp} z| = 2$. In this case lemma 8 may be applied. This completes the proof. □

4 Representation of System Functions

System functions map each component state into the set $\mathcal{S}$ of possible system states. Classical theory restricts the class of considered reliability systems to so–called maximal path systems. A *maximal path* $\boldsymbol{v} \in \{0,1\}^m$ is a component state with $\varphi(\boldsymbol{v}) = 0$ (remember 0 is the working state) such that

$$
\nexists \boldsymbol{x} \in \{0,1\}^m : \qquad \varphi(\boldsymbol{x}) = 0 \text{ and } \boldsymbol{x} \geq \boldsymbol{v} \quad .
$$

Thus a maximal path may be interpreted as a worst component state which the system operates at. (The notion "maximal path" is motivated by block diagrams, a graphical tool used to construct classical reliability systems.)

Lateron these principles have been extended to various multi–state reliability systems (see [7]). In most extensions each system state is associated with its own maximal path set. We will take up this idea in the case of continuous component sets.

Definition 10 Let $u \in U$ be a valence function. A system function $\varphi : \mathcal{Z} \to \mathcal{S}$ specifies a *maximal path system*, iff there exists for each system state $\kappa \in \mathcal{S}$ a finite set $\mathcal{V}_\kappa$ of u–subconcave functions such that

$$\forall z \in \mathcal{Z} : \qquad \varphi(z) \leq \kappa \iff \exists v \in \mathcal{V}_\kappa : \ z \preceq_u v \quad .$$

Without restriction we assume that the sets $\mathcal{V}_\kappa$ are minimal, i.e.

$$\nexists z_1, z_2 \in \mathcal{V}_\kappa, \ z_1 \neq z_2 : \qquad z_1 \preceq_u z_2$$

for all $\kappa \in \mathcal{S}$. Then each element of $\mathcal{V}_\kappa$ is called a *maximal path at system state* κ.

Example 7 We consider the isolated wire of example 4 and assume $\mathcal{C} = [0,1]$. The wire is said to be in a good state if there are only few and small failures and if the failure positions do not accumulate. In terms of kernel estimation this means that the realisation of the kernel estimator does not exceed a certain threshold. This situation can be described by a maximal path system with the following structure. Each system state $\kappa \in \mathcal{S}$ is associated with exactly one u–subconcave function f_κ. In the most simple system we have a constant threshold $\forall x \in \mathcal{C} : \ f_\kappa(x) = \kappa$. Inbedded in an inhomogenous environment (e.g. varying humidity) a component state $z \in \mathcal{Z}$ and its shifted state $z(\cdot + h) \in \mathcal{Z}$, $h \in \mathbb{R}$, may yield different system states. These situations require non–constant maximal paths f_κ, e.g.

$$\forall x \in \mathcal{C} : \quad f_\kappa(x) = \left(1 + \frac{1}{1+x}\right)\kappa \quad .$$

Alternative restrictions for each system state can be achieved using more than one maximal path per system state. A system that behaves critically if the failures accumulate on both ends of the wire is decribed by the maximal paths $\mathcal{V}_\kappa = \{f_\kappa^1, f_\kappa^2\}$ with

$$\forall x \in \mathcal{C} : \quad f_\kappa^1(x) = \left(1 + \frac{1}{1+x}\right)\kappa, \quad f_\kappa^2(x) = \left(1 + \frac{1}{2-x}\right)\kappa \quad .$$

Example 8 We consider the cooled pipeline of example 2. The pipeline is assumed to work better if the temperature in the pipeline (or, more precicely, the valuation function based on the temperature function) does not exceed a certain threshold. Then we have the same situation as in example 8.

5 Random Component States and System Reliabilities

We now consider the probabilistic behaviour of the reliability systems defined in the last section. Throughout this section we assume $u \in U$ and $\varphi : \mathcal{Z} \to \mathcal{S}$ to be its valence ant system function, respectively. Since deterministic component states have been modeled by discrete measures it is reasonable to construct random component states using random measures.

Let N be an nonnegative integer random variable and $(X_i, Y_i)_{i \in \mathbb{N}}$ be an i.i.d. sequence of $\mathcal{C} \times \mathbb{R}_+$–valued random variables. N and every pair (X_i, Y_i) are independent of each other (a dependency between X_i and Y_i is allowed). Let P_X be the distribution of X_1 with density function f_X, let $P_{Y|X}$ be the (conditional) distribution of Y_1 and let $p_N(n) = P(N = n)$, $n \in \mathbb{N}_0$, be the elementary probabilities of N.

Definition 11 The discrete random measure ζ defined by

$$\zeta = \sum_{i=1}^{N} Y_i \delta_{X_i} \quad .$$

is called a *random component state.*

Observe that the random failure positions $(X_i)_{i=1,\ldots,N}$ are a.s. pairwise distinct.

We now evaluate the *system realiability* $p_S(\kappa)$ at the system state $\kappa \in \mathcal{S}$. Let $\mathcal{V}_\kappa := \{f^1, \ldots, f^r\}$ be the maximal path set at the system state κ. By the definition of maximal path systems we obtain

$$\begin{aligned} p_S(\kappa) &:= P(\varphi(\zeta) \leq \kappa) \\ &= P(\exists f \in \mathcal{V}_\kappa : \zeta \preceq_u f) \\ &= \sum_{i=1}^{r} P(\zeta \preceq_u f^i) - \sum_{\substack{i,j=1 \\ i<j}}^{r} P(\zeta \preceq_u f^i \wedge f^j) \\ &\quad + \sum_{\substack{i,j,k=1 \\ i<j<k}}^{r} P(\zeta \preceq_u f^i \wedge f^j \wedge f^k) + \ldots \end{aligned}$$

Since the infimum of u–subconcave functions is u–subconcave it is sufficient to consider the evaluation of probabilities of the form $P(\zeta \preceq_u f)$ for an u–subconcave function f. Using theorem 9 we get

$$P(\zeta \preceq_u f) = P\left(\sum_{i=1}^{N} Y_i \delta_{X_i} \preceq_u f\right)$$

$$= P\left(\forall x \in \mathcal{C} : \sum_{i=1}^{N} Y_i u(|X_i - x|) \leq f(x)\right)$$

$$= P\left(\forall k \in \{1,\ldots,N\} : \sum_{i=1}^{N} Y_i u(|X_i - X_k|) \leq f(X_k)\right) \quad (4)$$

The resulting probability is (for bounded N) a finite–dimensional integral that may be computed by standard techniques.

Example 9 Let $\forall x \in \mathbb{R}_+ : \; u(x) = e^{-\frac{x}{5}}$. We consider the system of example 9 with $\forall \kappa \in \mathbb{R}_+ : \; \mathcal{V}_\kappa = \{f_\kappa\}$ and $\forall x \in \mathcal{C} : \; f_\kappa(x) = \kappa$. Let the failure degree distribution $P_{Y|X}$ be given by $P_{Y|X}([0,y]|\{x\}) = 1 - e^{-y}$ and let the distribution P_X of the failure position be a uniform distribution on $\mathcal{C}$. The elementary probabilities of the number of failures are assumed to be $p_N(0) = p_N(1) = p_N(2) = \frac{1}{3}$. By numerical quadrature of equation (4) the following results for the system reliabilities of the system states 0,1 and 2 are obtained.

$$\begin{aligned} p_S(0) &= 0.3333 \\ p_S(1) &= 0.6603 \\ p_S(2) &= 0.8538 \quad . \end{aligned}$$

References

[1] R. Barlow, F. Proschan (1965): *Statistical Theory of Reliability.* John Wiley & Sons, New York.

[2] R. Barlow, A. Wu (1978): *Coherent Systems with Multistate Components.* Math. Oper. Res. **3**, 275–281.

[3] L. Baxter (1984): *Continuum Structures I.* J. Appl. Prob. **21**, 802–815.

[4] L. Baxter (1986): *Continuum Structures II.* Prec. Camb. Phil. Soc. **99**, 331–338.

[5] W. Griffith (1980): *Multistate Reliability Theory.* J. Appl. Prob. **17**, 735–744.

[6] B. Natvig (1982): *Two Suggestions of How to Define a Multistate Cohent System.* Adv. Appl. Prob. **14**, 434–455.

[7] F. Ohi, T. Nishida (1984): *On Multistate Coherent Systems.* IEEE Trans. Rel. **R–33**, 284–287.

[8] T. Rolski, R. Szekli (1991): *Stochastic Ordering and Thinning of Point Processes.* Stochastic Processes and their Applications **37**, 299–312.

[9] S. Ross (1979): *Multivalued State Component Systems.* The Annals of Probability **7.2**, 379–383.

[10] A. Schöttl (1994): *Comparison of Discrete Measures.* TU Munich, Technical Report IAMS1994.6TUM.

[11] D. Stoyan (1983): *Comparison Methods for Queues and Other Stochastic Models.* John Wiley & Sons, New York.

[12] W. Wertz (1978): *Statistical Density Estimation: A Survey.* Venadenhoeck & Rupprecht, Göttingen.

Automatic Differentiation: A Structure-Exploiting Forward Mode with Almost Optimal Complexity for Kantorovič Trees

MICHAEL ULBRICH AND STEFAN ULBRICH

Institut für Angewandte Mathematik und Statistik
Technische Universität München

Abstract. A structure-exploiting forward mode is discussed that achieves almost optimal complexity for functions given by Kantorovič trees. It is based on approriate representations of the gradient and the Hessian. After a brief exposition of the forward and reverse mode of automatic differentiation for derivatives up to second order and compact proofs of their complexities, the new forward mode is presented and analyzed. It is shown that in the case of functions $f : \mathbb{R}^n \to \mathbb{R}$ with a tree as Kantorovič graph the algorithm is only $O(\ln(n))$ times as expensive as the reverse mode. Except for the fact that the new method is a very efficient implementation of the forward mode, it can be used to significantly reduce the length of characterizing sequences before applying the memory expensive reverse mode. For the Hessian all discussed algorithms are shown to be efficiently parallelizable. Some numerical examples confirm the advantages of the new forward mode.

Keywords. automatic differentiation, characterizing sequence, code list, forward mode, reverse mode, Kantorovič graph, Kantorovič tree, time complexity, parallelization.

1. Introduction

The paper is organized as follows. In Section 2 we introduce the basic definition of a characterizing sequence (or code list), which describes how a function $f : D \subset \mathbb{R}^n \longrightarrow \mathbb{R}$ can be evaluated at the point x by applying atomic unary and binary operations. The characterizing sequence can be converted into a corresponding Kantorovič graph. In this paper, characterizing sequences with a tree as Kantorovič graph are of special interest. In Section 3 we establish notations of computational complexity and collect some results concerning the complexity of partial derivatives. Section 4 introduces the basic idea of automatic differentiation of characterizing sequences and describes the forward mode for first and second order derivatives. The complexity of forward differentiation is treated in Section 5. Sections 6 and 7 deal with the reverse mode and include a new compact proof of its complexity. In Section 8 we describe the reverse mode for the Hessian and proof a complexity result. Section 9 contains the main part of this paper. Here we introduce a new structure-exploiting forward mode and show that it achieves almost optimal complexity for characterizing sequences with a tree as Kantorovič graph. The complexity considerations base on the observation that for Kantorovič trees, there exists a crucial relation between the total number of atomic operations necessary to compute $f(x)$, $\nabla f(x)$ (and $\nabla^2 f(x)$) and the total number of 'non-trivial created' nonzero components in temporary gradients, which is shown to grow with order $O(\ln(n)\#(f,S))$. Here $\#(f,S)$ denotes the computational costs to calculate $f(x)$ by means of the characterizing sequence S. In Section 10 some applications of the new method are proposed. A data structure for vectors and matrices tailored to the new algorithm is presented in Section 11. Section 12 briefly discusses how the methods of automatic differentiation can be efficiently parallelized. Some numerical results that show the advantages of the new forward mode can be found in Section 13.

2. Characterizing Sequences and Kantorovič Graphs

Let $f : D \subset \mathbb{R}^n \longrightarrow \mathbb{R}$, $n \in \mathbb{N}$, be a twice differentiable function on an open set D. For $x \in D$, we denote by $\nabla f(x) \in \mathbb{R}^n$ the gradient (a column vector) and by $\nabla^2 f(x) \in \mathbb{R}^{n\times n}$ the Hessian of f. In the following we will deal with functions f, which can be expressed as composition of elementary unary operations $\varphi \in \mathcal{U}$, binary operations $\beta \in \mathcal{B}$, the canonical projections $\pi_i : x \longmapsto x_i$ and constant functions $x \longmapsto c \in \mathcal{C} = \mathbb{R}$. Functions of the above type can be evaluated in an algorithmic manner by characterizing sequences.

Definition 2.1 (characterizing sequence, code list)

Set $I = \{1,\ldots,m\}$, $J = \{1,\ldots,n+m-1\}$. The vector

$$S = \left((\omega_i, k_i, l_i)\right)_{i\in I} \in \left((\mathcal{C}\times\{0\}^2)\cup(\mathcal{U}\times J\times\{0\})\cup(\mathcal{B}\times J^2)\right)^m,$$

$m \in \mathbb{N}$, is called a (connected) *charcterizing sequence* (code list) for f, iff

a) $k_i < i+n,\ l_i < i+n,\ \bigcup_{i=1}^{m}\{k_i,l_i\} \supset \{n+1,\ldots,n+m-1\}$

b) $f(x) = f_{m+n}$ for $x \in D$ and

$$f_j = x_j,\quad j = 1,\ldots,n$$

$$f_{i+n} = \begin{cases} \omega_i & ,\ \omega_i \in \mathcal{C} \\ \omega_i(f_{k_i}) & ,\ \omega_i \in \mathcal{U} \\ \omega_i(f_{k_i}, f_{l_i}) & ,\ \omega_i \in \mathcal{B} \end{cases} \quad ,\quad i = 1,\ldots,m$$

□

Each characterizing sequence describes a specific way to evaluate the corresponding *Kantorovič graph*. This directed labeled graph $G = (N,E)$ can be obtained as follows:

The set of nodes is given by

$$N = \bigcup_{i=1}^{m} \{k_i, l_i\} \cup \{n+m\} \setminus \{0\} .$$

Each node $j \in N$ is labeled by a labeling function λ:

$$\lambda(j) = \begin{cases} x_j & , \quad 1 \le j \le n \\ \omega_{j-n} & , \quad n < j \le n+m \end{cases}$$

The set $E \subset N^2 \times \{1,2\}$ of directed labeled edges is given by

$$E = \{(k_i, i+n, 1)\,;\; k_i > 0,\ 1 \le i \le m\} \cup \{(l_i, i+n, 2)\,;\; l_i > 0,\ 1 \le i \le m\} .$$

The edge $e = (j_1, j_2, \mu) \in E$ is directed from node j_1 to j_2. The label μ distinguishes between first and second operand of a binary operation. e is called *in-edge* of node j_2 and *out-edge* of j_1. A node without in-edge is called *leaf.*

Remark 2.2

- Leaves are always labeled by x_i or $c \in \mathcal{C}$.
- G is connected.
- G is cycle-free, because each edge $(j_1, j_2, \mu) \in E$ satisfies $j_1 < j_2$ and thus the nodes along a directed path are strictly increasing.
- In general, different characteristic sequences may lead to equivalent Kantorovič graphs.

□

If $r = n + m$ is the only node without out-edge and all nodes with more than one out-edge are leaves, then G is called *Kantorovič tree* with *root* r. In fact, a tree in the usual sense can be obtained by splitting a leaf with k out-edges into k different leaves with the same label and one out-edge. In this case, each temporary result f_i, $i > n$, is used for the calculation of exactly one f_j, $j > i$.

3. Complexity of Partial Derivatives

In the following, the binary operations are given by

$$\mathcal{B} = \{\beta : D_\beta \subset \mathbb{R}^2 \longrightarrow \mathbb{R}\,;\; \beta(a,b) = a \otimes b,\ \otimes \in \{+, -, \cdot\,, /\}\} \qquad (1)$$

For the analysis of time complexity we assume that all unary operations $\omega \in \mathcal{U}$, their derivatives ω', ω'' and the binary operations are *atomic* and can be executed in one time unit. Evaluations of constant functions and their vanishing derivatives of arbitrary order are free. Further, we don't count sign changes. Hence, the time complexity for computing $f(x)$ is proportional to the number of atomic operations.

Definition 3.1 Let S be a characterizing sequence for f of length m, then

$$\#(f,S) := |\{i\,;\; 1 \le i \le m,\; \omega_i \notin C\}|$$

denotes the number of atomic operations to evaluate S for a given x. The minimal operation count is denoted by $\#(f)$:

$$\#(f) := \min\{\#(f,S)\,;\; S \text{ is a characterizing sequence for } f\}\,.$$

If A is an algorithm to compute $f(x)$ and $\nabla^k f(x)$, $k = 1,\dots,s$, by means of S, $\#(f,\nabla f,\dots,\nabla^s f,S,A)$ denotes the number of necessary atomic operations. Finally, we set

$$\#(f,\nabla f,\dots,\nabla^s f,S) := \min\{\#(f,\nabla f,\dots,\nabla^s f,S,A)\,; \quad A \text{ computes } f(x),\dots,\nabla^s f(x) \text{ for given } x,S\}\,,$$

$$\#(f,\nabla f,\dots,\nabla^s f) := \min\{\#(f,\nabla f,\dots,\nabla^s f,S)\,;\; S \text{ char. sequence for } f\}\,.$$

□

The following basic result gives a bound for $\#(f,\nabla f)$ in terms of $\#(f)$.

Theorem 3.2 (W. Baur, V. Strassen, 1983)
Let $f : D \subseteq \mathbb{R}^n \longrightarrow \mathbb{R}$ be a rational function. Then the following inequality holds:

$$\#(f,\nabla f) \;\le\; 4\,\#(f)$$

Proof: See [1].

A similar result can be found in [3]: H. Fischer and H. Warsitz propose an algorithm $RM.G$ with the following property:

Theorem 3.3 (H. Fischer, H. Warsitz, 1995)
Let $f : D \subseteq \mathbb{R}^n \longrightarrow \mathbb{R}$ be a rational function and S an arbitrary characterizing sequence for f. Then:

$$\#(f,\nabla f,S,RM.G) \le 4\,\#(f,S)$$

Remark 3.4 In Section 6 we will introduce a simple alternative to algorithm $RM.G$, which also works for $\mathcal{U} \neq \emptyset$, and give a proof of the above theorem.

□

4. Automatic Differentiation

For a representative overview of the subject see [2], [5], [7], and the references therein. The main idea consists in differentiating the characterizing sequence. Let $f : D \subset \mathbb{R}^n \longrightarrow \mathbb{R}$ be a twice differentiable function given by S.

Characterizing sequence for the function f:

$j = 1, \dots, n$: $\quad f_j = x_j$

$i = 1, \dots, m$:

$$f_{i+n} = \begin{cases} \omega_i & , \ \omega_i \in \mathcal{C} \\ \omega_i(f_{k_i}) & , \ \omega_i \in \mathcal{U} \\ \omega_i(f_{k_i}, f_{l_i}) & , \ \omega_i \in \mathcal{B} \end{cases}$$

Sequence for the gradient ∇f:

$j = 1, \dots, n$: $\quad g_j = e_j = (\delta_{jk})_{1 \le k \le n}$

$i = 1, \dots, m$:

$$g_{i+n} = \begin{cases} 0 & , \ \omega_i \in \mathcal{C} \\ \omega_i'(f_{k_i})\, g_{k_i} & , \ \omega_i \in \mathcal{U} \\ \partial_a \omega_i(f_{k_i}, f_{l_i})\, g_{k_i} + \partial_b \omega_i(f_{k_i}, f_{l_i})\, g_{l_i} & , \ \omega_i \in \mathcal{B} \end{cases}$$

Sequence for the Hessian $\nabla^2 f$:

$j = 1, \dots, n$: $\quad H_j = 0$

$i = 1, \dots, m$:

$$H_{i+n} = \begin{cases} 0 & , \ \omega_i \in \mathcal{C} \\ \omega_i'(f_{k_i})\, H_{k_i} + \omega_i''(f_{k_i})\, g_{k_i} g_{k_i}^T & , \ \omega_i \in \mathcal{U} \\ \partial_a \omega_i(f_{k_i}, f_{l_i})\, H_{k_i} + \partial_b \omega_i(f_{k_i}, f_{l_i})\, H_{l_i} & \\ \quad + (g_{k_i}, g_{l_i})\, \nabla^2 \omega_i(f_{k_i}, f_{l_i})\, (g_{k_i}, g_{l_i})^T & , \ \omega_i \in \mathcal{B} \end{cases}$$

Now the values $f(x) = f_{n+m}$, $\nabla f(x) = g_{n+m}$ and $\nabla^2 f(x) = H_{n+m}$ can be obtained in a straightforward way by evaluating the above sequences in the order

$$f_j, g_j, H_j, \quad j = 1, \dots, n+m\,.$$

This method is called *forward mode*.

5. Complexity of the Forward Mode

The following tables show the number of *additional* atomic operations for the calculation of g_{i+n} and H_{i+n}, respectively. For the Hessian, we use the symmetry and results already computed for the gradient. The resulting algorithm for computing $f(x)$, $\nabla f(x)$ (and $\nabla^2 f(x)$) will be called F_g (F_H). We set $j = i + n$.

ω_i	Function	Gradient	Ops. for g_j
$\varphi(\cdot)$	$f_j = \varphi(f_{k_i})$	$g_j = \varphi'(f_{k_i})\, g_{k_i}$	$n+1$
$\pm$	$f_j = f_{k_i} \pm f_{l_i}$	$g_j = g_{k_i} \pm g_{l_i}$	n
$\cdot$	$f_j = f_{k_i} \cdot f_{l_i}$	$g_j = f_{l_i}\, g_{k_i} + f_{k_i}\, g_{l_i}$	$3n$
$/$	$f_j = f_{k_i} / f_{l_i}$	$g_j = (g_{k_i} - f_j\, g_{l_i}) / f_{l_i}$	$3n$

Table 1: Number of atomic operations for the calculation of g_{i+n}

ω_i	Hessian	Ops. for H_j
$\varphi(\cdot)$	$H_j = \varphi'(f_{k_i})\, H_{k_i} + \varphi''(f_{k_i})\, g_{k_i}\, g_{k_i}^T$	$(3n+2)\,(n+1)/2$
$\pm$	$H_j = H_{k_i} \pm H_{l_i}$	$n\,(n+1)/2$
$\cdot$	$H_j = f_{l_i}\, H_{k_i} + f_{k_i}\, H_{l_i}$ $+ g_{k_i}\, g_{l_i}^T + g_{l_i}\, g_{k_i}^T$	$n\,(7n+5)/2$
$/$	$H_j = (H_{k_i} - f_j\, H_{l_i}$ $- g_j\, g_{l_i}^T - g_{l_i}\, g_j^T) / f_{l_i}$	$n\,(7n+5)/2$

Table 2: Number of atomic operations for the calculation of H_{i+n}

Theorem 5.1 *Let S be a characterizing sequence for the twice differentiable function $f : D \subset \mathbb{R}^n \longrightarrow \mathbb{R}$. Then the following inequalities hold:*

$$(n+1)\,\#(f,S) \le \quad \#(f,\nabla f, S, F_g) \quad \le (3n+1)\,\#(f,S)$$

$$\frac{n^2+3n+2}{2}\,\#(f,S) \le \#(f,\nabla f,\nabla^2 f, S, F_H) \le \frac{7n^2+11n+2}{2}\,\#(f,S)$$

Proof: According to Table 1, the additional computation of ∇f_{i+n} in step i, $i = 1,\dots,m$, requires at least n and at most $3n$ operations. To calculate also the Hessian H_{i+n}, $n(n+1)/2$ to $n(7n+5)/2$ additional operations are necessary (see Table 2).

□

6. The Reverse Mode

Let S be a characterizing sequence for f of length m. Based on S, we define the sequence of functions $F_1, \dots, F_m$ as follows:

$i = 1, \ldots, m-1$:

$$F_i : D_i \subset \mathbb{R}^{n+i-1} \longrightarrow \mathbb{R}^{n+i}, \quad F_i(y_1, \ldots, y_{n+i-1}) = \begin{pmatrix} y_1 \\ \vdots \\ y_{n+i-1} \\ \gamma_i \end{pmatrix}$$

$$F_m : D_m \subset \mathbb{R}^{n+m-1} \longrightarrow \mathbb{R}, \quad F_m(y_1, \ldots, y_{n+m-1}) = \gamma_m$$

with $D_i = \{y \in \mathbb{R}^{n+i-1};\ y_{k_i} \in D_{\omega_i} \text{ if } \omega_i \in \mathcal{U},\ (y_{k_i}, y_{l_i}) \in D_{\omega_i} \text{ if } \omega_i \in \mathcal{B}\}$,

$$\gamma_i = \begin{cases} \omega_i & , \ \omega_i \in \mathcal{C} \\ \omega_i(y_{k_i}) & , \ \omega_i \in \mathcal{U} \\ \omega_i(y_{k_i}, y_{l_i}) & , \ \omega_i \in \mathcal{B} \end{cases} , \quad i = 1, \ldots, m .$$

We set

$$G_0 := x\ , \quad G_i^T := (f_1, \ldots, f_{n+i})^T = F_i \circ F_{i-1} \circ \cdots \circ F_1(x)\ ,\ i = 1, \ldots, m-1.$$

Differentiating the identity

$$f(x) = f_{n+m} = F_m \circ F_{m-1} \circ \cdots \circ F_1(x)$$

leads to

$$\nabla f(x)^T = DF_m(G_{m-1})\, DF_{m-1}(G_{m-2}) \cdots DF_1(x)\,, \tag{2}$$

where DF denotes the Jacobian of F. The forward mode corresponds to the evaluation of (2) *from the right to the left.* On the other hand, if $f_1, \ldots, f_{n+m}$ are already computed, (2) may also be evaluated *from the left to the right.* This method is called *reverse mode.*

7. Complexity of the Reverse Mode

In detail, (2) has the form

$$\nabla f(x)^T = \begin{pmatrix} \underset{\substack{\uparrow \\ k_m}}{\kappa_m} & \underset{\substack{\uparrow \\ l_m}}{\lambda_m} \end{pmatrix} \left(\begin{array}{cc} \multicolumn{2}{c}{I_{n+m-2}} \\ \hline \underset{\substack{\uparrow \\ k_{m-1}}}{\kappa_{m-1}} & \underset{\substack{\uparrow \\ l_{m-1}}}{\lambda_{m-1}} \end{array} \right) \cdots \left(\begin{array}{cc} \multicolumn{2}{c}{I_n} \\ \hline \underset{\substack{\uparrow \\ k_1}}{\kappa_1} & \underset{\substack{\uparrow \\ l_1}}{\lambda_1} \end{array} \right)$$

If $k_i = 0$ or $l_i = 0$, ignore the corresponding entry in the last row. The reverse mode is carried out in m steps:

$$v^m := DF_m(G_{m-1})\ , \quad v^i := v^{i+1} DF_i(G_{i-1})\ , \quad i = m-1, \ldots, 1$$

Denoting by v_j^i the jth component of v^i, v^i originates from v^{i+1} by adding $\kappa_i v_{i+n}^{i+1}$ (if $k_i \neq 0$) to the k_i^{th}, $\lambda_i v_{i+n}^{i+1}$ (if $l_i \neq 0$) to the l_i^{th} and deleting the last component. To obtain minimal complexity, we omit the additions, if $v_{k_i}^{i+1} = 0$ or $v_{l_i}^{i+1} = 0$, respectively. Table 3 collects the number of atomic operations to compute v^i from v^{i+1}; we set $j = i + n$ and define $\sigma : \mathbb{R} \longrightarrow \{0, 1\}$ by $\sigma(0) = 0$, $\sigma(\xi) = 1$ for $\xi \neq 0$.

ω_i	Function	κ_i	λ_i	Ops. κ_i, λ_i	Ops. $v^{i+1} \to v^i$
c	$f_j = c$	–	–	0	0
$\varphi(\cdot)$	$f_j = \varphi(f_{k_i})$	$\varphi'(f_{k_i})$	–	1	$2 + \sigma(v_{k_i}^{i+1})$
$\pm$	$f_j = f_{k_i} \pm f_{l_i}$	1	± 1	0	$\sigma(v_{k_i}^{i+1}) + \sigma(v_{l_i}^{i+1})$
$\cdot$	$f_j = f_{k_i} \cdot f_{l_i}$	f_{l_i}	f_{k_i}	0	$2 + \sigma(v_{k_i}^{i+1}) + \sigma(v_{l_i}^{i+1})$
$/$	$f_j = f_{k_i} / f_{l_i}$	$1/f_{l_i}$	$-f_j\,\kappa_i$	2	$2 + \sigma(v_{k_i}^{i+1}) + \sigma(v_{l_i}^{i+1})$

Table 3: Operation counts for a step of the reverse mode

The resulting algorithm is called R_g.

Theorem 7.1 *Let $f : D \subseteq \mathbb{R}^n \longrightarrow \mathbb{R}$ be a differentiable non-constant function and S an arbitrary characterizing sequence for f of length m. Then:*

$$\#(f, \nabla f, S, R_g) \leq 5\#(f, S) - \left| \bigcup_{i=1}^{m} \{k_i, l_i\} \setminus \{0\} \right| \leq 5\#(f, S) - m \leq 4\#(f, S)$$

Proof: From $f \neq$ const. and Definition 2.1 a) follows $m > 1$, $\omega_m \notin C$ and (note the dependence on at least one x_j)

$$N := \left| \bigcup_{i=1}^{m} \{k_i, l_i\} \setminus \{0\} \right| \geq m \,. \tag{3}$$

For simplicity, we set $v_0^i = 0$, $i = 1, \ldots, m$. According to Table 3, each of the $\#(f, S) - 1$ steps $v^{i+1} \to v^i$ with $\omega_i \notin C$ costs $\leq 2 + \sigma(v_{k_i}^{i+1}) + \sigma(v_{l_i}^{i+1})$ operations, those with $\omega_i \in C$ are free. The computation of v^m requires ≤ 2 operations. Thus, we have

$$\begin{aligned} \#(f, \nabla f, S, R_g) &\leq \#(f, S) + 2 + \sum_{\substack{1 \leq i < m \\ \omega_i \notin C}} (2 + \sigma(v_{k_i}^{i+1}) + \sigma(v_{l_i}^{i+1})) \\ &= 3\#(f, S) + \sum_{\substack{1 \leq i < m \\ \omega_i \notin C}} (\sigma(v_{k_i}^{i+1}) + \sigma(v_{l_i}^{i+1})) \end{aligned} \tag{4}$$

Since in step $v^{i+1} \to v^i$ additional non-zero elements can only arise at position k_i and l_i (if > 0), we have

$$\{j\,;\ 1 \le j \le n+i-1,\ v^i_j \ne 0\} \subset \left(\bigcup_{j=i}^{m} \{k_j, l_j\} \right) \cap \{1, \ldots, n+i-1\}\,.$$

With (3) and (4) follows the assertion:

$$\begin{aligned} \#(f, \nabla f, S, R_g) &\le 3\#(f,S) + \Big(2(\#(f,S)-1) - (N-2)\Big) = 5\#(f,S) - N \\ &\le 5\#(f,S) - m \le 4\#(f,S) \end{aligned}$$

□

Remark 7.2 If f is constant, then the theorem remains true for all sequences S which contain a k_i or l_i in $\{1, \ldots, n\}$.

□

8. A Reverse Mode of Optimal Order for the Hessian

The following example demonstrates that there exist functions f with

$$\#(f, \nabla f, \nabla^2 f) \ge cn\#(f)\ , \quad c \ge \frac{1}{2} \text{ independent of } n.$$

Example 8.1 Let $f(x) = \left(\prod_{i=1}^{n} x_i \right)^{-1}$. Then

$$\frac{\partial f(x)}{\partial x_i} = -\frac{f(x)}{x_i}\,, \quad \frac{\partial^2 f(x)}{\partial x_i \partial x_j} = \begin{cases} \dfrac{f(x)}{x_i x_j} & , \ i \ne j \\ \dfrac{2f(x)}{x_i^2} & , \ i = j \end{cases} \quad , \ i,j = 1, \ldots, n$$

and

$$\#(f) = n\ , \quad \#(f, \nabla f, \nabla^2 f) = 3n + \frac{n(n+1)}{2} > \frac{n+1}{2}\#(f)\,.$$

□

If S is a characterizing sequence of an arbitrary twice differentiable function f, then $f(x) = f_{n+m}$ and the partial derivative $\partial_{x_k} f(x) = h(x, e_k) = h_{n+m}$ can be computed simultanously using the forward mode:

$j = 1, \ldots, n$: $\quad f_j = x_j,\ h_j = y_j$

$i = 1, \ldots, m$:

$$f_{i+n} = \begin{cases} \omega_i & , \ \omega_i \in \mathcal{C} \\ \omega_i(f_{k_i}) & , \ \omega_i \in \mathcal{U} \\ \omega_i(f_{k_i}, f_{l_i}) & , \ \omega_i \in \mathcal{B} \end{cases}$$

$$h_{i+n} = \begin{cases} 0 & , \ \omega_i \in \mathcal{C} \\ \omega_i'(f_{k_i})\, h_{k_i} & , \ \omega_i \in \mathcal{U} \\ \partial_a\omega_i(f_{k_i}, f_{l_i})\, h_{k_i} + \partial_b\omega_i(f_{k_i}, f_{l_i})\, h_{l_i} & , \ \omega_i \in \mathcal{B} \end{cases}$$

Obviously, the expressions for h_j consist of at most three atomic operations (see Table 1) if we replace $\mathcal{U}$ by

$$\mathcal{U} \cup \{\varphi' ;\ \varphi \in \mathcal{U}\}\,.$$

Thus, after deleting unused temporary results f_j (all h_j are needed), we obtain a characterizing sequence S_h of length $m_h \leq 4m$ for $h(x,y)$ and application of the reverse mode to S_h at $(x,y) = (x, e_k)$ yields the kth row of the Hessian. We call the resulting algorithm R_H and get the following result:

Theorem 8.2 *Let $f : D \subseteq \mathbb{R}^n \longrightarrow \mathbb{R}$ be a twice differentiable non-constant function and S an arbitrary characterizing sequence for f. Then:*

$$\#(f, \nabla f, \nabla^2 f, S, R_H) \leq 16\, n\, \#(f, S)$$

Proof: If the computation of f_j requires l operations, then we need $\leq 3l$ operations for h_j. Thus, the computation of $f(x)$ and $\partial_{x_k} f(x) = h(x, e_k)$ by the above forward sequence costs $\leq 4\,\#(f,S)$ operations and consequently

$$\#(h, S_h) \leq 4\,\#(f, S)\,.$$

Since S contains at least one x_j-dependence, S_h contains a y_j-dependence and application of Theorem 7.1 in connection with Remark 7.2 leads to

$$\#(f, \nabla f, \nabla^2 f, S, R_H) \leq 16\, n\, \#(f, S)$$

□

Remark 8.3 Since the f_j are independent of y, they can be computed only once instead of n times. This improved algorithm $\tilde{R}_H$ satisfies

$$\#(f, \nabla f, \nabla^2 f, S, \tilde{R}_H) \leq (15n + 1)\,\#(f, S)\,.$$

□

Remark 8.4 The result of Remark 8.3 can also be found in [2]. In [6] M. IRI mentions an algorithm A_H with

$$\#(f, \nabla f, \nabla^2 f, S, A_H) \leq (10n + 4)\, \#(f, S)$$

□

9. An Efficient New Forward Mode

The applicability of the reverse mode is severely restricted by the fact that the required memory grows linearly with $\#(f, S)$. A. GRIEWANK overcomes this difficulty by formulating a recursive reverse mode (see [4]) that achieves time-complexity $O\left(\left(c_1 + c_2 \log_2\left(\#(f, S)\right)\right) \#(f, S)\right)$ and reduces the space-complexity to $O\left(c_3 \log_2\left(\#(f, S)\right)\right)$. In the following, we will propose new (to the knowledge of the authors) structure-exploiting forward algorithms SF_g and SF_H with complexities

$$\#(f, \nabla f, S, SF_g) \leq 7\#(f, S) - b + (\#(f, S) + 1) \ln\left(\max\{b, 1\}\right)$$

and

$$\#(f, \nabla f, \nabla^2 f, S, SF_H) \leq$$
$$\leq 5(b+1)\left(7\#(f, S) - \frac{3}{2}b + (\#(f, S) + 1) \ln\left(\max\{b, 1\}\right)\right) + 3b$$

for characterizing sequences with a tree as Kantorovič graph. Here b denotes the number of x_k's on which f depends. To this end, we improve the forward mode by the following modifications:

- For the temporary gradients g_j and Hessians H_j use the representations

  $$g_j = \alpha_j v_j \ , \quad H_j = \beta_j v_j v_j^T + \gamma_j w_j w_j^T + \delta_j M_j$$

  with $\alpha_j, \beta_j, \gamma_j, \delta_j \in \mathbb{R}$, $\alpha_j, \delta_j \neq 0$, vectors v_j, w_j and matrices M_j.
- Restrict the computations to non-zero elements.
- When two temporary results are combined by a binary atomic operation, obtain the new gradient- and Hessian-representation by copying and modifying the one with the less sparse v_j. Don't copy non-leaves with one out-edge, since they are no more needed and can be overwritten.
- When applying a unary atomic operation to a non-leaf with one out-edge, obtain the new temporary result by modifying the old one.

Definition 9.1 A characterizing sequence S is called *T-sequence* iff the corresponding Kantorovič graph is a Kantorovič tree.

□

Let S be a characterizing sequence for $f : D \subset \mathbb{R}^n \longrightarrow \mathbb{R}$ of length m. Obviously, the characterizing subsequence $S_i = (\omega_s, k_s, l_s)_{s \in N_i}$ for the calculation of f_{i+n}, $1 \leq i \leq m$, can be constructed by setting

$$N_i := \bigcup_{k=1}^{i} N_i^k \tag{5}$$

with

$$N_i^1 := \{i\} \ , \quad N_i^{k+1} := \bigcup_{s \in N_i^k} \{k_s - n, l_s - n\} \setminus \{-n, \ldots, 0\} \ , \ k = 1, \ldots, i-1 \, .$$

Denote by b_i the number of x_k's used in S_i, $1 \leq i \leq m$, i.e.

$$b_i := |B_i| \ \text{ with } \ B_i := \begin{cases} \bigcup_{s \in N_i} \{k_s, l_s\} \cap \{1, \ldots, n\} & , \ 1 \leq i \leq m \\ \emptyset & , \ i \leq 0 \end{cases} \tag{6}$$

and define

$$\begin{aligned} q_i &:= b_i - \max\{b_{k_i - n}, b_{l_i - n}\} \\ p_i &:= \begin{cases} q_i & , \ \{k_i, l_i\} \cap \{1, \ldots, n\} = \emptyset \\ 0 & , \ \text{otherwise} \end{cases} \end{aligned} \tag{7}$$

q_i can be seen as the number of fill-ins, when v_{i+n} is calculated by merging the sparser of v_{k_i}, v_{l_i} into the other. p_i counts only fill-ins, if both operands aren't x_k's.

Algorithm SF_g:

$j = 1, \ldots, n$:

$\alpha_j = 1$, $v_j = e_j$

$i = 1, \ldots, m$:

$j := i + n$

I) $\omega_i \in C$:

$f_j = \omega_i, \ g_j = 0 = \alpha_j v_j$

with

$\alpha_j = 1$, $v_j = 0$.

II) $\omega_i \in \mathcal{U}$:

$$f_j = \omega_i(f_{k_i}), \quad g_j = \omega_i'(f_{k_i})g_{k_i} = \omega_i'(f_{k_i})\alpha_{k_i}v_{k_i} = \alpha_j v_j$$

with

1) $k_i \in \{1,\dots,n\}: \quad \alpha_j = 1,\ v_j = \omega_i'(x_{k_i})\, e_{k_i}.$

2) $k_i \notin \{1,\dots,n\},\ \omega_i'(f_{k_i}) \neq 0: \quad \alpha_j = \omega_i'(f_{k_i})\alpha_{k_i},\ v_j = v_{k_i}.$

3) $k_i \notin \{1,\dots,n\},\ \omega_i'(f_{k_i}) = 0: \quad \alpha_j = 1,\ v_j = 0.$

III) $\omega_i \in \mathcal{B},\ \omega_i(a,b) = a \pm b$:

$$f_j = f_{k_i} \pm f_{l_i}, \quad g_j = g_{k_i} \pm g_{l_i} = \alpha_{k_i}v_{k_i} \pm \alpha_{l_i}v_{l_i} = \alpha_j v_j$$

with

1) $k_i, l_i \in \{1,\dots,n\}: \quad \alpha_j = 1,\ v_j = e_{k_i} \pm e_{l_i}.$

2) $k_i \notin \{1,\dots,n\},\ l_i \in \{1,\dots,n\}: \quad \alpha_j = \alpha_{k_i},\ v_j = v_{k_i} \pm \dfrac{1}{\alpha_{k_i}} e_{l_i}.$

3) $l_i \notin \{1,\dots,n\},\ k_i \in \{1,\dots,n\}:$ Analogous to 2).

4) $k_i, l_i \notin \{1,\dots,n\},\ b_{k_i-n} \geq b_{l_i-n}$: $\alpha_j = \alpha_{k_i},\ v_j = v_{k_i} \pm \dfrac{\alpha_{l_i}}{\alpha_{k_i}} v_{l_i}.$

5) $k_i, l_i \notin \{1,\dots,n\},\ b_{k_i-n} < b_{l_i-n}$: Analogous to 4).

IV) $\omega_i \in \mathcal{B},\ \omega_i(a,b) = a \cdot b$:

$$f_j = f_{k_i} f_{l_i}, \quad g_j = f_{l_i} g_{k_i} + f_{k_i} g_{l_i} = f_{l_i}\alpha_{k_i}v_{k_i} + f_{k_i}\alpha_{l_i}v_{l_i} = \alpha_j v_j$$

with

1) $k_i, l_i \in \{1,\dots,n\}: \quad \alpha_j = 1,\ v_j = x_{l_i} e_{k_i} + x_{k_i} e_{l_i}.$

2) $k_i \notin \{1,\dots,n\},\ l_i \in \{1,\dots,n\},\ x_{l_i} \neq 0: \quad \alpha_j = x_{l_i}\alpha_{k_i},\ v_j = v_{k_i} + \dfrac{f_{k_i}}{\alpha_j} e_{l_i}.$

3) $k_i \notin \{1,\dots,n\},\ l_i \in \{1,\dots,n\},\ x_{l_i} = 0: \quad \alpha_j = 1,\ v_j = f_{k_i} e_{l_i}.$

4) $l_i \notin \{1,\dots,n\},\ k_i \in \{1,\dots,n\},\ x_{k_i} \neq 0:$ Analogous to 2).

5) $l_i \notin \{1,\dots,n\},\ k_i \in \{1,\dots,n\},\ x_{k_i} = 0:$ Analogous to 3).

6) $k_i, l_i \notin \{1,\dots,n\},\ b_{k_i-n} \geq b_{l_i-n},\ f_{l_i} \neq 0$: $\alpha_j = f_{l_i}\alpha_{k_i},$
$$v_j = v_{k_i} + \frac{f_{k_i}\alpha_{l_i}}{\alpha_j} v_{l_i}.$$

7) $k_i, l_i \notin \{1,\dots,n\},\ b_{k_i-n} \geq b_{l_i-n},\ f_{l_i} = 0$: $\alpha_j = 1,\ v_j = f_{k_i}\alpha_{l_i}v_{l_i}.$

8) $k_i, l_i \notin \{1,\ldots,n\}$, $b_{k_i-n} < b_{l_i-n}$, $f_{k_i} \neq 0$: Analogous to 6).

9) $k_i, l_i \notin \{1,\ldots,n\}$, $b_{k_i-n} < b_{l_i-n}$, $f_{k_i} = 0$: Analogous to 7).

V) $\omega_i \in \mathcal{B}$, $\omega_i(a,b) = a/b$:

$$f_j = \frac{f_{k_i}}{f_{l_i}}, \; g_j = \frac{1}{f_{l_i}}(g_{k_i} - f_j g_{l_i}) = \frac{\alpha_{k_i}}{f_{l_i}} v_{k_i} - \frac{f_j \alpha_{l_i}}{f_{l_i}} v_{l_i} = \alpha_j v_j$$

with

1) $k_i, l_i \in \{1,\ldots,n\}: \quad \alpha_j = \dfrac{1}{x_{l_i}}, \; v_j = e_{k_i} - f_j e_{l_i}.$

2) $k_i \notin \{1,\ldots,n\}, \; l_i \in \{1,\ldots,n\}: \quad \alpha_j = \dfrac{\alpha_{k_i}}{x_{l_i}}, \; v_j = v_{k_i} - \dfrac{f_j}{\alpha_{k_i}} e_{l_i}.$

3) $l_i \notin \{1,\ldots,n\}, \; k_i \in \{1,\ldots,n\}, \; x_{k_i} \neq 0: \quad \alpha_j = -\dfrac{f_j \alpha_{l_i}}{f_{l_i}},$

$$v_j = v_{l_i} - \frac{1}{\alpha_{l_i} f_j} e_{k_i}.$$

4) $l_i \notin \{1,\ldots,n\}, \; k_i \in \{1,\ldots,n\}, \; x_{k_i} = 0: \quad \alpha_j = 1, \; v_j = \dfrac{1}{f_{l_i}} e_{k_i}.$

5) $k_i, l_i \notin \{1,\ldots,n\}, \; b_{k_i-n} \geq b_{l_i-n}: \quad \alpha_j = \dfrac{\alpha_{k_i}}{f_{l_i}}, \; v_j = v_{k_i} - \dfrac{\alpha_{l_i} f_j}{\alpha_{k_i}} v_{l_i}.$

6) $k_i, l_i \notin \{1,\ldots,n\}, \; b_{k_i-n} < b_{l_i-n}, \; f_{k_i} \neq 0: \quad \alpha_j = -\dfrac{f_j \alpha_{l_i}}{f_{l_i}},$

$$v_j = v_{l_i} - \frac{\alpha_{k_i}}{\alpha_{l_i} f_j} v_{k_i}.$$

7) $k_i, l_i \notin \{1,\ldots,n\}, \; b_{k_i-n} < b_{l_i-n}, \; f_{k_i} = 0: \quad \alpha_j = \dfrac{\alpha_{k_i}}{f_{l_i}}, \; v_j = v_{k_i}.$

Algorithm SF_H can be found in the Appendix. Table 4 contains the number of atomic operations that are required in each step of algorithm SF_g (SF_H). We turn to the analysis of time complexity:

Lemma 9.2 *If* $0 < y_1 \leq z_1$, $0 < y_2 \leq z_2$ *and* $0 \leq y \leq \min\{y_1, y_2\}$, *then*

$$z_1 \ln y_1 + z_2 \ln y_2 + y \leq (z_1 + z_2) \ln (y + \max\{y_1, y_2\})$$

Proof: Without loss of generality, assume $y_1 \leq y_2$. We have to show

$$\psi(y) := (z_1 + z_2) \ln (y + y_2) - z_1 \ln y_1 - z_2 \ln y_2 - y \geq 0 \quad \forall y \in [0, y_1].$$

This follows immediately from

$$\psi(0) = z_1 (\ln y_2 - \ln y_1) \geq 0$$

and

$$\psi'(y) = \frac{z_1 + z_2}{y + y_2} - 1 \geq \frac{z_1 + z_2}{y_1 + y_2} - 1 \geq 0 \quad \forall y \in [0, y_1] .$$

□

Case	Atomic operations for f_j	α_j, v_j	$\beta_j, \gamma_j, \delta_j, w_j, M_j$	Rem.
I)	0	0	0	
II.1)	1	1	1	
II.2)	1	3	7	
II.3)	1	2	3	
III.1)	1	0	0	
III.2)	1	$2 - q_i$	$\leq 2b_s + 5$	
III.3)	1	$2 - q_i$	$\leq 2b_t + 5$	
III.4)	1	$2b_t - q_i + 1$	$\leq (3b_t + 2b_s + 7)b_t + 7$	$b_s \geq b_t$
III.5)	1	$2b_s - q_i + 1$	$\leq (3b_s + 2b_t + 7)b_s + 7$	$b_s < b_t$
IV.1)	1	$2 - q_i$	0	
IV.2)	1	$4 - q_i$	$\leq 2b_s + 11$	
IV.3)	1	1	1	
IV.4)	1	$4 - q_i$	$\leq 2b_t + 11$	
IV.5)	1	1	1	
IV.6)	1	$2b_t - q_i + 4$	$\leq (3b_t + 2b_s + 7)b_t + 15$	$b_s \geq b_t$
IV.7)	1	$b_t + 2$	$\leq (2b_t + b_s + 5)b_t + 4$	$b_s \geq b_t$
IV.8)	1	$2b_s - q_i + 4$	$\leq (3b_s + 2b_t + 7)b_s + 15$	$b_s < b_t$
IV.9)	1	$b_s + 2$	$\leq (2b_s + b_t + 5)b_s + 4$	$b_s < b_t$
V.1)	1	$3 - q_i$	3	
V.2)	1	$3 - q_i$	$\leq 2b_s + 12$	
V.3)	1	$5 - q_i$	$\leq 2b_t + 16$	
V.4)	1	2	$\leq b_t + 3$	
V.5)	1	$2b_t - q_i + 3$	$\leq (3b_t + 2b_s + 7)b_t + 18$	$b_s \geq b_t$
V.6)	1	$2b_s - q_i + 4$	$\leq (3b_s + 2b_t + 7)b_s + 18$	$b_s < b_t$
V.7)	1	2	$\leq (2b_t + 2)b_s$	$b_s < b_t$

Table 4: Number of atomic operations in step i of algorithm SF_g (SF_H), $s = k_i - n$, $t = l_i - n$, $j = i + n$

Lemma 9.3 *Let S be a characterizing sequence for $f : D \subset \mathbb{R}^n \longrightarrow \mathbb{R}$ of length m. If S is a T-sequence, then*

$$b_m \leq \#(f, S) + 1 , \tag{8}$$

$$\sum_{i=1}^{m} p_i \leq (\#(f, S) + 1) \ln (\max \{b_m, 1\}) \tag{9}$$

$$\overline{\#}(f, \nabla f, S, SF_g) \leq 7\#(f, S) - 2b_m + \sum_{i=1}^{m} p_i\,, \tag{10}$$

$$\overline{\#}(f, \nabla f, \nabla^2 f, S, SF_H) \leq 5(b_m + 1)\left(7\#(f, S) - 2b_m + \sum_{i=1}^{m} p_i\right) \tag{11}$$

with b_m, p_i according to (6), (7). *$\overline{\#}(f, \nabla f, S, SF_g)$ denotes the number of atomic operations needed to calculate f_{m+n} and g_{m+n} in the factorized form $g_{m+n} = \alpha_{m+n} v_{m+n}$ with algorithm SF_g. $\overline{\#}(f, \nabla f, \nabla^2 f, S, SF_H)$ is defined analogously with factorization*

$$H_{m+n} = \beta_{m+n} v_{m+n} v_{m+n}^T + \gamma_{m+n} w_{m+n} w_{m+n}^T + \delta_{m+n} M_{m+n}\,.$$

Proof: We use the notations (5), (6) and (7). We proof the assertions by induction with respect to the length m of T-sequences:
If $m = 1$, then the following cases may occur:

$\underline{\omega_1 \in \mathcal{C}:}$

$b_1 = p_1 = 0$, $\#(f, S) = \overline{\#}(f, \nabla f, S, SF_g) = \overline{\#}(f, \nabla f, \nabla^2 f, S, SF_H) = 0$.

$\underline{\omega_1 \in \mathcal{U},\ k_1 \in \{1, \ldots, n\}:}$

$b_1 = 1$, $p_1 = 0$, $\#(f, S) = 1$,
$\overline{\#}(f, \nabla f, S, SF_g) = 2$, $\overline{\#}(f, \nabla f, \nabla^2 f, S, SF_H) = 3$.

$\underline{\omega_1 \in \mathcal{B},\ k_1, l_1 \in \{1, \ldots, n\}:}$

$b_1 = q_1 \leq 2$, $p_1 = 0$, $\#(f, S) = 1$,
$\overline{\#}(f, \nabla f, S, SF_g) \leq 4 - q_1$, $\overline{\#}(f, \nabla f, \nabla^2 f, S, SF_H) = 7 - q_1$.

In all cases (8), (9), (10) and (11) are fulfilled.

Now assume that the assertions hold for all T-sequences of length $\leq m$ and let S be a T-sequence of length $m + 1$. We have to consider the following cases:

$\underline{\omega_{m+1} \in \mathcal{U}:}$

Then $k_{m+1} = m + n$ and S_m is a T-sequence of length m defining a function $\tilde{f}$. By induction, we have

$$b_m \leq \#(\tilde{f}, S_m) + 1\,, \tag{12}$$

$$\sum_{i=1}^{m} p_i \leq \left(\#(\tilde{f}, S_m) + 1\right) \ln\left(\max\{b_m, 1\}\right)\,, \tag{13}$$

$$\overline{\#}(\tilde{f}, \nabla\tilde{f}, S_m, SF_g) \leq 7\,\#(\tilde{f}, S_m) - 2b_m + \sum_{i=1}^{m} p_i \tag{14}$$

and

$$\overline{\#}(\tilde{f}, \nabla\tilde{f}, \nabla^2\tilde{f}, S_m, SF_H) \leq 5(b_m + 1)\left(7\,\#(\tilde{f}, S_m) - 2b_m + \sum_{i=1}^{m} p_i\right). \quad (15)$$

Moreover,

$$\#(f, S) = \#(\tilde{f}, S_m) + 1,\ \overline{\#}(f, \nabla f, S, SF_g) \leq \overline{\#}(\tilde{f}, \nabla\tilde{f}, S_m, SF_g) + 4,$$
$$\overline{\#}(f, \nabla f, \nabla^2 f, S, SF_H) \leq \overline{\#}(\tilde{f}, \nabla\tilde{f}, \nabla^2\tilde{f}, S_m, SF_H) + 11$$
$$b_{m+1} = b_m,\ p_{m+1} = 0,$$

and therefore

$$b_{m+1} = b_m \leq \#(f, S)\,,$$
$$\sum_{i=1}^{m+1} p_i = \sum_{i=1}^{m} p_i \leq \#(f, S)\ln\left(\max\{b_{m+1}, 1\}\right)\,,$$
$$\begin{aligned}\overline{\#}(f, \nabla f, S, SF_g) &\leq 7(\#(f, S) - 1) - 2b_{m+1} + \sum_{i=1}^{m+1} p_i + 4\\ &\leq 7\#(f, S) - 2b_{m+1} + \sum_{i=1}^{m+1} p_i\,,\end{aligned}$$
$$\begin{aligned}\overline{\#}(f, \nabla f, \nabla^2 f, S, SF_H) &\leq 5(b_m + 1)\left(7\#(f, S) - 7 - 2b_{m+1} + \sum_{i=1}^{m+1} p_i\right) + 11\\ &\leq 5(b_{m+1} + 1)\left(7\#(f, S) - 2b_{m+1} + \sum_{i=1}^{m+1} p_i\right).\end{aligned}$$

$\omega_{m+1} \in \mathcal{B}$, $\{k_i, l_i\} \cap \{1, \ldots, n\} \neq \emptyset$:

Then $k_{m+1} = m + n$ or $l_{m+1} = m + n$ and S_m is a T-sequence of length m defining a function $\tilde{f}$. As above, (12), (13), (14) and (15) hold, and we have

$$\#(f, S) = \#(\tilde{f}, S_m) + 1,\ \overline{\#}(f, \nabla f, S, SF_g) \leq \overline{\#}(\tilde{f}, \nabla\tilde{f}, S_m, SF_g) + 6 - q_{m+1},$$
$$\overline{\#}(f, \nabla f, \nabla^2 f, S, SF_H) \leq \overline{\#}(\tilde{f}, \nabla\tilde{f}, \nabla^2\tilde{f}, S_m, SF_H) + 2b_m + 22 - q_{m+1}$$
$$b_{m+1} = b_m + q_{m+1},\ 0 \leq q_{m+1} \leq 1,\ p_{m+1} = 0\,.$$

Hence,

$$b_{m+1} \leq b_m + 1 \leq \#(f, S) + 1\,,$$
$$\sum_{i=1}^{m+1} p_i = \sum_{i=1}^{m} p_i \leq \#(f, S)\ln\left(\max\{b_{m+1}, 1\}\right)\,,$$

$$\overline{\#}(f,\nabla f,S,SF_g) \le 7(\#(f,S)-1) - 2(b_{m+1}-q_{m+1}) + \sum_{i=1}^{m+1} p_i + 6 - q_{m+1}$$
$$\le 7\#(f,S) - 2b_{m+1} + \sum_{i=1}^{m+1} p_i \,.$$

With

$$5(b_{m+1}+1)(6-q_{m+1}) \ge 25(b_m+1) \ge 2b_m + 22 - q_{m+1}$$

we obtain

$$\overline{\#}(f,\nabla f,\nabla^2 f,S,SF_H) \le 5(b_{m+1}+1)\left(7\#(f,S) - 2b_{m+1} + \sum_{i=1}^{m+1} p_i\right) .$$

<u>$\omega_{m+1} \in \mathcal{B}$, $k_i, l_i \in \{n+1,\dots,n+m\}$:</u>

Then $\tilde{S} := S_{k_{m+1}-n}$ and $\hat{S} := S_{l_{m+1}-n}$ are T-sequences of length $1 \le \tilde{m}, \hat{m} \le m$ for functions $\tilde{f}$ and $\hat{f}$. By induction, with $s = k_{m+1}-n$ and $t = l_{m+1}-n$ we have

$$b_s \le \#(\tilde{f},\tilde{S}) + 1$$
$$b_t \le \#(\hat{f},\hat{S}) + 1$$
$$\sum_{i\in N_s} p_i \le \left(\#(\tilde{f},\tilde{S}) + 1\right) \ln\left(\max\{1,b_s\}\right)$$
$$\sum_{i\in N_t} p_i \le \left(\#(\hat{f},\hat{S}) + 1\right) \ln\left(\max\{1,b_t\}\right)$$
$$\overline{\#}(\tilde{f},\nabla\tilde{f},\tilde{S},SF_g) \le 7\#(\tilde{f},\tilde{S}) - 2b_s + \sum_{i\in N_s} p_i$$
$$\overline{\#}(\hat{f},\nabla\hat{f},\hat{S},SF_g) \le 7\#(\hat{f},\hat{S}) - 2b_t + \sum_{i\in N_t} p_i \,.$$
$$\overline{\#}(\tilde{f},\nabla\tilde{f},\nabla^2\tilde{f},\tilde{S},SF_H) \le 5(b_s+1)\left(7\#(\tilde{f},\tilde{S}) - 2b_s + \sum_{i\in N_s} p_i\right)$$
$$\overline{\#}(\hat{f},\nabla\hat{f},\nabla^2\hat{f},\hat{S},SF_H) \le 5(b_t+1)\left(7\#(\hat{f},\hat{S}) - 2b_t + \sum_{i\in N_t} p_i\right) .$$

In addition, $m = \tilde{m} + \hat{m}$, $\#(f,S) = \#(\tilde{f},\tilde{S}) + \#(\hat{f},\hat{S}) + 1$,

$$\overline{\#}(f,\nabla f,S,SF_g) \le \overline{\#}(\tilde{f},\nabla\tilde{f},\tilde{S},SF_g) + \overline{\#}(\hat{f},\nabla\hat{f},\hat{S},SF_g) + 5$$
$$+2\min\{b_s,b_t\} - q_{m+1} \,,$$

$$\overline{\#}(f,\nabla f,\nabla^2 f,S,SF_H) \le \overline{\#}(\tilde{f},\nabla\tilde{f},\nabla^2\tilde{f},\tilde{S},SF_H) + \overline{\#}(\hat{f},\nabla\hat{f},\nabla^2\hat{f},\hat{S},SF_H)$$
$$+23 + (5b_{m+1}+9)\min\{b_s,b_t\} - q_{m+1}$$

and

$$b_{m+1} = \max\{b_s,\ b_t\} + q_{m+1},\ p_{m+1} = q_{m+1} = b_{m+1} - \max\{b_s,b_t\} \,.$$

Hence,

$$b_{m+1} = |B_{m+1}| = |B_s \cup B_t| \leq b_s + b_t \leq \#(f,S) + 1\,.$$

Moreover,

$$p_{m+1} \leq b_s + b_t - \max\{b_s, b_t\} = \min\{b_s, b_t\}\ .$$

Thus, setting

$$y := p_{m+1},\ y_1 := \max\{1, b_s\}\,,\ y_2 := \max\{1, b_t\}\,,$$

$$z_1 := \#(\tilde{f}, \tilde{S}) + 1,\ z_2 := \#(\hat{f}, \hat{S}) + 1\,,$$

we may apply Lemma 9.2:

$$\begin{aligned}
\sum_{i=1}^{m+1} p_i &= \sum_{i\in N_s} p_i + \sum_{i\in N_t} p_i + p_{m+1} \leq z_1 \ln y_1 + z_2 \ln y_2 + y \\
&\leq (z_1 + z_2) \ln (y + \max\{y_1, y_2\}) \\
&= (\#(f,S) + 1) \ln (p_{m+1} + \max\{1, b_s, b_t\}) \\
&= (\#(f,S) + 1) \ln (\max\{1, b_{m+1}\})\ .
\end{aligned}$$

In the last step we have used that $b_s = b_t = 0$ implies $b_{m+1} = 0$. Finally,

$$\begin{aligned}
&\overline{\#}(f, \nabla f, S, SF_g) \leq \\
&\quad \leq 7\#(\tilde{f}, \tilde{S}) - 2b_s + \sum_{i\in N_s} p_i + 7\#(\hat{f}, \hat{S}) - 2b_t + \sum_{i\in N_t} p_i + 5 \\
&\qquad +2\min\{b_s, b_t\} - q_{m+1} \\
&\quad = 7\#(f,S) - 7 + \sum_{i=1}^{m+1} p_i - 2q_{m+1} - 2\,(b_s + b_t - \min\{b_s, b_t\}) + 5 \\
&\quad = 7\#(f,S) - 2\,(\max\{b_s, b_t\} + q_{m+1}) + \sum_{i=1}^{m+1} p_i - 2 \\
&\quad < 7\#(f,S) - 2b_{m+1} + \sum_{i=1}^{m+1} p_i
\end{aligned}$$

The estimate

$$\begin{aligned}
&5(b_{m+1} + 1)(5 + 2\min\{b_s, b_t\} - q_{m+1}) \geq 5(b_{m+1} + 1)(5 + \min\{b_s, b_t\}) \\
&= 25b_{m+1} + 5b_{m+1} \min\{b_s, b_t\} + 5\min\{b_s, b_t\} + 25 \\
&\geq 23 + (5b_{m+1} + 9)\min\{b_s, b_t\} - q_{m+1}
\end{aligned}$$

yields

$$\overline{\#}(f,\nabla f,\nabla^2 f,S,SF_H) \le 5(b_{m+1}+1)\Big(7\#(\tilde{f},\tilde{S}) - 2b_s + \sum_{i\in N_s} p_i + 7\#(\hat{f},\hat{S})$$
$$-2b_t + \sum_{i\in N_t} p_i + 5 + 2\min\{b_s,b_t\} - q_{m+1}\Big)$$
$$\le 5(b_{m+1}+1)\left(7\#(f,S) - 2b_{m+1} + \sum_{i=1}^{m+1} p_i\right)$$

□

By induction, one easily verifies:

Lemma 9.4 *Let S be a characterizing sequence for $f : D \subset \mathbb{R}^n \longrightarrow \mathbb{R}$ of length m and B_i defined according to* (6). *Then, during the application of algorithm SF_g, all s^{th} components of the vector v_{i+n} with $s \notin B_i$ vanish. Moreover, during the application of algorithm SF_H, all s^{th} components, $s \notin B_i$, of the vectors v_{i+n} and w_{i+n} vanish and so do all (s,t)-components of M_{i+n} with $(s,t) \notin B_i \times B_i$.*

Now we are in a position to state the main theorem:

Theorem 9.5 *Let S be a characterizing sequence for $f : D \subset \mathbb{R}^n \longrightarrow \mathbb{R}$ of length m. If S is a T-sequence, then*

$$\#(f,\nabla f,S,SF_g) \le 7\#(f,S) - b + (\#(f,S)+1)\ln(\max\{b,1\})$$

and

$$\#(f,\nabla f,\nabla^2 f,S,SF_H) \le 5(b+1)\Big(7\#(f,S) - \frac{3}{2}b$$
$$+ (\#(f,S)+1)\ln(\max\{b,1\})\Big) + 3b$$

Here $b = \left|\bigcup_{i=1}^{m}\{k_i,l_i\} \cap \{1,\ldots,n\}\right|$ *denotes the number of x_k's on which f depends.*

Proof: Due to Lemma 9.4, the computation of $\nabla f(x) = \alpha_{m+n}v_{m+n}$ costs at most $b = b_m$ multiplications. Thus,

$$\#(f,\nabla f,S,SF_g) \le \overline{\#}(f,\nabla f,S,SF_g) + b\,.$$

Computing

$$\nabla^2 f(x) = \beta_{m+n}v_{m+n}v_{m+n}^T + \gamma_{m+n}w_{m+n}w_{m+n}^T + \delta_{m+n}M_{m+n}$$

requires $\leq 5\frac{b(b+1)}{2} + 2b$ operations. Now the assertions follow immediately by combining the results of Lemma 9.3.

□

Remark 9.6 In many cases one would like to compute the directonal derivative $\nabla^2 f(x)d$, $d \in \mathbb{R}^n \setminus \{0\}$, instead of the whole Hessian. The representations

$$g_j = \alpha_j v_j \ , \ \alpha_j \neq 0 \ , \quad H_j d = \beta_j v_j + \gamma_j w_j + \delta_j z_j \ , \ \delta_j \neq 0 \ , \quad g_j^T d = \varepsilon_j \, .$$

allow efficient update formulas for the scalars $\alpha_j, \beta_j, \gamma_j, \delta_j, \varepsilon_j$ and the vectors v_j, w_j, z_j similar to those in algorithm SF_H.

□

10. Applications

In general, algorithm SF_g (and SF_H) develops its full efficiency only if it is applied to a T-sequence. We will put forward some arguments why it is nevertheless valuable. First of all, the results of Theorem 9.5 are interesting from a theoretical point of view, because they show that in special cases the forward mode is almost as fast as the reverse mode. There are classes of functions (e.g. Horner schemes) that are evaluable by T-sequences without additional costs and would require too much workspace when a reverse mode is applied. The algorithms SF_g, SF_H are directly applicable with almost optimal complexities $O(\ln(n)\#(f,S))$ and $O(\ln(n)n\#(f,S))$, which are in general better than the complexities $O((1+\ln(\#(f,S)))\#(f,S))$ and $O((1+\ln(\#(f,S)))n\#(f,S))$ of Griewank's recursive reverse mode. The numerical results in Section 13 even show typical costs of $\leq 4\#(f,S)$ and $\leq 2\,n\#(f,S)$, respectively. Further, the SF-algorithms are very efficient general purpose implementations of the forward mode and achieve the complexity of Theorem 9.5 in all subtrees of the Kantorovič graph.

Another application of the new algorithms consists in shortening characterizing sequences before applying the reverse mode. Every characterizing sequence for a function $f : D \subset \mathbb{R}^n \longrightarrow \mathbb{R}$ can be split up as follows:

$$t_1 = \tau_1(x) \ , \quad i = 2, \ldots, l : \ t_i = \tau_i(x, t_1, \ldots, t_{i-1})$$

where the functions

$$\tau_i : D_i \subset \mathbb{R}^{n+i-1} \longrightarrow \mathbb{R}$$

are given by T-sequences S_{τ_i}. In most cases, the function f is already formulated in this fashion. Analogously to Section 6, we define

$$\Phi_i : y \in D_i \longmapsto \begin{pmatrix} y \\ \tau_i(y) \end{pmatrix} \in \mathbb{R}^{n+i}, \ i = 1, \ldots, l-1 \, , \quad \Phi_l : y \in D_l \longmapsto \tau_l(y) \in \mathbb{R}$$

and

$$T_0 := x \ , \quad T_i := \begin{pmatrix} T_{i-1} \\ t_i \end{pmatrix} = \Phi_i \circ \cdots \circ \Phi_1(x) \ , \quad i = 1, \ldots, l-1 \, .$$

Then

$$\nabla f(x)^T = D\Phi_l(T_{l-1})\, D\Phi_{l-1}(T_{l-2}) \cdots D\Phi_1(x)\,.$$

Each Jacobian $D\Phi_i$ has the structure

$$D\Phi_i(y) = \begin{pmatrix} I_{n+i-1} \\ \nabla\tau_i(y)^T \end{pmatrix}.$$

Now we may apply a *reduced reverse mode* RR_g as follows:

1) forward calculation of the t_i, $i = 1, \ldots, l-1$:

   $t_1 := \tau_1(x)$, $i = 2, \ldots, l-1:\ t_i = \tau_i(x, t_1, \ldots, t_{i-1})$

   Here, the functions τ_i are evaluated using S_{τ_i}.

2) Compute $f(x) = t_l = \tau_l(T_{l-1})$ and $v^l := \nabla\tau_l(T_{l-1})$ by applying algorithm SF_g to S_{τ_l}.

3) Reverse steps:

   $i = l-1, \ldots, 1$:

   a) Compute $\nabla\tau_i(T_{i-1})$ by applying algorithm SF_g to S_{τ_i}.

   b) Compute $v^i := v^{i+1} \begin{pmatrix} I_{n+i-1} \\ \nabla\tau_i(T_{i-1})^T \end{pmatrix}$

4) $\nabla f(x)^T = v^1$.

Note that the computation of v^i in 3 b) requires at most $2\, b_{\tau_i}$ operations, where b_{τ_i} denotes the number of non-zeroes in $\nabla\tau_i(T_{i-1})$. Since Lemma 9.3 and 9.4 imply $b_{\tau_i} \le \#(\tau_i, S_{\tau_i}) + 1$, we get with

$$\#(f, S) = \sum_{i=1}^{l} \#(\tau_i, S_{\tau_i})$$

in the case $\#(\tau_i, S_{\tau_i}) \ge 1$, $i = 1, \ldots, l$, the estimate

$$\begin{aligned} \#(f, \nabla f, S, RR_g) &\le \sum_{i=1}^{l-1} \#(\tau_i, S_{\tau_i}) + \sum_{i=1}^{l} \#(\tau_i, \nabla\tau_i, S_{\tau_i}, SF_g) + 2\sum_{i=1}^{l-1} b_{\tau_i} \\ &\le 3\sum_{i=1}^{l-1} \#(\tau_i, S_{\tau_i}) + \sum_{i=1}^{l} \#(\tau_i, \nabla\tau_i, S_{\tau_i}, SF_g) + 2(l-1) \\ &\le \left(3 + \max_{1\le i\le l} \frac{\#(\tau_i, \nabla\tau_i, S_{\tau_i}, SF_g)}{\#(\tau_i, S_{\tau_i})}\right) \#(f, S) + 2(l-1)\,. \end{aligned}$$

Theorem 9.5 gives an upper bound for the maximum in the last expression. An algorithm for the computation of the kth row of the Hessian can be obtained by considering the sequence

$$t_1 = \tau_1(x)\ ,\quad t_1^k = \tau_1^k(x)$$

$$i = 2,\dots,l:\quad t_i = \tau_i(x, t_1, \dots, t_{i-1})\ ,\quad t_i^k = \tau_i^k(x, t_1, \dots, t_{i-1}, t_1^k, \dots, t_{i-1}^k)$$

with $\tau_1^k(x) = \partial_{x_k}\tau_1(x)$ and

$$\tau_i^k : (x, y, z) \in \mathbb{R}^n \times \mathbb{R}^{i-1} \times \mathbb{R}^{i-1} \longmapsto \nabla\tau_i(x, y)^T \begin{pmatrix} e_k \\ z \end{pmatrix}\ ,\quad i = 2, \dots, l\,,$$

which simultanously computes $f(x) = t_l$ and $\partial_{x_k} f(x) = t_l^k$. To carry out a reverse mode, we need in each step $i+1 \to i$ the gradients $\nabla\tau_i(T_{i-1})$ and $\nabla\tau_i^k(T_{i-1}, T_{i-1}^k)$, where $T_i^k = \left(t_1^k, \dots, t_i^k\right)^T$. They can be computed by applying a modification of algorithm SF_H in the sense of Remark 9.6 to S_{τ_i} and by using the identity

$$\nabla\tau_i^k(x, y, z) = \begin{pmatrix} \nabla^2\tau_i(x, y)\begin{pmatrix} e_k \\ z \end{pmatrix} \\ \nabla_y\tau_i(x, y) \end{pmatrix}\,.$$

11. Technical Details

An efficient implementation of algorithm SF_g and SF_H requires a data structure for vectors that allows fast execution of the following operations:

- Set to zero.
- Determine the number of nonzero components.
- Check, whether a component is nonzero.
- Access all nonzero components.
- Create a new nonzero component.

By nonzero we mean the possibility of being nonzero, because the component was attached during the algorithm.

A vector $v \in \mathbb{R}^d$ is represented by

$$b^v, m^v \in \mathbb{N}\,,\ c^v \in \mathbb{N}^d,\ d^v \in \mathbb{N}^d,\ v^v \in \mathbb{R}^d\,.$$

b^v is the number of nonzeros, $\left(c_j^v\right)_{1\le j\le b^v}$ contains the indices of possibly nonzero components in arbitrary order and $d_i^v = m^v$ means that v_i is possibly nonzero, i.e. $i \in \left\{c_j^v\,;\ 1 \le j \le b^v\right\}$, with $v_i = v_i^v$.

A workspace of those vectors v is managed by a free-list. At the beginning, all vectors d^v and the b^v's are set to zero, the marks m^v are initialized to one. Then the above mentioned operations can be implemented as follows:

Set to zero: $m^v := m^v + 1$, $b^v := 0$

Number of nonzeros: b^v

Check whether component i is nonzero: $d_i^v \stackrel{?}{=} m^v$

Access to all nonzero components: $v_{c_j^v}^v$, $j = 1, \ldots, b^v$

Create a component $v_j = a$: $b_v := b_v + 1$, $c_{b^v}^v := j$, $v_j^v := a$, $d_j^v := m^v$

The matrix M can be represented in the same way. If the implementations of algorithm SF_g and SF_H are based on the above data structures and the assumptions of Theorem 9.5 are met, a dominating part of the runtime is spent to perform atomic operations. This is confirmed by the numerical results presented in Section 13.

12. Parallelization

The key to an efficient parallelization of the forward mode F_g is the observation that in each step $i-1 \to i$ the components of g_{i+n} can be computed concurrently if f_{k_i}, f_{l_i} and f_{i+n} are known. So, on a multiprocessor machine each processor can be entrusted with the evaluation of $f(x)$ and some partial derivatives $\partial_{x_k} f(x)$. Similarily, in algorithm F_H the components of H_{i+n} can be computed concurrently from f_{k_i}, f_{l_i}, f_{i+n} and g_{k_i}, g_{l_i}, g_{i+n}. Thus, each processor calculates $f(x)$, $\nabla f(x)$ and a part of the Hessian $\nabla^2 f(x)$. This method has been successfully implemented on a transputer workstation.
If one doesn't exploit the eventual concurrencies in the Kantorovič graph, for methods with better complexity only the parallelization of the Hessian seems to be promising. The reverse mode R_H consists of n characterizing sequences for $f(x)$ and $\partial_{x_k} f(x)$, which can be interpreted and backward-differentiated fully in parallel. Algorithm SF_H can be parallelized in the same way as F_H.

13. Numerical Results

In the following examples $T(f)$ denotes the execution time to compute $f(x)$ on a HP 712/80 workstation. $T(f, \nabla f)$ and $T(f, \nabla f, \nabla^2 f)$ denote the execution times of algorithm SF_g and SF_H, respectively.

Example 13.1 $f : A \in \mathbb{R}^{d\times d} \longrightarrow \det A$, $d \geq 2$, evaluated by Laplace's expansion rule, i.e. $\#(f, S) \in [5n!/2 - 2, en! - 2[$.

$n = d^2$	Algorithm SF_g			Algorithm SF_H	
d	$T(f)$	$T(f,\nabla f)$	$\frac{T(f,\nabla f)}{T(f)}$	$T(f,\nabla f,\nabla^2 f)$	$\frac{T(f,\nabla f,\nabla^2 f)}{T(f)}$
6	2.67	8.41	3.15	26.7	$10.0 \approx 0.28n$
7	19.3	58.5	3.03	196	$10.2 \approx 0.21n$
8	150	472	3.15	1550	$10.3 \approx 0.16n$
9	1350	4250	3.15	14400	$10.7 \approx 0.13n$
10	13500	42900	3.18	143000	$10.6 \approx 0.11n$

Table 5: Execution times for Example 13.1 in 10^{-3} seconds

Example 13.2

$$f(x)=\sum_{k=1}^{K}\left(\frac{y_k-\left(1+\sum_{j=1}^{n_1}u_j^kx_j\right)\left(\sum_{j=n_1+1}^{n_2}u_j^kx_j\right)}{\sum_{j=n_2+1}^{n}u_j^kx_j}\right)^2$$

with given data y_k, u_j^k.

K=500			Algorithm SF_g			Algorithm SF_H	
n_1	n_2	n	$T(f)$	$T(f,\nabla f)$	$\frac{T(f,\nabla f)}{T(f)}$	$T(f,\nabla f,\nabla^2 f)$	$\frac{T(f,\nabla f,\nabla^2 f)}{T(f)}$
5	7	9	0.016	0.046	2.88	0.19	$11.9\approx 1.32n$
8	14	18	0.027	0.077	2.85	0.49	$18.1\approx 1.01n$
16	28	36	0.050	0.14	2.80	1.50	$30.0\approx 0.83n$
32	56	72	0.094	0.27	2.87	5.39	$57.3\approx 0.80n$
64	112	144	0.18	0.53	2.94	21.5	$119\approx 0.83n$
128	224	288	0.36	1.05	2.92	89.6	$249\approx 0.86n$
256	448	576	0.72	2.09	2.90	354	$492\approx 0.85n$
512	896	1152	1.37	4.22	3.08	1402	$1023\approx 0.89n$
1024	1792	2304	2.66	8.73	3.28	–	–
4096	7168	9216	11.0	36.2	3.29	–	–
16384	28672	36864	43.2	144.1	3.34	–	–

Table 6: Execution times for Example 13.2 in seconds

Example 13.3 A worst case example can be obtained by evaluating a balanced Kantorovič tree of depth d with cyclic distribution of the n independent variables to the leaves and divisions as binary operations. The characterizing sequence is obtained by recursive descent. In the second part of the table, the constant increment of $T(f,\nabla f)$ by about 0.5 seconds when doubling n shows the logarithmic dependence on n.

		Algorithm SF_g			Algorithm SF_H	
d	n	$T(f)$	$T(f,\nabla f)$	$\frac{T(f,\nabla f)}{T(f)}$	$T(f,\nabla f,\nabla^2 f)$	$\frac{T(f,\nabla f,\nabla^2 f)}{T(f)}$
17	64	0.212	0.757	3.57	22.19	$105\approx 1.64n$
17	128	0.212	0.814	3.84	45.4	$214\approx 1.67n$
17	256	0.212	0.823	3.88	92.5	$436\approx 1.70n$
17	512	0.212	0.890	4.20	183	$863\approx 1.69n$
20	256	1.66	6.55	3.95	–	–
20	512	1.66	7.06	4.25	–	–
20	1024	1.66	7.56	4.55	–	–
20	2048	1.66	8.16	4.92	–	–
20	4096	1.66	8.67	5.22	–	–
20	8192	1.66	9.17	5.52	–	–

Table 7: Execution times for Example 13.3 in seconds

Acknowledgments

We wish to thank Dr. Herbert Fischer for his helpful suggestions and Michael Rempter for proof reading.

Appendix

Algorithm SF_H:

$j = 1, \dots, n$:

$$\alpha_j = 1,\ v_j = e_j,$$
$$\beta_j = \gamma_j = 0,\ \delta_j = 1,\ w_j = 0,\ M_j = 0.$$

$i = 1, \dots, m:$

$j := i + n$

I) $\omega_i \in \mathcal{C}:$

$$f_j = \omega_i,\ \ g_j = 0 = \alpha_j v_j,\ \ H_j = 0 = \beta_j v_j v_j^T + \gamma_j w_j w_j^T + \delta_j M_j$$

with

$$\alpha_j = 1,\ v_j = 0,$$
$$\beta_j = \gamma_j = 0,\ \delta_j = 1,\ w_j = 0,\ M_j = 0.$$

II) $\omega_i \in \mathcal{U}$:

$$\begin{aligned} f_j &= \omega_i(f_{k_i}),\quad g_j = \omega_i'(f_{k_i}) g_{k_i} = \omega_i'(f_{k_i})\alpha_{k_i} v_{k_i} = \alpha_j v_j \\ H_j &= \omega_i'(f_{k_i}) H_{k_i} + \omega_i''(f_{k_i}) g_{k_i} g_{k_i}^T \\ &= (\omega_i''(f_{k_i})\alpha_{k_i}^2 + \omega_i'(f_{k_i})\beta_{k_i}) v_{k_i} v_{k_i}^T + \omega_i'(f_{k_i})\gamma_{k_i} w_{k_i} w_{k_i}^T + \omega_i'(f_{k_i})\delta_{k_i} M_{k_i} \\ &= \beta_j v_j v_j^T + \gamma_j w_j w_j^T + \delta_j M_j \end{aligned}$$

with

1) $k_i \in \{1, \dots, n\}:$

$$\alpha_j = 1,\ v_j = \omega_i'(x_{k_i})\, e_{k_i},$$
$$\beta_j = 0,\ \delta_j = 1,\ \gamma_j = \omega_i''(x_{k_i}),\ w_j = e_{k_i},\ M_j = 0.$$

2) $k_i \notin \{1, \dots, n\},\ \omega_i'(f_{k_i}) \neq 0:$

$$\alpha_j = \omega_i'(f_{k_i})\alpha_{k_i},\ v_j = v_{k_i},$$
$$\beta_j = (\omega_i''(f_{k_i})\alpha_{k_i}^2 + \omega_i'(f_{k_i})\beta_{k_i}),\ \gamma_j = \omega_i'(f_{k_i})\gamma_{k_i},\ \delta_j = \omega_i'(f_{k_i})\delta_{k_i},$$

$w_j = w_{k_i},\ M_j = M_{k_i}.$

3) $k_i \notin \{1,\dots,n\},\ \omega_i'(f_{k_i}) = 0:$

$\alpha_j = 1,\ v_j = 0,$

$\beta_j = 0,\ \delta_j = 1,\ \gamma_j = \omega_i''(f_{k_i})\alpha_{k_i}^2,\ w_j = v_{k_i},\ M_j = 0.$

III) $\omega_i \in \mathcal{B},\ \omega_i(a,b) = a \pm b$:

$$\begin{aligned} f_j &= f_{k_i} \pm f_{l_i}, \quad g_j = g_{k_i} \pm g_{l_i} = \alpha_{k_i} v_{k_i} \pm \alpha_{l_i} v_{l_i} = \alpha_j v_j \\ H_j &= H_{k_i} \pm H_{l_i} = \beta_{k_i} v_{k_i} v_{k_i}^T + \gamma_{k_i} w_{k_i} w_{k_i}^T + \delta_{k_i} M_{k_i} \\ &\quad \pm \left(\beta_{l_i} v_{l_i} v_{l_i}^T + \gamma_{l_i} w_{l_i} w_{l_i}^T + \delta_{l_i} M_{l_i}\right) = \beta_j v_j v_j^T + \gamma_j w_j w_j^T + \delta_j M_j \end{aligned}$$

with

1) $k_i, l_i \in \{1,\dots,n\}:$

$\alpha_j = 1,\ v_j = e_{k_i} \pm e_{l_i},$

$\beta_j = \gamma_j = 0,\ \delta_j = 1,\ w_j = 0,\ M_j = 0.$

2) $k_i \notin \{1,\dots,n\},\ l_i \in \{1,\dots,n\}:$

$\alpha_j = \alpha_{k_i},\ v_j = v_{k_i} \pm \dfrac{1}{\alpha_{k_i}} e_{l_i},$

$\beta_j = \beta_{k_i},\ \gamma_j = \gamma_{k_i},\ \delta_j = \delta_{k_i},\ w_j = w_{k_i},$

$M_j = M_{k_i} - \dfrac{\beta_{k_i}}{\alpha_{k_i}\delta_{k_i}} \left(\dfrac{1}{\alpha_{k_i}} e_{l_i} e_{l_i}^T \pm \left(v_{k_i} e_{l_i}^T + e_{l_i} v_{k_i}^T \right) \right)$

3) $l_i \notin \{1,\dots,n\},\ k_i \in \{1,\dots,n\}:$

Analogous to 2).

4) $k_i, l_i \notin \{1,\dots,n\},\ b_{k_i - n} \geq b_{l_i - n}$:

$\alpha_j = \alpha_{k_i},\ v_j = v_{k_i} \pm \dfrac{\alpha_{l_i}}{\alpha_{k_i}} v_{l_i},$

$\beta_j = \beta_{k_i},\ \gamma_j = \gamma_{k_i},\ \delta_j = \delta_{k_i},\ w_j = w_{k_i},$

$$\begin{aligned} M_j = M_{k_i} \pm \frac{1}{\delta_{k_i}} \Bigg(&\left(\beta_{l_i} \mp \beta_{k_i} \left(\frac{\alpha_{l_i}}{\alpha_{k_i}} \right)^2 \right) v_{l_i} v_{l_i}^T + \gamma_{l_i} w_{l_i} w_{l_i}^T + \delta_{l_i} M_{l_i} \\ &- \beta_{k_i} \frac{\alpha_{l_i}}{\alpha_{k_i}} \left(v_{k_i} v_{l_i}^T + v_{l_i} v_{k_i}^T \right) \Bigg) \end{aligned}$$

5) $k_i, l_i \notin \{1,\dots,n\},\ b_{k_i - n} < b_{l_i - n}$:

Analogous to 4).

IV) $\omega_i \in \mathcal{B}$, $\omega_i(a,b) = a \cdot b$:

$$\begin{aligned} f_j &= f_{k_i} f_{l_i}, \quad g_j = f_{l_i} g_{k_i} + f_{k_i} g_{l_i} = f_{l_i} \alpha_{k_i} v_{k_i} + f_{k_i} \alpha_{l_i} v_{l_i} = \alpha_j v_j \\ H_j &= f_{l_i} H_{k_i} + f_{k_i} H_{l_i} + g_{k_i} g_{l_i}^T + g_{l_i} g_{k_i}^T \\ &= f_{l_i} \beta_{k_i} v_{k_i} v_{k_i}^T + f_{l_i} \gamma_{k_i} w_{k_i} w_{k_i}^T + f_{l_i} \delta_{k_i} M_{k_i} + f_{k_i} \beta_{l_i} v_{l_i} v_{l_i}^T + f_{k_i} \gamma_{l_i} w_{l_i} w_{l_i}^T \\ &\quad + f_{k_i} \delta_{l_i} M_{l_i} + \alpha_{k_i} \alpha_{l_i} \left(v_{k_i} v_{l_i}^T + v_{l_i} v_{k_i}^T \right) = \beta_j v_j v_j^T + \gamma_j w_j w_j^T + \delta_j M_j \end{aligned}$$

with

1) $k_i, l_i \in \{1, \ldots, n\}$:

$\alpha_j = 1$, $v_j = x_{l_i} e_{k_i} + x_{k_i} e_{l_i}$,

$\beta_j = \gamma_j = 0$, $\delta_j = 1$, $w_j = 0$, $M_j = e_{k_i} e_{l_i}^T + e_{l_i} e_{k_i}^T$.

2) $k_i \notin \{1, \ldots, n\}$, $l_i \in \{1, \ldots, n\}$, $x_{l_i} \neq 0$:

$\alpha_j = x_{l_i} \alpha_{k_i}$, $v_j = v_{k_i} + \dfrac{f_{k_i}}{\alpha_j} e_{l_i}$,

$\beta_j = x_{l_i} \beta_{k_i}$, $\gamma_j = x_{l_i} \gamma_{k_i}$, $\delta_j = x_{l_i} \delta_{k_i}$, $w_j = w_{k_i}$,

$$M_j = M_{k_i} + \frac{1}{\delta_j} \left(\left(\alpha_{k_i} - \beta_j \frac{f_{k_i}}{\alpha_j} \right) \left(v_{k_i} e_{l_i}^T + e_{l_i} v_{k_i}^T \right) - \beta_j \left(\frac{f_{k_i}}{\alpha_j} \right)^2 e_{l_i} e_{l_i}^T \right)$$

3) $k_i \notin \{1, \ldots, n\}$, $l_i \in \{1, \ldots, n\}$, $x_{l_i} = 0$:

$\alpha_j = 1$, $v_j = f_{k_i} e_{l_i}$,

$\beta_j = \gamma_j = 0$, $\delta_j = \alpha_{k_i}$, $w_j = 0$, $M_j = v_{k_i} e_{l_i}^T + e_{l_i} v_{k_i}^T$

4) $l_i \notin \{1, \ldots, n\}$, $k_i \in \{1, \ldots, n\}$, $x_{k_i} \neq 0$:

Analogous to 2).

5) $l_i \notin \{1, \ldots, n\}$, $k_i \in \{1, \ldots, n\}$, $x_{k_i} = 0$:

Analogous to 3).

6) $k_i, l_i \notin \{1, \ldots, n\}$, $b_{k_i - n} \geq b_{l_i - n}$, $f_{l_i} \neq 0$:

$\alpha_j = f_{l_i} \alpha_{k_i}$, $v_j = v_{k_i} + \dfrac{f_{k_i} \alpha_{l_i}}{\alpha_j} v_{l_i}$,

$\beta_j = f_{l_i} \beta_{k_i}$, $\gamma_j = f_{l_i} \gamma_{k_i}$, $\delta_j = f_{l_i} \delta_{k_i}$, $w_j = w_{k_i}$,

$$\begin{aligned} M_j = M_{k_i} + \frac{1}{\delta_j} \Bigg(& \left(f_{k_i} \beta_{l_i} - \beta_j \left(\frac{f_{k_i} \alpha_{l_i}}{\alpha_j} \right)^2 \right) v_{l_i} v_{l_i}^T + f_{k_i} \gamma_{l_i} w_{l_i} w_{l_i}^T \\ & + f_{k_i} \delta_{l_i} M_{l_i} + \left(\alpha_{k_i} \alpha_{l_i} - \beta_j \frac{f_{k_i} \alpha_{l_i}}{\alpha_j} \right) \left(v_{k_i} v_{l_i}^T + v_{l_i} v_{k_i}^T \right) \Bigg) \end{aligned}$$

7) $k_i, l_i \notin \{1,\ldots,n\}$, $b_{k_i-n} \geq b_{l_i-n}$, $f_{l_i} = 0$:

$\alpha_j = 1$, $v_j = f_{k_i}\alpha_{l_i}v_{l_i}$,

$\beta_j = 0$, $\gamma_j = f_{k_i}\gamma_{l_i}$, $\delta_j = 1$, $w_j = w_{l_i}$,

$M_j = f_{k_i}\beta_{l_i}v_{l_i}v_{l_i}^T + f_{k_i}\delta_{l_i}M_{l_i} + \alpha_{k_i}\alpha_{l_i}\left(v_{k_i}v_{l_i}^T + v_{l_i}v_{k_i}^T\right)$

8) $k_i, l_i \notin \{1,\ldots,n\}$, $b_{k_i-n} < b_{l_i-n}$, $f_{k_i} \neq 0$:

Analogous to 6).

9) $k_i, l_i \notin \{1,\ldots,n\}$, $b_{k_i-n} < b_{l_i-n}$, $f_{k_i} = 0$:

Analogous to 7).

V) $\omega_i \in \mathcal{B}$, $\omega_i(a,b) = a/b$:

$$\begin{aligned}
f_j &= \frac{f_{k_i}}{f_{l_i}}, \quad g_j = \frac{1}{f_{l_i}}\left(g_{k_i} - f_j g_{l_i}\right) = \frac{\alpha_{k_i}}{f_{l_i}}v_{k_i} - \frac{f_j\alpha_{l_i}}{f_{l_i}}v_{l_i} = \alpha_j v_j \\
H_j &= \frac{1}{f_{l_i}}\left(H_{k_i} - f_j H_{l_i} - g_j g_{l_i}^T - g_{l_i}g_j^T\right) \\
&= \frac{\beta_{k_i}}{f_{l_i}}v_{k_i}v_{k_i}^T + \frac{\gamma_{k_i}}{f_{l_i}}w_{k_i}w_{k_i}^T + \frac{\delta_{k_i}}{f_{l_i}}M_{k_i} - \frac{f_j\beta_{l_i}}{f_{l_i}}v_{l_i}v_{l_i}^T - \frac{f_j\gamma_{l_i}}{f_{l_i}}w_{l_i}w_{l_i}^T \\
&\quad -\frac{f_j\delta_{l_i}}{f_{l_i}}M_{l_i} - \frac{\alpha_{l_i}\alpha_j}{f_{l_i}}\left(v_j v_{l_i}^T + v_{l_i}v_j^T\right) = \beta_j v_j v_j^T + \gamma_j w_j w_j^T + \delta_j M_j
\end{aligned}$$

with

1) $k_i, l_i \in \{1,\ldots,n\}$:

$\alpha_j = \dfrac{1}{x_{l_i}}$, $v_j = e_{k_i} - f_j e_{l_i}$,

$\beta_j = \gamma_j = 0$, $\delta_j = \alpha_j^2$, $w_j = 0$, $M_j = -e_{k_i}e_{l_i}^T - e_{l_i}e_{k_i}^T + 2f_j e_{l_i}e_{l_i}^T$.

2) $k_i \notin \{1,\ldots,n\}$, $l_i \in \{1,\ldots,n\}$:

$\alpha_j = \dfrac{\alpha_{k_i}}{x_{l_i}}$, $v_j = v_{k_i} - \dfrac{f_j}{\alpha_{k_i}}e_{l_i}$,

$\beta_j = \dfrac{\beta_{k_i}}{x_{l_i}}$, $\gamma_j = \dfrac{\gamma_{k_i}}{x_{l_i}}$, $\delta_j = \dfrac{\delta_{k_i}}{x_{l_i}}$, $w_j = w_{k_i}$,

$$\begin{aligned}
M_j = M_{k_i} + \frac{1}{\delta_j}\Bigg(&\left(\beta_j\frac{f_j}{\alpha_{k_i}} - \frac{\alpha_j}{x_{l_i}}\right)\left(v_{k_i}e_{l_i}^T + e_{l_i}v_{k_i}^T\right) \\
&-\frac{f_j}{\alpha_{k_i}}\left(\beta_j\frac{f_j}{\alpha_{k_i}} - 2\frac{\alpha_j}{x_{l_i}}\right)e_{l_i}e_{l_i}^T\Bigg)
\end{aligned}$$

3) $l_i \notin \{1,\ldots,n\}$, $k_i \in \{1,\ldots,n\}$, $x_{k_i} \neq 0$:

$\alpha_j = -\dfrac{f_j\alpha_{l_i}}{f_{l_i}}$, $v_j = v_{l_i} - \dfrac{1}{\alpha_{l_i}f_j}e_{k_i}$,

$$\beta_j = -2\frac{\alpha_{l_i}\alpha_j}{f_{l_i}} - \frac{f_j\beta_{l_i}}{f_{l_i}},\ \gamma_j = -\frac{f_j\gamma_{l_i}}{f_{l_i}},\ \delta_j = -\frac{f_j\delta_{l_i}}{f_{l_i}},\ w_j = w_{l_i},$$

$$M_j = M_{l_i} + \frac{1}{\delta_j}\left(\frac{1}{\alpha_{l_i} f_j}\left(\beta_j + \frac{\alpha_{l_i}\alpha_j}{f_{l_i}}\right)(e_{k_i}v_{l_i}^T + v_{l_i}e_{k_i}^T) - \left(\frac{1}{\alpha_{l_i} f_j}\right)^2 \beta_j e_{k_i}e_{k_i}^T\right)$$

4) $l_i \notin \{1,\ldots,n\}$, $k_i \in \{1,\ldots,n\}$, $x_{k_i} = 0$:

$$\alpha_j = 1,\ v_j = \frac{1}{f_{l_i}}e_{k_i},$$

$$\beta_j = \gamma_j = 0,\ \delta_j = -\frac{\alpha_{l_i}}{f_{l_i}^2},\ w_j = 0,\ M_j = e_{k_i}v_{l_i}^T + v_{l_i}e_{k_i}^T$$

5) $k_i, l_i \notin \{1,\ldots,n\}$, $b_{k_i-n} \geq b_{l_i-n}$:

$$\alpha_j = \frac{\alpha_{k_i}}{f_{l_i}},\ v_j = v_{k_i} - \frac{\alpha_{l_i} f_j}{\alpha_{k_i}}v_{l_i},$$

$$\beta_j = \frac{\beta_{k_i}}{f_{l_i}},\ \gamma_j = \frac{\gamma_{k_i}}{f_{l_i}},\ \delta_j = \frac{\delta_{k_i}}{f_{l_i}},\ w_j = w_{k_i},$$

$$M_j = M_{k_i} + \frac{1}{\delta_j}\left(\left(-\frac{f_j\beta_{l_i}}{f_{l_i}} - \frac{\alpha_{l_i} f_j}{\alpha_{k_i}}\left(\beta_j\frac{\alpha_{l_i} f_j}{\alpha_{k_i}} - 2\frac{\alpha_{l_i}\alpha_j}{f_{l_i}}\right)\right)v_{l_i}v_{l_i}^T - \frac{f_j\gamma_{l_i}}{f_{l_i}}w_{l_i}w_{l_i}^T - \frac{f_j\delta_{l_i}}{f_{l_i}}M_{l_i} + \left(\beta_j\frac{\alpha_{l_i} f_j}{\alpha_{k_i}} - \frac{\alpha_{l_i}\alpha_j}{f_{l_i}}\right)(v_{k_i}v_{l_i}^T + v_{l_i}v_{k_i}^T)\right)$$

6) $k_i, l_i \notin \{1,\ldots,n\}$, $b_{k_i-n} < b_{l_i-n}$, $f_{k_i} \neq 0$:

$$\alpha_j = -\frac{f_j\alpha_{l_i}}{f_{l_i}},\ v_j = v_{l_i} - \frac{\alpha_{k_i}}{\alpha_{l_i} f_j}v_{k_i},$$

$$\beta_j = -\frac{f_j\beta_{l_i}}{f_{l_i}} - 2\frac{\alpha_{l_i}\alpha_j}{f_{l_i}},\ \gamma_j = -\frac{f_j\gamma_{l_i}}{f_{l_i}},\ \delta_j = -\frac{f_j\delta_{l_i}}{f_{l_i}},\ w_j = w_{l_i},$$

$$M_j = M_{l_i} + \frac{1}{\delta_j}\left(\left(\frac{\beta_{k_i}}{f_{l_i}} - \beta_j\left(\frac{\alpha_{k_i}}{\alpha_{l_i} f_j}\right)^2\right)v_{k_i}v_{k_i}^T + \frac{\gamma_{k_i}}{f_{l_i}}w_{k_i}w_{k_i}^T + \frac{\delta_{k_i}}{f_{l_i}}M_{k_i} + \frac{\alpha_{k_i}}{\alpha_{l_i} f_j}\left(\beta_j + \frac{\alpha_{l_i}\alpha_j}{f_{l_i}}\right)(v_{k_i}v_{l_i}^T + v_{l_i}v_{k_i}^T)\right)$$

7) $k_i, l_i \notin \{1,\ldots,n\}$, $b_{k_i-n} < b_{l_i-n}$, $f_{k_i} = 0$:

$$\alpha_j = \frac{\alpha_{k_i}}{f_{l_i}},\ v_j = v_{k_i},$$

$$\beta_j = \frac{\beta_{k_i}}{f_{l_i}},\ \gamma_j = \frac{\gamma_{k_i}}{f_{l_i}},\ \delta_j = \frac{\delta_{k_i}}{f_{l_i}},\ w_j = w_{k_i},$$

$$M_j = M_{k_i} - \frac{\alpha_{l_i}\alpha_j}{\delta_j f_{l_i}}(v_{k_i}v_{l_i}^T + v_{l_i}v_{k_i}^T)$$

References

[1] W. BAUR, V. STRASSEN: *The Complexity of Partial Derivatives.* Theor. Comp. Sci. 22, 1983, 317-330.

[2] H. FISCHER: *Automatisches Differenzieren.* In: Wissenschaftliches Rechnen (J. Herzberger ed.), Akademie Verlag, Berlin, 1995.

[3] H. FISCHER, H. WARSITZ: *Complexity Investigations Concerning the Derivative of a Rational Function.* Technical Report 138, Inst. f. Angew. Math. u. Stat., TU München, 1995.

[4] A. GRIEWANK: *Achieving Logarithmic Growth of Temporal and Spacial Complexity in Reverse Automatic Differentiation.* Optimiz. Meth. Softw. 1, 1992, 35-54.

[5] A. GRIEWANK, G. F. CORLISS (EDS.): *Automatic Differentiation of Algorithms: Theory, Implementations and Applications.* SIAM, Philadelphia, 1991.

[6] M. IRI: *History of Automatic Differentiation and Rounding Error Estimation.* In: [5].

[7] L. B. RALL: *Automatic Differentiation: Techniques and Applications.* Lect. Notes Comp. Sci. 120, Springer-Verlag, Berlin, 1981.

APPROXIMATE STRUCTURED OPTIMIZATION BY CYCLIC BLOCK-COORDINATE DESCENT*

Jorge Villavicencio and Michael D. Grigoriadis
Rutgers University, New Brunswick, NJ 08903, USA.

Abstract. A uniform randomized exponential-potential block-coordinate descent method for the approximate solution of block-angular convex resource-sharing programs was analyzed in [5] and for the linear case in [14]. The former method is rendered deterministic by replacing its random block selection by arbitrary sweeps of its block coordinates, akin to classical implementations of Gauss-Seidel relaxation and coordinate descent in unconstrained optimization, recently used in concurrent network flows [15]. The general block-angular model consists of K disjoint convex compact sets ("blocks") and M nonnegative convex block-separable inequalities ("coupling constraints"). It is shown that for linear coupling constraints and for a given but arbitrary relative accuracy $\varepsilon \in (0,1]$, the proposed derandomized algorithm runs in $O(K \ln M(\varepsilon^{-2} + \ln \min\{K, M\})$ coordination steps or block optimizations, which is lower than all other existing bounds. It is also shown that this bound on coordination steps also applies to a reformulation of the above general nonlinear problem.

Key words: block angular problem, exponential potential, fully polynomial time approximation scheme, linear programming, multicommodity flow, sharing problem, structured optimization

AMS subject classification 68Q25, 90C05, 90C30

1. Introduction

This paper deals with *block-angular resource-sharing* problems of the form:

$$(\mathcal{P}) \qquad \lambda^* \doteq \lambda(x^*) = \min\{\, \lambda \mid f(x) \le \lambda e,\ x \in B \,\},$$

where the set B is the product of K given nonempty disjoint convex compact sets B^k, $k = 1, \ldots, K$, called *blocks*, and $x = (x^1, x^2, \ldots, x^k)$ is the vector of K *block variables* $x^k \in B^k$. The $M \ge 2$ inequalities $f(x) = \sum_{k=1}^K f^k(x^k) \le \lambda e$ are block-separable componentwise convex *coupling constraints* such that $f^k(x^k) \ge 0$ for all $x^k \in B^k$. The symbol e denotes the vector of all ones. We shall assume linear $f^k(x^k) = A^k x^k$, where A^k are given real matrices, and denote $f^k \doteq f^k(x^k)$, $f \doteq \sum_{k=1}^K f^k$ and $\lambda(x) = \min\{f_1(x), \ldots, f_M(x)\}$. The feasibility problem of computing an $x \in B$ such that $f(x) \le b$ for any positive b is included in $\mathcal{P}$.

---

* Research supported by the National Science Foundation under grant CCR-9208539.

Although there are several classical decomposition and partitioning methods for block angular programs, our interest here is in fast approximation schemes for solving large instances of this model, e.g., [5], [6], [14], and several of its special cases, e.g., [11],[15],[16]. These methods typically compute an *ε-approximate solution* of $\mathcal{P}$ by solving the approximation problem

$$(\mathcal{P}_\varepsilon) \qquad \text{find } x \in B \text{ such that } f(x) \leq (1+\varepsilon)\lambda^* e.$$

where $\varepsilon \in (0,1]$ is a given *relative accuracy*.

We shall base our analysis on the exponential-potential restricted Lagrangian decomposition algorithm described in [5]. The *deterministic* variant of this block-coordinate descent algorithm solves the general convex programming case of $\mathcal{P}_\varepsilon$ in

$$(1.1) \qquad N(\varepsilon) = O\left(K(\ln M)(\varepsilon^{-2} + \ln \min\{K, M\})\right)$$

coordination steps, each of which computes a nonnegative *price vector p* for the coupling constraints, issues K parallel calls to "block solvers" to compute block-coordinate potential-function descent directions by minimizing a convex combination of the M components of f^k with weights p, selects the best direction among these, and then moves to a new point with a lower potential value. The *randomized* variant is similar but its block selection is simpler. At each coordination step, it uniformly selects *one* random block-coordinate to process, and performs a move if it is descending. The expression (1.1) then bounds the *expected* number of coordination steps to solve $\mathcal{P}_\varepsilon$. For an arbitrary $\varepsilon \in (0,1]$, these iteration bounds are the best overall, better by a factor of $\ln(M/\varepsilon)/\ln M$ than those for the all-linear case [14] and for the special case of concurrent network flows [4]. The randomized bound of [11] is substantially improved by (1.1).

More elaborate block selection strategies in decomposition algorithms have been tried in the past, including those based on ideas borrowed from coordinate sweep strategies for Gauss-Seidel relaxation and for cyclic coordinate-descent schemes for unconstrained nonlinear optimization (see e.g. [2], [18], [19]). Recently, Radzik [15] showed that round-robin processing of the commodities in the concurrent network flow algorithms of [11] and [4] results in a bound that matches their expected time complexities.

In this paper, we extend this result to the more general case of block-angular programs with linear coupling constraints and K disjoint convex compact blocks. We present a simple modification that derandomizes Algorithm $\mathcal{R}$ of [5], and provide a straightforward analysis to show that (1.1) bounds the coordination complexity, or equivalently, the number of single-block solver calls. The new algorithm produces an approximate dual price vector in addition to a solution of $\mathcal{P}_\varepsilon$ and cuts down by a factor of K the number of single block optimizations required by the previous best deterministic algorithm [5]. We further show that our algorithm solves a reformulation of *nonlinear* problems $\mathcal{P}$ augmented by additional variables, with no increase in its coordination complexity.

In Section 2 we define the exponential potential function, review some of its properties and state our Algorithm $\mathcal{D}$ which implements cyclic block-coordinate descent. Section 3 contains our key results, namely, proofs of correctness and derivations of iteration bounds for solving $\mathcal{P}$ to a prescribed accuracy. The last section discusses a number of simple applications of the proposed method.

2. Block-coordinate potential-reduction algorithm

Although exponential potential functions have been widely used in science and and engineering, their application to systems of linear inequalities was first proposed by Motzkin [12]. Here we shall work with the following *exponential potential function*:

$$\Phi_t(x) = \ln \sum_{m=1}^{M} \exp \frac{f_m(x)}{t}, \tag{2.1}$$

where $t > 0$ is a small parameter. It is easy to see from this definition that

$$t\Phi_t(x) \geq \lambda(x) \geq t\Phi_t(x) - t\ln M, \tag{2.2}$$

which implies:

$$\lambda(x) \leq \lambda(y) + t\ln M \quad \text{for any } x, y \in B \quad \text{such that} \quad \Phi_t(x) \leq \Phi_t(y). \tag{2.3}$$

Since $\lambda(x)$ is sandwiched between $t\Phi_t(x)$ and $t\Phi_t(x) - t\ln M$ for all $x \in B$, the potential $\Phi_t(x)$ is a convenient smooth (w.r.t. f) substitute for the nondifferentiable convex function $\lambda(x)$ over $x \in B$. In fact, setting $t = \sigma/\ell \ln M$ for a given absolute accuracy $\sigma > 0$ and a scalar $\ell > 1$, the problem of minimizing $\lambda(x)$ over $x \in B$ to an absolute accuracy of $\sigma > 0$ can be replaced by the minimization of $\Phi_t(x)$ over $x \in B$ to an absolute accuracy of $(\ell - 1)\ln M$. (We shall later set $\ell = 7$.) Algorithm $\mathcal{D}$ below attempts to solve this potential minimization problem instead of the sharing problem $\mathcal{P}$.

Another useful property of $\Phi_t(x)$ is that the difference of the potential at any two points $x, y \in B$ can be written as:

$$\Phi_t(y) - \Phi_t(x) = \ln \sum_{m=1}^{M} p_m \exp\left\{\frac{f_m(y) - f_m(x)}{t}\right\}, \tag{2.4}$$

where

$$p_m \doteq p_m(x) = \frac{\exp\{f_m(x)/t\}}{\sum_{\mu=1}^{M} \exp\{f_\mu(x)/t\}}, \quad m = 1, \ldots, M. \tag{2.5}$$

This *price vector* $p \in P = \{p \in \mathbb{R}^M \mid e^T p = 1,\ p \geq 0\}$ is simply the gradient of the potential function with respect to f at $f(x)$. It is related to $\lambda(x)$ by the following inequality (Lemma 2 in [5]):

$$\lambda(x) \leq p^T f(x) + t\ln M \text{ for any } x \in B. \tag{2.6}$$

That is, the p-weighted convex combination of f provides a close approximation to $\lambda(x)$ for a sufficiently small $t > 0$. Indeed, by standard Lagrangian duality arguments, the *block problems*

$$(\mathcal{P}^k) \qquad g^k(p) = \min\{\, p^T f^k(x^k) \mid f^k(x^k) \le qe,\ x^k \in B^k \,\},\ k = 1, \ldots, K,$$

for some fixed $q > 1$, provide the relation

$$\lambda^* = \max_{p \in P} g(p) = \sum_{k=1}^{K} g^k(p). \tag{2.7}$$

We shall only need an approximate solution of the block problems $\mathcal{P}^k$. This is not only expedient but often the only realistic choice, depending on the generality of the sets B^k. Specifically, for a given relative *optimization* tolerance $\varepsilon' > 0$ and a *feasibility* tolerance $q > 1$, we define an (ε', q)-*good block solver* as a subroutine which computes a block solution $\hat{x}^k \in B^k$ such that $f^k(\hat{x}^k) \le qe$ and $p^T f^k(\hat{x}^k) \le (1+\varepsilon')\min\{\, p^T f^k(x^k) \mid f^k(x^k) \le qe, x^k \in B^k \,\}$.

We are now ready to state our algorithm for solving $\mathcal{P}$ to a given absolute accuracy of $\sigma \in (0, 1/3]$. We shall assume that an initial point $x_0 \in B$ such that $f(x_0) \le e$ is available and that a feasibility tolerance $q > 1$ is given. We shall denote by $p^{(k)}$ the contemporaneous price vector with which the block solver solves $\mathcal{P}^k$ in Step $\mathcal{D}$2-1 below.

Algorithm $\mathcal{D}(f, x_0, \sigma, q)$:

Step $\mathcal{D}0$. (Initialize.) Select $t = \sigma/7 \ln M$, set $x := x_0$ and compute p from (2.5). Let π be an arbitrary permutation of the integers $1, 2, \ldots, K$.

Step $\mathcal{D}1$. (Sweep.) For $k = \pi(1), \pi(2), \ldots, \pi(K)$ do steps $\mathcal{D}$1-$\mathcal{D}$2 below. When the list is exhausted go to Step $\mathcal{D}3$.

Step $\mathcal{D}2$. (Iteration.) Perform the following two steps:

Step $\mathcal{D}$2-1. (Block optimization.) Use a $(\sigma/7, q)$-good block solver to compute a block solution $\hat{x}^k \in B^k$.

Step $\mathcal{D}$2-2. (Coordination.) For a fixed *step length* $\tau^k \doteq \tau \doteq \sigma t/11q$, compute the point $y(k, \tau^k)$, which is the current iterate x with its kth block coordinate x^k replaced by $(1-\tau^k)x^k + \tau^k \hat{x}^k$. Compute $\Delta^k\Phi = \Phi_t(y(k, \tau^k)) - \Phi_t(x)$. If $\Delta^k\Phi < 0$, replace x^k by $y(k, \tau^k)$ to obtain the new iterate x, and recompute p from (2.5).

Step $\mathcal{D}3$. (Stopping criterion.) If $\sum_{k=1}^{K} \Delta^k\Phi > -\tau\sigma/4t$, report x and stop. Otherwise, reset or change π and go to Step $\mathcal{D}1$. □

Algorithm $\mathcal{D}$ is a simple deterministic variant of Algorithm $\mathcal{R}$, the three-step ($\mathcal{R}1$, $\mathcal{R}2$, $\mathcal{R}3$) uniform randomized block-selection algorithm of [5]. We replaced the $(K+1)$-sided fair coin toss of its Step $\mathcal{R}1$ by our Step $\mathcal{D}1$, K invocations of which constitute a *sweep* of an arbitrary permutation π of the K block coordinates. Such strategies were proposed in the early fifties in the context of Gauss-Seidel type relaxation methods for solving systems of equations (see, e.g., [8]) and were later extended to unconstrained optimization (see, e.g., [2], [18], [19]) . We also simplified Step $\mathcal{R}3$ by dropping the (now redundant) requirement that, on the average, a full sweep after each K randomized iterations be performed; what remains of Step $\mathcal{R}3$ in Step $\mathcal{D}3$ is just the stopping criterion.

3. Analysis of Algorithm $\mathcal{D}$

In this section we will prove Algorithm $\mathcal{D}$ correct and analyze its performance. Our proof strategy differs from [5] in several important aspects. First, as mentioned above, Algorithm $\mathcal{R}$ in that paper mandates that every K random block-solver calls are followed, on the average, by one deterministic full sweep that solves all K block problems, selecting the best among them to move, a fact used in its proof. No such step is retained here.

Second, the componentwise decreases of the price vector p, which occur at each iteration of Algorithms $\mathcal{R}$ and $\mathcal{D}$, are shown here to be bounded below and can be thus controlled to be small as the algorithm progresses (Lemma 1). This property was first observed by Radzik [15] for the special case of concurrent flows, where coupling constraints are arc capacities and each block is a single-pair path polytope.

Third, and most important, the block solutions $\hat{x}^k$ are no longer related to λ^* in the manner defined by the Lagrangian duality relation (2.7) since, in our case, the prices p change at each coordination step. We relax this duality relation so that it pertains to block solutions with as many as K *different* price vectors $p^{(k)}$ produced by the algorithm, and so that it provides a useful lower bound on the optimal value λ^* (Lemma 2).

The correctness of Algorithm $\mathcal{D}$ and its iteration bound are shown in Theorem 1. Its proof relies on the properties of the potential function, the stopping criterion, and Lemmas 1 and 2 below. Theorem 1 also ensures that Algorithm $\mathcal{D}$ provides a good approximation to the optimal price vector.

LEMMA 1. *For any pair of blocks r, s such that r is processed before s within a sweep of* Algorithm $\mathcal{D}$,

$$p^{(s)} \geq (1 - 22\tau/21t)\, p^{(r)}.$$

Proof. Let x denote the iterate at the beginning of the sweep and $x^{(k)}$ the iterate obtained in Step $\mathcal{D}2$-2 for a block "k" selected in Step $\mathcal{D}1$. Similarly, denote $f^{(k)} \equiv Ax^{(k)} = \sum_{\ell=1}^{K} A^\ell x^{\ell(k)}$. As before, x^k denotes the kth block variable of problem $\mathcal{P}$.

Consider the sequence of blocks $r = k_1, k_2, \ldots, k_q = s$ selected within a sweep and denote $S = \{k_2, \ldots, k_q\}$. Clearly, no block index repeats in S. Then, by the linearity and the nonnegativity of the f^k,

$$f^{(s)} - f^{(r)} = \tau \sum_{k \in S} f^k(\hat{x}^k - x^k) \geq -\tau \sum_{k \in S} f^k(x^k) \geq -\tau f^{(r)}.$$

Divide the above by t, exponentiate both sides, and normalize by $\sum_{m=1}^{M} \exp(f_m^{(s)}/t)$. This yields

$$p_m^{(s)} \geq \frac{\exp(f_m^{(r)}/t)}{\sum_{\mu=1}^{M} \exp(f_\mu^{(s)}/t)} \exp\left(-\frac{\tau}{t} f_m^{(r)}\right) \geq \frac{\exp(f_m^{(r)}/t)}{\sum_{\mu=1}^{M} \exp(f_\mu^{(r)}/t)} \exp\left(-\frac{\tau}{t} f_m^{(r)}\right)$$

where the last inequality follows from $\Phi(x^{(s)}) \leq \Phi(x^{(r)})$. The definition (2.5) for p and the standard inequality $\exp(-\beta) \geq 1 - \beta$ for any β provide:

$$p_m^{(s)} \geq \left(1 - \frac{\tau}{t} f_m^{(r)}\right) p_m^{(r)}. \tag{3.1}$$

For the point $x^{(r)}$, where necessarily $\Phi(x^{(r)}) \leq \Phi(x_0)$, property (2.3) implies that $f_m^{(r)} \leq \lambda(x_0) + t\ln(M) \leq 1 + \sigma/7$ for all m, and $f_m^{(r)} \leq 22/21$ since $\sigma \leq 1/3$. Using this bound in the right term of (3.1) proves the claim. □

Next we generalize the Lagrangian duality relation (2.7) so that it pertains to block solutions with as many as K *different* price vectors $p^{(k)}$ produced in the course of a single sweep of Algorithm $\mathcal{D}$.

LEMMA 2.

$$\sum_{k=1}^{K} p^{(k)T} \hat{f}^k \leq \left(1 + \frac{\sigma}{4}\right) \lambda^*$$

Proof. Let s denote the last block in a sweep, $\bar{x}^k$ denote an exact solution of $\mathcal{P}^k$ and $\bar{f}^k \doteq f^k(\bar{x}^k)$. Recall that $\hat{x}^k$ is the solution computed by the $(\sigma/7, q)$-good block solver in Step $\mathcal{D}$2-1 of the algorithm. Then,

$$p^{(k)T} \hat{f}^k \leq (1 + \sigma/7)\ p^{(k)T} \bar{f}^k \leq \frac{1 + \sigma/7}{1 - 22\tau/21t}\ p^{(s)T} \bar{f}^k \leq (1 + \sigma/4) p^{(s)T} \bar{f}^k,$$

where the second inequality is due to Lemma 1 and the last inequality follows from the definition of τ. Summing these inequalities over an entire sweep gives

$$\begin{aligned} \sum_{k=1}^{K} p^{(k)T} \hat{f}^k &\leq (1 + \sigma/4)\ p^{(s)T} \sum_{k=1}^{K} \bar{f}^k \\ &= (1 + \sigma/4)\ g(p^{(s)}) \\ &\leq (1 + \sigma/4)\lambda^*. \quad \square \end{aligned} \tag{3.2}$$

We are now ready to prove Algorithm $\mathcal{A}$ correct and to develop its coordination complexity. We shall first bound the potential decrease in a single iteration, add all such decreases over an entire sweep and then refine these estimates so that $\lambda(x) \leq \lambda^* + \sigma$ for the $x \in B$ produced by Algorithm $\mathcal{D}$. The stopping criterion and Lemmas 1 and 2 are used to show that the last price vector is optimal to within an absolute accuracy σ.

THEOREM 1. Algorithm $\mathcal{D}$ *halts in*

$$N_D = O\left(qK \ln M\left(\frac{\lambda(x)-\lambda^*}{\sigma^3}+\frac{1}{\sigma^2}\right)\right)$$

coordination steps with an $x \in B$ *such that* $\lambda(x) \leq \lambda^* + \sigma$. *Moreover,* $g(p^{(s)}) \geq \lambda^* - 0.85\sigma$ *for the last price vector* $p^{(s)}$ *computed by (2.5).*

Proof. Let $\Delta^k\Phi$ be the potential decrease achieved in a single iteration of the algorithm, say for block k. Then,

$$\Delta^k\Phi \leq \sum_{m=1}^{M} p_m^{(k)}\left(\exp\left\{-\tfrac{\tau}{t}(f_m^k - \hat{f}_m^k)\right\}-1\right), \tag{3.3}$$

where $f^k = A^k x^k$ and $\hat{f}^k = A^k \hat{x}^k$. This can be verified from (2.4) by using the elementary inequality $\ln(z) \leq z-1$. In view of the linearity of f^k, it is easy to see by induction on the number of iterations that

$$f_m^k, \hat{f}_m^k \in [0, q]. \tag{3.4}$$

For our values of $\tau = \sigma t/11q$ and $\sigma \leq 1/3$, this expression implies that $\left|-(\tau/t)(f_m^k - \hat{f}_m^k)\right| \leq \sigma/11 \leq 1/18$ for all m. Using the inequality $\exp\{u\} - 1 \leq u + (6/11)u^2$ for all $|u| \leq 1/18$, (3.3) can be further bounded as follows:

$$\Delta^k\Phi \leq -\frac{\tau}{t}\sum_{m=1}^{M} p_m^{(k)}(f_m^k - \hat{f}_m^k) + \frac{6\tau^2}{11t^2}\sum_{m=1}^{M} p_m^{(k)}(f_m^k - \hat{f}_m^k)^2. \tag{3.5}$$

On the other hand, it is easy to see from (3.4) that $(f_m^k - \hat{f}_m^k)^2 \leq (f_m^k)^2 + (\hat{f}_m^k)^2 \leq q(f_m^k + \hat{f}_m^k)$. A useful form of the single-iteration bound is then obtained by substituting this inequality in (3.5):

$$\Delta^k\Phi \leq -\frac{\tau}{t}\left(1-\frac{6q\tau}{11t}\right)p^{(k)T}(f^k - \hat{f}^k) + \frac{12q\tau^2}{11t^2}p^{(k)T}\hat{f}^k.$$

The total decrease of the potential over an entire sweep can be bounded by simply summing the above single-iteration bounds:

$$\sum_{k=1}^{K}\Delta^k\Phi \leq -\frac{\tau}{t}\left(1-\frac{6q\tau}{11t}\right)\sum_{k=1}^{K} p^{(k)T}(f^k - \hat{f}^k) + \frac{12q\tau^2}{11t^2}\sum_{k=1}^{K} p^{(k)T}\hat{f}^k.$$

By using our value for τ, the stopping criterion, the fact that $\lambda^* \leq 1$ and Lemma 2, the above inequality can be simplified to:

$$-\frac{\sigma\tau}{4qt} \leq -\frac{119\tau}{121t}\sum_{k=1}^{K} p^{(k)T}(f^k - \hat{f}^k) + \frac{13q\tau^2}{11t^2}$$

which clearly implies that

$$(3.6)\qquad \sum_{k=1}^{K} p^{(k)T}(f^k - \hat{f}^k) \le \frac{4\sigma}{11}\,.$$

We shall refine this estimate as follows. Denote by $p^{(r)}$, $x^{(r)}$ the initial price and solution vectors for this sweep. Use Lemmas 1 and 2 to write

$$\sum_{k=1}^{K} p^{(k)T}(f^k - \hat{f}^k) \ge \left(1 - \frac{22}{21}\frac{\tau}{t}\right) \sum_{k=1}^{K} p^{(r)T} f^k - \lambda^* - \frac{\sigma}{4}.$$

For the initial and current iterates $x^{(r)}$ and x, respectively, $\sum_{k=1}^{K} p^{(r)T} f^k \ge \lambda(x^{(r)}) - t \ln M$ by property (2.6), and $\lambda(x^{(r)}) \ge \lambda(x) - t \ln M$ by property (2.3). Adding these two inequalities gives $\sum_{k=1}^{K} p^{(r)T} f^k \ge \lambda(x) - 2t \ln M$. Hence,

$$(3.7)\qquad \sum_{k=1}^{K} p^{(k)T}(f^k - \hat{f}^k) \ge \left(1 - \frac{22}{21}\frac{\tau}{t}\right) (\lambda(x) - 2t \ln M) - \lambda^* - \frac{\sigma}{4},$$

and since $\tau = 6t/11q$ and $\lambda(x) \le 1 + t \ln M$,

$$(3.8)\qquad \left(1 - \frac{22}{21}\frac{\tau}{t}\right) (\lambda(x) - 2t \ln M) \ge \lambda(x) - \frac{8\sigma}{21}.$$

Now combine (3.7) and (3.8) to obtain

$$\sum_{k=1}^{K} p^{(k)T}(f^k - \hat{f}^k) \ge \lambda(x) - \lambda^* - \frac{7\sigma}{11},$$

which, along with (3.6), yields $\lambda(x) \le \lambda^* + \sigma$, as claimed.

To show the approximate optimality of the last price vector $p^{(s)}$, where s be the last block index in a sweep. Denote by $\bar{x}^k$ the (unknown) exact solution of $\mathcal{P}^k$ and let $\bar{f}^k \doteq f^k(\bar{x}^k)$. Since the block solution $\hat{x}^k$ is computed approximately by a $(\sigma/7, q)$-good block solver in Step $\mathcal{D}$2-1 of the algorithm, inequalities (3.2) and (3.6) provide:

$$(3.9)\qquad \left(1 + \frac{\sigma}{4}\right)\; g(p^{(s)}) \ge \sum_{k=1}^{K} p^{(k)T} f^k - \frac{4\sigma}{11}.$$

On the other hand, from Lemma 1 we have

$$\sum_{k=1}^{K} p^{(k)T} f^k \ge \left(1 - \frac{22}{21}\frac{\tau}{t}\right) \sum_{k=1}^{K} p^{(r)T} f^k.$$

Clearly, $\lambda(x^{(r)}) \geq \lambda^*$ for the initial iterate $x^{(r)}$ and $\sum_{k=1}^{K} p^{(r)T} f^k \geq \lambda(x^{(r)}) - t\ln M$ from property (2.6). Hence the above inequality can be further bounded below as follows:

$$\sum_{k=1}^{K} p^{(k)T} f^k \geq \left(1 - \frac{22}{21}\frac{\tau}{t}\right)(\lambda^* - t\ln M) \geq \lambda^* - \frac{22\tau}{21t}\lambda^* - \frac{\sigma}{7} \geq \lambda^* - \frac{5\sigma}{21}.$$

The claim is obtained by combining this last inequality and (3.9):

$$g(p^{(s)}) \geq \frac{\lambda^*}{(1+\sigma/4)} - \frac{139\sigma}{231(1+\sigma/4)} \geq \frac{\lambda^*}{(1+\sigma/4)} - 0.6\sigma$$
$$\geq \lambda^* - 0.25\sigma - 0.6\sigma = \lambda^* - 0.85\sigma.$$

Finally, the stopping criterion (Step $\mathcal{D}3$) implies that the potential function decreases by at least $\Phi(x^{(s)}) - \Phi(x^{(r)}) = \sum_{k=1}^{K} \Delta^k \Phi \leq -\sigma^2/44q$ during this sweep, that is, $\Phi(x^*) - \Phi(x_0) \leq -N'\sigma^2/44q$ in N' sweeps. By (2.2), this translates into $\lambda(x^*)/t - \lambda(x_0)/t - \ln M \leq -N'\sigma^2/44q$. Since $t = \sigma/7\ln M$, we compute

$$N' = O\left(q\ln M\left(\frac{\lambda(x_0) - \lambda^*}{\sigma^3} + \frac{1}{\sigma^2}\right)\right),$$

which is N_D/K since each sweep consists of K coordination steps. □

Our overall polynomial time approximation scheme for $\mathcal{P}_\varepsilon$ consists of a *ternary search* procedure $\mathcal{TS}$ which uses Algorithm $\mathcal{D}$ as a subroutine. Starting with any $x_0 \in B$, $x = x_0$, $\underline{\lambda} = 0$, $\mathcal{TS}$ maintains the interval $[\underline{\lambda}, \lambda(x)]$, its length $\delta = \lambda(x) - \underline{\lambda}$, and the parameters $\lambda_1 = \underline{\lambda} + \delta/3$, $\lambda_2 = \underline{\lambda} + 2\delta/3$ and $\nu = 1/(2\underline{\lambda} + \lambda(x))$. For an absolute tolerance $\sigma = \nu\delta/3$, $\mathcal{TS}$ proceeds as follows.

while $\lambda(x) > \varepsilon\underline{\lambda}$ **do**

Run $\mathcal{D}(\nu f, x, \sigma, q)$ to compute a point y such that $\lambda(y) \leq \lambda^* + \delta/3$.

if $\lambda(y) \leq \lambda_2$ **then** $x := y$ and $\lambda(x) := \lambda(y)$ **else** $\underline{\lambda} := \lambda_1$ **end**

end

As shown in [5], $\mathcal{TS}$ solves $\mathcal{P}_\varepsilon$ in

$$N = O\left(qK(\ln M)\left(\varepsilon^{-2} + \ln\left(\lambda(x_0)/\lambda^*\right)\right)\right) \tag{3.10}$$

coordination steps, or that many block optimizations. The data-dependent logarithmic term in this expression can be bounded further by an appropriate initialization. The simplest way to do this is to solve each block problem for $p = e/M$, resulting in an initial point x_0 such that $\ln(\lambda(x_0)/\lambda^*) = O(\ln M)$, see [6]. The *greedy* initialization described in [5] calls the block solver an additional $O(\ln M)$ times to refine the initial x_0 so that $\ln(\lambda(x_0)/\lambda^*) = O(\ln K)$. Thus, the expression (3.10) for $q = O(1)$ bounds the number coordination steps (1.1) as claimed in the Introduction.

At present, extending these coordination complexity bounds to problems $\mathcal{P}$ with nonnegative convex coupling functions f^k entails the considerable cost of appending new variables y^k for each block $k = 1, \ldots, K$. This results in the augmented problem:

$$(\mathcal{P}_y) \qquad \lambda^* = \min \left\{ \lambda \;\middle|\; \sum_{k=1}^{K} y^k \leq \lambda e,\ f(x^k) \leq y^k,\ x \in B \right\}.$$

This problem can then be reduced to the form required by our Algorithm $\mathcal{D}$ by defining the kth block problem over the set $B_y^k = \{\, x^k \in B^k, y^k \in \mathbb{R}^M \mid f(x^k) \leq y^k\}$, which does not satisfy our compactness assumption. At any rate, it is more convenient to compute an approximate $\hat{x}^k$ by solving the block problem $\min\{p^T f^k(x^k) \mid f(x^k) \leq qe,\ x^k \in B^k\}$, since $\hat{y}^k$ can be obtained as $\hat{y}^k = f(\hat{x}^k)$. Clearly, the coordination complexity (1.1) of our algorithm remains valid for the augmented problem $\mathcal{P}_y$ and thus matches the randomized complexity bound of Algorithm $\mathcal{R}$ of [5]. In [7] this reduction is further extended to solve the block-angular convex *optimization* variant of $\mathcal{P}$ with the same coordination complexity. On the other hand, it is important to note that our deterministic algorithm applied to $\mathcal{P}_y$ cannot be considered a derandomized variant of Algorithm $\mathcal{R}$ due to the presence of these additional y^k variables. The question of whether there is a block selection scheme that formally derandomizes Algorithm $\mathcal{R}$, with iterates in the epigraph of the convex coupling constraints remains open.

We summarize the above discussion as follows.

THEOREM 2. There is a deterministic approximation algorithm that solves problem $\mathcal{P}$ with nonnegative block-separable convex coupling functions $f(x)$ to a given relative accuracy $\varepsilon \in (0, 1]$ in

$$N(\varepsilon) = O\left(qK(\ln M)(\varepsilon^{-2} + \ln \min\{K, M\})\right)$$

coordination steps, or equivalently in that many block optimizations, using (ε, q)-good block solvers. □

As before, for $q \approx 1$ or even $q = O(1)$, this expression bounds the number of block optimizations as claimed in the Introduction.

We note in closing that although arbitrary cyclic block coordinate selection provides a convenient mechanism for improving the theoretical bound on block optimizations for exponential-potential restricted Lagrangian decomposition, our analysis also implies that the resulting schemes are short-step methods. Several issues related to the existence of long-step deterministic variants of comparable coordination complexity, possibly allowing line searches in block coordinates, or even K-search (jointly in several block coordinates, see, e.g., [5], [17]) remain open, as is the more practical question of whether block-coordinate sweep strategies provide any appreciable improvement over other block selection strategies.

4. Applications

The above iteration bounds can often be translated to running times by considering the complexity of the block solvers for $\mathcal{P}^k$ and the tradeoffs between the feasibility tolerance $q > 1$ expected of the block solver, its running time and the coordination complexity (1.1). It is clearly advantageous if $q > 1$ can fixed to a number that does not affect the running time of the block solver. Such is the case for the applications we discuss below. For a fixed $\varepsilon \in (0, 1]$ and ignoring logarithmic factors, an ε-approximate solution of all of these models can be computed in $O(KM)$ time.

First, consider the M-source, K-sink *generalized transportation sharing problem with destination budgets*, which is to minimize λ subject to source constraints $\sum_{k=1}^{K} D^k x^k \leq \lambda a$, and the constraints $e^T x^k = b^k$, $c^{kT} x^k \leq C^k$ for each sink $k = 1, \ldots, K$. The data for this problem consist of an M-vector $a > 0$ which specifies supply production proportions at the sources, the demands $b^1, b^2, \ldots, b^K > 0$, the procurement costs $c^k \geq 0$ at sink k against its fixed budget $C^k > 0$, an M-vector of arc-flow multipliers d^k for the incident arcs at k, which define $D^k = \text{diag}(d^k)$ for each sink k. After scaling, the problem can be written as

$$\min \left\{ \lambda \;\middle|\; \sum_{k=1}^{K} D^k x^k \leq \lambda e,\; x^k \in B^k,\; k = 1, \ldots, K \right\}$$

where $B^k = \{\, x^k \mid c^k x^k \leq 1,\; e^T x^k = 1,\; x^k \geq 0 \,\}$ are the blocks, one per sink. Our algorithm uses $O\left(K \ln M (\varepsilon^{-2} + \ln \min\{M, K\})\right)$ block solver calls. Each block problem $\mathcal{P}^k$ is a linear program with two constraints and upper bounds which can be solved $O(M)$ time [13]. Thus the overall time for this application is $O\left(KM \ln M (\varepsilon^{-2} + \ln \min\{K, M\})\right)$.

Second, if the fixed budget requirement at each sink is replaced by an *equitable budget* allocation constraint, the problem becomes:

$$\min \left\{ \lambda \;\middle|\; \sum_{k=1}^{K} D^k x^k \leq \lambda e,\; c^{kT} x^k \leq \lambda,\; x^k \in B^k,\; k = 1, \ldots, K \right\} \tag{4.1}$$

where $B^k = \{x^k \mid e^T x^k = 1,\; x^k \geq 0\}$ are the normalized sink demands. The coupling constraints now have an additional constraint per block, or a total of $M + K$ constraints. Nevertheless, each block problem $\mathcal{P}^k$: $\min\{p^T D^k x^k + p_{M+1} c^{kT} x^k \mid D^k x^k \leq qe,\; c^{kT} x^k \leq q,\; x^k \in B^k\}$ exhibits essentially the same structure as that of the previous example, and it is solved in $O(M)$ time as before, leading to $O(KM \ln(M + K)(\varepsilon^{-2} + \ln K))$ time.

A noteworthy application of (4.1) is the *scheduling of K jobs on M << K unrelated parallel machines with preemption* so that the longest completion time λ is minimized [9]. Here the M-vectors $d^k > 0$ denote the given processing times for job k and furthermore $c^k = d^k$ for each $k = 1, \ldots, K$. Hence, the bounds $D^k x^k \le qe$ in $\mathcal{P}^k$ are redundant as they are implied by the inequality $c^{kT} x^k \le q$, and each $\mathcal{P}^k$ then becomes a two-constraint knapsack problem which can be solved in $O(M)$ time [3]. Our algorithm provides an ε-approximate nonintegral schedule in $O\left(KM \ln K(\varepsilon^{-2} + \ln K)\right)$ time. For a small but fixed $\varepsilon \in (0, 1]$, our $O(KM \ln K)$ time beats the best known deterministic time by a factor of K and it is a factor of $O(\ln M)$ faster than the corresponding randomized time. To compute an integer schedule the continuous approximate solution x of minmax length λ we obtain can be rounded by the deterministic rounding procedure described in [14]. This results in a schedule of minmax time of at most $(1 + \varepsilon)\lambda$.

Third, consider again the problem (4.1), but this time *with no budget constraints*, i.e., with $B^k = \{x^k \mid e^T x^k = 1,\ x^k \ge 0\}$. This can be viewed as a generalized supply sharing transportation network, a concurrent flow problem with uniform capacities, or as an LP relaxation of the minmax version of the generalized assignment problem. Each $\mathcal{P}^k$ is now a bounded knapsack problem, solvable in $O(M)$ time [1], resulting in $O\left(KM \ln M(\varepsilon^{-2} + \ln \min\{M, K\})\right)$ overall time.

A popular variant of the latter model is the problem of *scheduling K independent jobs on M unrelated machines* so as to minimize the longest completion time λ. This model is defined in [10] with the additional nonconvex constraints $x^k_m = 0$ if $d^k_m > \lambda$, $m = 1, \ldots, M$. The smallest *feasible* $\lambda = \lambda^*$ can be computed in $O(\ln M)$ iterations of bisection search in the interval $(0, M\lambda^*]$. Each iteration consists of a "preprocessing" step that invokes the above additional nonconvex constraints to delete some variables for the current value of λ, and the solution of the feasibility problem by our algorithm. Since the deletion of these variables implies that the upper bounds $d^k_m x^k_m \le q$ for the remaining variables are redundant in $\mathcal{P}^k$, each such $\mathcal{P}^k$ reduces to the minimization of a linear function on the $(M - 1)$-dimensional unit simplex, in $O(M)$ time. Thus, the total complexity of this scheme is $O(KM \ln^2 M(\varepsilon^{-2} + \ln M))$. For a small but fixed $\varepsilon > 0$, our time of $O(KM \ln^2 M)$ is a factor of K faster than the best deterministic time and slightly faster than corresponding randomized time. The approximate LP-relaxed solution x^* of minmax value λ^* we obtain can then be transformed to an integer schedule of value at most $2\lambda^*$, by first converting it to a *basic* solution using positive-flow cycle elimination [14] and then by applying the rounding procedure of [10].

Finally, our algorithm leads to fast approximation schemes for solving several specially-structured linear programs, including minimum-cost K-commodity flow problems in N-node, M-arc networks which can thus be solved in $O(\varepsilon^{-2} KNM)$ time, disregarding logarithmic factors [7]. This is the fastest known approximation scheme for minimum-cost multicommodity flows.

5. References

[1] E. BALAS AND E. ZEMEL, *An algorithm for large zero-one knapsack problems*, Operations Research 28 (1980), 1131-1154.

[2] M. S. BAZARAA, H. D. SHERALI AND C.M. SHETTY, *Nonlinear Programming: Theory and Algorithms*, John Wiley & Sons, New York, 1993.

[3] M.E. DYER, *An $O(n)$ algorithm for the multiple-choice knapsack linear program*, Mathematical Programming 29 (1984), 57-63.

[4] A.V. GOLDBERG, *A natural randomization strategy for multicommodity flow and related problems*, Information Processing Let. 42 (1992), 249-256.

[5] M. D. GRIGORIADIS AND L. G. KHACHIYAN, *Fast approximation schemes for convex programs with many blocks and coupling constraints*, SIAM¡ J. Optimization 4 (1994), 86-107.

[6] ——, *Coordination complexity of parallel price-directive decomposition*, TR 94-19, DIMACS, Rutgers Univ., New Brunswick, NJ, April 1994. To appear in Mathematics of Operations Research.

[7] ——, *Approximate minimum-cost multicommodity flows in $\tilde{O}(\varepsilon^{-2}KNM)$ time*, TR 95-13, DIMACS, Rutgers Univ., New Brunswick, NJ, May, 1995. To appear in Mathematical Programming.

[8] E. ISAACSON AND H.B. KELLER, *Analysis of Numerical Methods*, John Wiley and Sons, NY, 1966.

[9] E.L. LAWLER AND J. LABETOULLE, *On preemptive scheduling of unrelated parallel processors by linear programming*, J. Assoc. Comput. Mach., 25 (1978), 612-619.

[10] J.K. LENSTRA, D.B. SHMOYS AND E. TARDOS, *Approximation algorithms for scheduling unrelated parallel machines*, Mathematical Programming, 24 (1990), 259-272.

[11] T. LEIGHTON, F. MAKEDON, S. PLOTKIN, C. STEIN, E. TARDOS AND S. TRAGOUDAS, *Fast approximation algorithms for multicommodity flow problems*, Journal of Computer System Sciences, 50 (1995), 228 -243.

[12] T.S. MOTZKIN, *New technique for linear inequalities and optimization*, in Project SCOOP Symp. on Linear Inequalities and Programming, Planning Res. Div., U.S. Air Force, Washington, D.C., 1952.

[13] N. MEGIDDO AND A. TAMIR, *Linear time algorithms for some separable quadratic programming problems*, Operations Research Letters 13 (1993), 203-211.

[14] S.A. PLOTKIN, D.B. SHMOYS AND E. TARDOS, *Fast approximation algorithms for fractional packing and covering problems*, Mathematics of Operations Research, 20 (1995), 257 -301.

[15] T. RADZIK, *Fast deterministic approximation for the multicommodity flow problem*, Proc. 6th ACM-SIAM Symp. on Discrete Algorithms (1995), 486-492.

[16] F. SHAHROKHI AND D.W. MATULA, *The maximum concurrent flow problem*, J. of the ACM 37 (1990) 318-334.

[17] G.L. SCHULTZ AND R.R. MEYER, *An interior point method for block angular optimization*, SIAM J. Optimization 1 (1991), 583-602.

[18] N. ZADEH, *A note on cyclic coordinate ascent method*, Management Science 3 (1970) 643-644.

[19] W.I. ZANGWILL, *Nonlinear Programming: A Unified Approach*, Prentice-Hall, Englewood Cliffs, NJ, 1969.

References

[1] [illegible] R.K. and [illegible] "Algorithms for [illegible] capacitated [illegible]", Operations Research [illegible] (19[illegible]).

[2] [illegible] D. [illegible] and G.M. [illegible] "[illegible] Programming [illegible]", [illegible] and [illegible], New York, 19[illegible].

[3] [illegible] "An [illegible] algorithm for the multi-[illegible] [illegible]", [illegible] Programming [illegible] (19[illegible]).

[4] [illegible] "[illegible] algorithm for [illegible] [illegible]", [illegible] (19[illegible]) [illegible].

[5] [illegible] and [illegible] HILLMAN "[illegible]", SIAM J. [illegible] (19[illegible]).

[6] [illegible] "[illegible] complexity of [illegible]", [illegible] Report, [illegible] (1[illegible]). [illegible] to appear in Mathematics of [illegible] Research.

[7] [illegible] "[illegible] and [illegible] problems", [illegible] TR [illegible], [illegible] University, [illegible].

[8] [illegible] "[illegible] of Numerical Methods", [illegible] and Sons, [illegible].

[9] [illegible] "The [illegible] of weighted [illegible]", [illegible] (19[illegible]) [illegible].

[10] [illegible] PARDOS [illegible] "[illegible] algorithms for [illegible]", Mathematical Programming [illegible] (19[illegible]) [illegible].

[11] [illegible] PARDOS [illegible] "[illegible] algorithms for the [illegible] flow problems", Journal of Computer and System Sciences [illegible] (19[illegible]) [illegible].

[12] [illegible] "[illegible] and [illegible]", in: [illegible] and Programming, [illegible].

[13] [illegible] "[illegible] for [illegible] separable [illegible]", [illegible] and Research [illegible] (19[illegible]) [illegible].

[14] [illegible] "[illegible] network [illegible]", [illegible].

[15] [illegible] "[illegible] for [illegible] flow problems", [illegible] (19[illegible]) [illegible].

[16] [illegible] "The [illegible] concurrent flow problem", [illegible] (19[illegible]) [illegible].

[17] [illegible] "[illegible] method [illegible]", [illegible] Optimization [illegible] (19[illegible]) [illegible].

[18] [illegible] (19[illegible]).

[19] [illegible] "[illegible] Programming: A Unified Approach", Prentice-Hall, [illegible], 19[illegible].

Author Index

Prof. Dr. Nikolaos Apostolatos
Department of Informatics, Section of Theoretical Informatics
University of Athens, Panepistemiopolis, G-15710 Athens, Greece

Prof. Miroslav D. Ašić
Department of Mathematics
The Ohio State University, Ohio, USA

OStR Ludwig Barnerßoi
Institut für Angewandte Mathematik und Statistik
Technische Universität München, 80290 München, Germany

Prof. Dr. Fred Alois Behringer
Institut für Angewandte Mathematik und Statistik
Technische Universität München, 80290 München, Germany
e-mail: behringe@statistik.tu-muenchen.de

Prof. Michael J. Best
Department of Combinatorics and Optimization
University of Waterloo, Waterloo, Ontario N2L 3G1, Canada
e-mail: mjbest@math.uwaterloo.ca

Ioannis T. Christou
Center for Parallel Optimization, Computer Sciences Department
University of Wisconsin, Madison, Wisconsin 53706-1685, USA

Prof. John E. Dennis
Department of Computational and Applied Mathematics
Rice University, Houston, Texas 77005-1892, USA
e-mail: dennis@caam.rice.edu

Prof. Dr. Werner Ehm
Institut für Angewandte Mathematik und Statistik
Technische Universität München, 80290 München, Germany
e-mail: ehm@statistik.tu-muenchen.de

Dr. Herbert Fischer
Institut für Angewandte Mathematik und Statistik
Technische Universität München, 80290 München, Germany
e-mail: fischer@statistik.tu-muenchen.de

Prof. Michael D. Grigoriadis
Department of Computer Science
Rutgers University, New Brunswick, NJ 08903, USA
e-mail: grigoria@cs.rutgers.edu

Anthony J. Kearsley
Department of Mathematics
University of Massachusetts, Dartmouth, Massachusetts, USA

Prof. Dr. Bernhard Korte
Forschungsinstitut für Diskrete Mathematik
Universität Bonn, Nassestr. 2, 53113 Bonn, Germany
e-mail: dm@or.uni-bonn.de

Prof. Vera V. Kovačević-Vujčić
Faculty of Organizational Sciences
University of Belgrade, Belgrade, Yugoslavia
e-mail: verakov@fon.fon.bg.ac.yu

Dr. Christian Kredler
Institut für Angewandte Mathematik und Statistik
Technische Universität München, 80290 München, Germany
e-mail: kredler@statistik.tu-muenchen.de

Prof. Olvi L. Mangasarian
Computer Sciences Department
University of Wisconsin, Madison, Wisconsin 53706-1685, USA
e-mail: olvi@cs.wisc.edu

Prof. Vladimir Mařik
AI Division, Department of Control Engineering
Czech Technical University, Prague, Czech Republic

Prof. Robert R. Meyer
Center for Parallel Optimization, Computer Sciences Department
University of Wisconsin, Madison, Wisconsin 53706-1685, USA
e-mail: rrm@cs.wisc.edu

Martin Pfingstl
Institut für Angewandte Mathematik und Statistik
Technische Universität München, 80290 München, Germany

Dr. Libor Přeučil
AI Division, Department of Control Engineering
Czech Technical University, Prague, Czech Republic
e-mail: preucil@labe.felk.cvut.cz

Prof. Louis B. Rall
Department of Mathematics
University of Wisconsin, Madison, Wisconsin 53706, USA
e-mail: rall@math.wisc.edu

Michael Rempter
Institut für Angewandte Mathematik und Statistik
Technische Universität München, 80290 München, Germany
e-mail: rempter@statistik.tu-muenchen.de

Dr. Bruno Riedmüller
Institut für Angewandte Mathematik und Statistik
Technische Universität München, 80290 München, Germany
e-mail: riedmuel@statistik.tu-muenchen.de

Prof. Stephen M. Robinson
Department of Industrial Engineering
University of Wisconsin, Madison, Wisconsin 53706-1572, USA
e-mail: smr@cs.wisc.edu

Priv.-Doz. Dr. Stefan Schäffler
Institut für Angewandte Mathematik und Statistik
Technische Universität München, 80290 München, Germany
e-mail: stefan@statistik.tu-muenchen.de

Dr. Walter Schlee
Institut für Angewandte Mathematik und Statistik
Technische Universität München, 80290 München, Germany
e-mail: schleew@statistik.tu-muenchen.de

Rolf Schöne
Institut für Angewandte Mathematik und Statistik
Technische Universität München, 80290 München, Germany
e-mail: schoene@statistik.tu-muenchen.de

Dr. Alfred Schöttl
Institut für Angewandte Mathematik und Statistik
Technische Universität München, 80290 München, Germany
e-mail: schoettl@statistik.tu-muenchen.de

Guillermina Schröder
Institut für Angewandte Mathematik und Statistik
Technische Universität München, 80290 München, Germany

Waldemar Schultz
Institut für Angewandte Mathematik und Statistik
Technische Universität München, 80290 München, Germany
e-mail: waldi@statistik.tu-muenchen.de

Petr Štěpán
AI Division, Department of Control Engineering
Czech Technical University, Prague, Czech Republic

Prof. Richard Tapia
Department of Computational and Applied Mathematics
Rice University, Houston, Texas 77005-1892, USA
e-mail: rat@masc18.rice.edu

Michael W. Trosset
Department of Computational and Applied Mathematics
Rice University, Houston, Texas 77005-1892, USA

Michael Ulbrich
Institut für Angewandte Mathematik und Statistik
Technische Universität München, 80290 München, Germany
e-mail: ulbrich@statistik.tu-muenchen.de

Stefan Ulbrich
Institut für Angewandte Mathematik und Statistik
Technische Universität München, 80290 München, Germany
e-mail: ulbrich@statistik.tu-muenchen.de

Luís N. Vicente
Department of Computational and Applied Mathematics
Rice University, Houston, Texas 77005-1892, USA

Jorge Villavicencio
Department of Computer Science
Rutgers University, New Brunswick, NJ 08903, USA